Chamomile
Medicinal, Biochemical, and Agricultural Aspects

Traditional Herbal Medicines for Modern Times

Each volume in this series provides academia, health sciences, and the herbal medicines industry with in-depth coverage of the herbal remedies for infectious diseases, certain medical conditions, or the plant medicines of a particular country.

Series Editor: Dr. Roland Hardman

Volume 1
Shengmai San, edited by Kam-Ming Ko

Volume 2
Rasayana: Ayurvedic Herbs for Rejuvenation and Longevity, by H.S. Puri

Volume 3
Sho-Saiko-To: (Xiao-Chai-Hu-Tang) Scientific Evaluation and Clinical Applications, by Yukio Ogihara and Masaki Aburada

Volume 4
Traditional Medicinal Plants and Malaria, edited by Merlin Willcox, Gerard Bodeker, and Philippe Rasoanaivo

Volume 5
Juzen-taiho-to (Shi-Quan-Da-Bu-Tang): Scientific Evaluation and Clinical Applications, edited by Haruki Yamada and Ikuo Saiki

Volume 6
Traditional Medicines for Modern Times: Antidiabetic Plants, edited by Amala Soumyanath

Volume 7
Bupleurum Species: Scientific Evaluation and Clinical Applications, edited by Sheng-Li Pan

Volume 8
Herbal Principles in Cosmetics: Properties and Mechanisms of Action, by Bruno Burlando, Luisella Verotta, Laura Cornara, and Elisa Bottini-Massa

Volume 9
Figs: The Genus Ficus, by Ephraim Philip Lansky and Helena Maaria Paavilainen

Volume 10
Phyllanthus Species: Scientific Evaluation and Medicinal Applications edited by Ramadasan Kuttan and K. B. Harikumar

Volume 11
Honey in Traditional and Modern Medicine, edited by Laïd Boukraâ

Volume 12
Caper: The Genus Capparis, Ephraim Philip Lansky, Helena Maaria Paavilainen, and Shifra Lansky

Volume 13
Chamomile: Medicinal, Biochemical, and Agricultural Aspects, Moumita Das

Traditional Herbal Medicines for Modern Times

Chamomile
Medicinal, Biochemical, and Agricultural Aspects

Moumita Das

CRC Press
Taylor & Francis Group
Boca Raton London New York

CRC Press is an imprint of the
Taylor & Francis Group, an **informa** business

CRC Press
Taylor & Francis Group
6000 Broken Sound Parkway NW, Suite 300
Boca Raton, FL 33487-2742

First issued in paperback 2019

© 2015 by Taylor & Francis Group, LLC
CRC Press is an imprint of Taylor & Francis Group, an Informa business

No claim to original U.S. Government works

ISBN-13: 978-1-4665-7759-6 (hbk)
ISBN-13: 978-0-367-37855-4 (pbk)

This book contains information obtained from authentic and highly regarded sources. Reasonable efforts have been made to publish reliable data and information, but the author and publisher cannot assume responsibility for the validity of all materials or the consequences of their use. The authors and publishers have attempted to trace the copyright holders of all material reproduced in this publication and apologize to copyright holders if permission to publish in this form has not been obtained. If any copyright material has not been acknowledged please write and let us know so we may rectify in any future reprint.

Except as permitted under U.S. Copyright Law, no part of this book may be reprinted, reproduced, transmitted, or utilized in any form by any electronic, mechanical, or other means, now known or hereafter invented, including photocopying, microfilming, and recording, or in any information storage or retrieval system, without written permission from the publishers.

For permission to photocopy or use material electronically from this work, please access www.copyright.com (http://www.copyright.com/) or contact the Copyright Clearance Center, Inc. (CCC), 222 Rosewood Drive, Danvers, MA 01923, 978-750-8400. CCC is a not-for-profit organization that provides licenses and registration for a variety of users. For organizations that have been granted a photocopy license by the CCC, a separate system of payment has been arranged.

Trademark Notice: Product or corporate names may be trademarks or registered trademarks, and are used only for identification and explanation without intent to infringe.

Library of Congress Cataloging-in-Publication Data

Das, Moumita.
  Chamomile : medicinal, biochemical, and agricultural aspects / author: Moumita Das.
    pages cm. -- (Traditional herbal medicines for modern times ; 13)
  Includes bibliographical references and index.
  ISBN 978-1-4665-7759-6 (alk. paper)
  1. German chamomile. 2. German chamomile--Therapeutic use. 3. Herbs--Therapeutic use. I. Title. II. Series: Traditional herbal medicines for modern times ; v. 13.

  RS165.C24D37 2014
  615.3'21--dc23   2014004189

Visit the Taylor & Francis Web site at
http://www.taylorandfrancis.com

and the CRC Press Web site at
http://www.crcpress.com

Contents

Series Preface xv
Preface xvii
Acknowledgments xix
Author xxi

Chapter 1 Introduction to Chamomile 1

    1.1 Introduction 1
    1.2 Chamomile as Food 3
    1.3 Chamomile as a Cosmetic Ingredient 4
    1.4 Other Uses of Chamomile 4
    1.5 Medicinal Uses of Chamomile 4
        1.5.1 Use of Chamomile in the Traditional System of Medicine 5
            1.5.1.1 Preparation of Tea 5
            1.5.1.2 Preparation of Tincture and Extract 5
            1.5.1.3 Essential Oil 5
            1.5.1.4 Disease Conditions in Which Chamomile Is Used 7
            1.5.1.5 Herbal Formulations of Chamomile 10
            1.5.1.6 Pharmacopoeias and Monographs 10
            1.5.1.7 Quality, Safety, and Efficacy Issues of Chamomile Herbal Formulations 16
            1.5.1.8 Methods to Ensure Quality, Safety, and Efficacy of Chamomile Drugs 18
            1.5.1.9 Good Manufacturing Practices for Herbal Formulations per World Health Organization Guidelines 20
            1.5.1.10 National Regulations for Herbal Formulations 21
        1.5.2 Homeopathy 23
            1.5.2.1 Disease Conditions 24
            1.5.2.2 Formulations in Homeopathy 25
            1.5.2.3 Homeopathic Pharmacopoeia 26
            1.5.2.4 Quality Issues of Chamomile Homeopathic Formulations 28
            1.5.2.5 Methods to Improve Quality, Safety, and Efficacy of Homeopathic Drugs 29
            1.5.2.6 Guidelines for Quality, Safety, and Efficacy of Homeopathic Drugs 30
            1.5.2.7 National Regulations on Homeopathy System of Medicine 32

		1.5.3	Unani System of Medicine	35
			1.5.3.1 Disease Conditions	35
			1.5.3.2 Formulations of Chamomile	35
			1.5.3.3 Unani Pharmacopoeia	38
			1.5.3.4 Quality Issues of Unani Formulations	39
			1.5.3.5 Quality Control of Unani Formulations	39
			1.5.3.6 National Legislation in India	41
	References			42

Chapter 2 Chamomile: Botany 49

2.1 Introduction 49
2.2 Common Names 49
2.3 Occurrence 49
2.4 Botanical Classification and Nomenclature of Chamomile 50
    2.4.1 Linnaeus on Chamomile 52
        2.4.1.1 *Hortus Cliffortianus* 54
        2.4.1.2 *Flora Suecica* (First Edition) 55
        2.4.1.3 *Species Plantarum* (First Edition) 55
        2.4.1.4 *Flora Suecica* (Second Edition) 56
        2.4.1.5 *Species Plantarum* (Second Edition) 57
        2.4.1.6 Differences in Classification 57
        2.4.1.7 English Translation of Linnaeus's Work 57
    2.4.2 Nomenclature 58
    2.4.3 *Flora Europea* 59
    2.4.4 Flora of Great Britain and Ireland 60
    2.4.5 Taxonomic Key 60
    2.4.6 Classification Based on New Taxonomic Tools 61
2.5 Anatomical Characteristics of Chamomile 61
    2.5.1 Stem 61
    2.5.2 Leaf 62
    2.5.3 Peduncle 63
    2.5.4 Inflorescence or Capitulum 63
    2.5.5 Roots 63
2.6 Reproductive Biology 64
    2.6.1 Stages of Flowering 64
2.7 Cytology and Ploidy Effects 64
2.8 Physiology 66
    2.8.1 Germination of Chamomile Seeds 67
    2.8.2 Effect of Light 67
    2.8.3 Effect of Temperature 67
    2.8.4 Effect of Photoperiod and Diurnal Rhythms of
        Secondary Metabolites 68
    2.8.5 Effect of Nitrogen, Potassium, and Phosphorus 68

Contents	vii

        2.8.6   Effect of Calcium, Sulfur, and Magnesium 68
        2.8.7   Effect of Indole Acetic Acid, Gibberellin, Kinetin, and Other Growth Regulators 69
        2.8.8   Water Stress 70
        2.8.9   Salt Stress 70
        2.8.10 Heat Stress 72
        2.8.11 Toxic Metals 72
        2.8.12 Effect of Inorganic and Organic Fertilizers and Mycorrhiza 74
        2.8.13 Effect of Weedicide 75
        2.8.14 Biotic Stress 75
        2.8.15 Effect of Biopesticides 75
  2.9   Biochemistry 78
        2.9.1   Enzymes 78
        2.9.2   Polysaccharides and Pectic Acid 79
        2.9.3   Chamazulene Synthesis 79
        2.9.4   Biosynthesis of Other Sesquiterpenes 79
        2.9.5   Biosynthesis of Coumarins 80
  2.10 Descriptors of Chamomile 80
        2.10.1 Need of Descriptors 80
        2.10.2 Definition and Types of Descriptor 80
        2.10.3 Chamomile Descriptors: Morphology, Yield of Flower and Oil, and Chemical Constituents of Oil 81
References 89

Chapter 3 Chamomile: Medicinal Properties 99

  3.1   Introduction 99
  3.2   Anti-Inflammatory Effects of Chamomile 100
        3.2.1   Anti-Inflammatory Effect of Chamomile Extract 101
        3.2.2   Anti-Inflammatory Effect of Chamazulene 102
        3.2.3   Anti-Inflammatory Effect of (−)-α-Bisabolol 103
        3.2.4   Anti-Inflammatory Effect of *Trans*-En-Yn-Dicycloether 104
        3.2.5   Anti-Inflammatory Effect of Flavonoids 104
                3.2.5.1   Anti-Inflammatory Effect of Apigenin 106
                3.2.5.2   Anti-Inflammatory Effect of Quercetin 106
                3.2.5.3   Anti-Inflammatory Effect of Luteolin 109
  3.3   Chamomile as an Antioxidant 111
        3.3.1   Antioxidant Property of Chamomile 111
                3.3.1.1   Chamazulene 111
                3.3.1.2   (−)-α-Bisabolol 112
                3.3.1.3   Chamomile Extract and Essential Oil 112
                3.3.1.4   Polysaccharides 113

3.4	Antinociceptive Effect of Chamomile		113
	3.4.1	Pain-Relieving Properties of Chamomile	113
3.5	Chamomile and Wound Healing		114
	3.5.1	Wound Healing and Ulcer-Protective Property of Chamomile	114
		3.5.1.1 Chamomile Extract	114
		3.5.1.2 (−)-α-Bisabolol	115
		3.5.1.3 Chamomile Oil	116
		3.5.1.4 Others	116
3.6	Chamomile as an Immunomodulator		117
	3.6.1	Immunomodulating Effect of Chamomile	117
3.7	Antimutagenic Effect of Chamomile		118
	3.7.1	Antigenotoxic and Tumoricidal Effect of Chamomile	118
3.8	Anticancer Effect of Chamomile		119
	3.8.1	Anticancer Property of Chamomile	119
3.9	Chamomile for Treating Insomnia		120
	3.9.1	Chamomile as a Sedative	121
3.10	Chamomile for Relieving Stress		122
	3.10.1	Anxiolytic and Stress-Reducing Properties of Chamomile	122
3.11	Neuroprotective Effect of Chamomile		123
	3.11.1	Neuroprotective Properties of Chamomile	123
3.12	Effect of Chamomile on Morphine Withdrawal Syndrome		124
	3.12.1	Effect of Chamomile Extract on Morphine Dependence and Abstinence	125
3.13	Anxiolytic Effect of Chamomile		125
	3.13.1	Anxiolytic Properties of Chamomile	126
3.14	Chamomile as Psychostimulant and Its Effect on Attention Deficit Hyperactivity Disorder		126
	3.14.1	Psychopharmacological Effect of Chamomile	126
3.15	Effect of Chamomile on Some Dermatological Problems		127
	3.15.1	Treatment of Dermatological Problems with Chamomile	127
3.16	Phytoestrogenic Effect of Chamomile		128
	3.16.1	Chamomile as Phytoestrogen	128
3.17	Chamomile and Cardiovascular Disease		128
	3.17.1	Effect of Chamomile on Some Cardiovascular Disease Conditions	129
3.18	Chamomile and Oral Health		129
	3.18.1	Effect of Chamomile on Oral Disease Conditions	129
3.19	Treatment of Colicky Infants with Chamomile		130
	3.19.1	Effect of Chamomile on Colicky Infants	130

Contents ix

        3.20 Spasmolytic Effect of Chamomile 130
                3.20.1 Antispasmodic Effects of Chamomile 130
        3.21 Hepatoprotective Effect of Chamomile 131
                3.21.1 Hepatoprotective Properties of Chamomile 131
        3.22 Nephroprotective Effect of Chamomile 132
                3.22.1 Effect of Chamomile on Nephrotoxicity 132
        3.23 Antileishmania Effect of Chamomile 133
                3.23.1 Inhibitory Effect of Chamomile on *Leishmania* 133
        3.24 Antibiotic Effect of Chamomile 133
                3.24.1 Antibiotic Effect of Chamomile Extracts 133
                3.24.2 Antibiotic Effect of Chamomile Essential Oil 134
        3.25 Effect of Chamomile against Anisakiasis 135
                3.25.1 Chamomile as a Potential Cure for Anisakiasis 135
        3.26 Anticulex Effect of Chamomile 135
                3.26.1 Inhibitory Effect of Chamomile against *Culex* 136
        3.27 Chamomile and Pregnancy 136
        3.28 Chamomile and Pest Management 136
        3.29 Chamomile Use in Poultry and Animal Husbandry 136
        3.30 Chamomile as Veterinary Medicine 137
        3.31 Toxicity and Allergenicity of Chamomile Compounds 138
        3.32 Pharmacokinetic Studies and In Vivo Skin Penetrations of
              Chamomile Compounds 139
        References 139

Chapter 4 Chamomile Oil and Extract: Localization, Chemical
Composition, Extraction, and Identification 155

        4.1 Introduction 155
        4.2 Localization and Content of the Chamomile Essential Oil 156
              4.2.1 Essential Oil Content in Flowers 156
                      4.2.1.1 Variation in Essential Oil Content in
                                 Different Locations 157
                      4.2.1.2 Variation in Essential Oil Content in
                                 Different Stages of Flowering 157
                      4.2.1.3 Variation in Essential Oil Content Due to
                                 Farming Practices 157
                      4.2.1.4 Variation in Composition of Organ-
                                 Specific Essential Oil 158
        4.3 Discovery of the Compounds in the Essential Oil 159
        4.4 Chemical Composition of the Flower Essential Oil 160
              4.4.1 Major Compounds 163
                      4.4.1.1 Chamazulene and Its Precursors 163
                      4.4.1.2 Bisabolols and Bisabolol Oxides 164

		4.4.1.3	(E)-β-Farnesene	165
		4.4.1.4	Spiroethers	165
	4.4.2	Minor Compounds		166
		4.4.2.1	Germacrene D	166
		4.4.2.2	(E)-Nerolidol	166
		4.4.2.3	Farnesol	166
		4.4.2.4	Spathulenol	167
		4.4.2.5	n-Hexanol	167
		4.4.2.6	Isoborneol	168
		4.4.2.7	Nerol	168
		4.4.2.8	Phytol	168
		4.4.2.9	Limonene	169
		4.4.2.10	Camphene	169
		4.4.2.11	α-Terpineol	170
		4.4.2.12	Chamomillol	170
		4.4.2.13	β-Elemene	170
		4.4.2.14	α-Humulene	171
		4.4.2.15	Hexadec-11-yn-13,15-diene	171
		4.4.2.16	β-Bourbonene	172
		4.4.2.17	τ-Cadinol	172
4.5	Physical Properties of the Flower Essential Oil			173
4.6	Chemical Composition of the Herb Essential Oil (Stem and Leaves)			173
4.7	Chemical Composition of the Root Essential Oil			173
4.8	Comparative Account of the Essential Oils of Various Parts of Chamomile Plant			174
4.9	Chemical Composition of Chamomile Flower Extract			174
	4.9.1	Major Compounds		185
		4.9.1.1	Apigenin	185
		4.9.1.2	Luteolin	185
		4.9.1.3	Quercetin	185
		4.9.1.4	Herniarin	185
		4.9.1.5	Umbelliferone	189
	4.9.2	Other Compounds		190
		4.9.2.1	Polysaccharides	190
		4.9.2.2	Lipidic and Ceraceous Substances	190
		4.9.2.3	Phenyl Carboxylic Acids	190
4.10	Extraction Methods			190
	4.10.1	Infusion		191
	4.10.2	Steam Distillation		191
	4.10.3	Solvent Extraction		192
	4.10.4	Supercritical Carbon Dioxide Fluid Extraction		193
	4.10.5	Vacuum Headspace		194

Contents

     4.10.6 Solid-Phase Extraction 194
     4.10.7 Solid-Phase Microextraction 195
  4.11 Identification Methods of Chemical Compounds in
     Chamomile Oil and Extract..195
     4.11.1 Thin-Layer Chromatography 195
     4.11.2 Thin-Layer Chromatography–Ultra Violet
         Spectrometry.. 196
     4.11.3 Spectrocolorimeter 196
     4.11.4 High-Performance Thin-Layer Chromatography 196
     4.11.5 High-Performance Liquid Chromatography.............. 196
     4.11.6 Overpressured-Layer Chromatography 197
     4.11.7 Ultra-Performance Liquid Chromatography 197
     4.11.8 Capillary Electrochromatography................................ 197
     4.11.9 Gas Chromatography-Mass Spectrometry................ 197
     4.11.10 Nuclear Magnetic Resonance Spectroscopy............. 198
  References...198

Chapter 5 Genetics and Breeding of Chamomile..209

  5.1 Introduction...209
  5.2 Variability in Chamomile Population.......................................209
     5.2.1 Variation in Oil Content... 211
     5.2.2 Variation in Oil Composition....................................... 211
     5.2.3 Chemical Types of Chamomile 215
     5.2.4 Chemotypes Based on Flower Extract........................ 217
     5.2.5 Variability Due to Ontogeny 218
     5.2.6 Variability Due to Environmental Conditions............. 218
  5.3 Genetics 220
     5.3.1 Genetics of Chamazulene and Bisaboloids 220
     5.3.2 Molecular Techniques to Detect Genetic Variation
        in Chamomile 221
     5.3.3 Patents on Cloned Genes 222
  5.4 Breeding..223
  5.5 Comparative Account of Tetraploid and Diploid Varieties
     of Chamomile 227
     5.5.1 Morphological Characters 227
     5.5.2 Determination of Ploidy Levels....................................228
     5.5.3 Compounds in the Essential Oil 229
     5.5.4 Compounds in the Extract 230
  5.6 Biometrics...230
  5.7 Micropropagation of Chamomile 232
     5.7.1 Essential Oil in Cultures...232
     5.7.2 Cryopreservation 233
  References...234

Chapter 6	Cultivation of Chamomile	243
	6.1 Introduction	243
	6.2 Habitat and Climatic Conditions	243
	6.3 Soil Conditions	245
	6.4 Seeds	246
	6.5 Sowing	247
	6.5.1 Direct Sowing	248
	6.5.2 Indirect Sowing	248
	6.5.3 Self-Sowing	248
	6.5.4 Field Preparation	249
	6.5.5 Germination	249
	6.5.6 Transplantation	249
	6.6 Fertilization	249
	6.6.1 Inorganic Fertilizers	249
	6.6.2 Vermicompost	251
	6.6.3 Combined Fertilizers	252
	6.6.4 Physiological Treatments	253
	6.6.5 Vascular Arbuscular Mycorrhiza	253
	6.6.6 Rhizobacteria	254
	6.7 Irrigation	254
	6.8 Weeding	254
	6.9 Herbicide Resistance	256
	6.10 Pest Control	256
	6.11 Harvesting	256
	6.12 Yield	258
	6.13 Postharvest Treatment	258
	6.14 Fitness for Rotation with Other Crops	261
	6.15 Hydroponic Culture	261
	References	261
Chapter 7	Chamomile: Patents and Products	269
	7.1 Chamomile Patents	269
	7.1.1 Oral Medicine Containing Chamomile	269
	7.1.2 Topical Use of Chamomile	270
	7.1.3 Medical Devices	273
	7.1.4 Oral Hygiene	274
	7.1.5 Diapers	274
	7.1.6 Feminine Hygiene	274
	7.1.7 Food	275
	7.1.7.1 Candies	275
	7.1.7.2 Flavoring	275
	7.1.8 Beverages	275
	7.1.9 Perfume	277

	7.1.10 Cosmetics	277
	7.1.11 Insecticide	279
	7.1.12 Trichomonadicidal	279
	7.1.13 Veterinary Medicine	279
	7.1.14 Animal Feed Supplement	280
	7.1.15 Chamomile Variety	280
	7.1.16 Equipment	280
7.2	Chamomile Products	280
References		283
Index		287

Series Preface

Chamomile, *Matricaria recutita* L., is considered to have originated in the Near East and south and east Europe. It is found almost all over Europe, in Western Siberia, Asia Minor, the Caucasus Mountains, Iran, Afghanistan, and India and cultivated in many of these countries and introduced for this purpose into North and South America, Australia, and New Zealand. This information and the medicinal uses are reviewed.

In the Unani system of medicine, chamomile (called Babuna) has been used since ancient times. Today, throughout the world the flowers are used in the form of a simple tea (tisane) as a gentle medicine for colicky babies and for adults with mild upset stomach or symptoms of mild stress. Extracts of the flowers and also the essential oil distilled from them provide other formulations that extend the range of medicinal benefits through antioxidant, anti-inflammatory, antifungal, and antibacterial activities. Several studies have indicated that chamomile has potential anticancer activity. Other uses are in cosmetics and as a flavoring agent in foods, beverages, bakery products, ice cream, and tobacco.

On the basis of the composition of the essential oil isolated from the flowers, there are now chemical types documented all over the world. Dr. Moumita Das has researched this volume in relation to her native country, India. She describes the chemical nature of the essential oils available from, and their distribution in, the leaves, stems, and roots. Of course, as a single plant it is very small and is usually cultivated in many countries by two sowings of seed in spring and autumn.

Dr. Das has concentrated on the genetic makeup of her Indian chamomile and also described the breeding that has been carried out for the development of high yielding varieties. The associated tissue culture methods and biotechnology required to generate the desired varieties are given in this book.

For temperate crops such as chamomile, India makes use of its Himalayan range providing lower slopes with an appropriate climate. Intercropping is often used throughout the country, to the benefit of two crops growing in proximity, and this too is sometimes applied to chamomile. Dr. Das includes cultivation and agronomic practices, such as layout of the field, choice of fertilizers, irrigation, harvesting and storage, and postharvest technology, for essential oil extraction. The selection of suitable cultivars for India has been researched by her and breeding programs have also been developed to maximize the crop's benefit to the Indian population.

Relevant inventions, classified according to usage, conclude this book.

Throughout the manuscript, Dr. Das has commented on the latest data and in doing so has supplied 1165 references.

I must thank Dr. Das for showing such admirable persistence and determination to complete her very good book on time.

Roland Hardman
Series Editor

Preface

Chamomile is a fascinating plant. It has been used as a medicinal plant since time immemorial in Europe and the countries of the Middle East. The traditional systems of medicine in these countries list chamomile as a very important plant and the spectrum of disease conditions in which it is used is simply mind-boggling. In the American, Australian, and Asian countries, the use of chamomile is comparatively recent, but growing. Apart from its medicinal properties, chamomile has aromatic properties and is extensively used in flavoring and perfumery. There is, without a doubt, a growing demand for this plant. In this scenario, it is only appropriate that there should be an updated ready reference for the researchers, cultivators, and entrepreneurs who wish to work with chamomile.

The genesis of this work lies in my doctoral work. I was working on the breeding of chamomile for my PhD and I had included a short monograph of this plant in my thesis in 1999. My interest in chamomile sustained for more than a decade after that, and I thought of transforming that monograph into a book, with the hope that it would be useful to its readers. In this book, I have made an attempt to collate the advances in the research on chamomile that have been carried out by innumerable botanists, taxonomists, chemists, biotechnologists, plant breeders, and other associated researchers. The contents of this book include the various uses of chamomile; botanical, chemical, biotechnological, and cultivation aspects; and patents and products of chamomile. This information, I hope, will be of interest to the readers.

Acknowledgments

This work has been possible because of the help, support, and encouragement of many friends and well-wishers, and I take this opportunity to express my gratitude to them.

At the outset, I thank Dr. Roland Hardman for his gentle mentoring since the inception of this book. He kept me motivated with a generous supply of references from time to time while I was writing.

I extend my thanks to Dr. Ramesh Kumar for his kind help with the information on the patents on chamomile.

I thank the editorial team of CRC Press/Taylor & Francis Group for their support. I also thank John Sulzycki, Barbara Norwitz, Katherine Everett, Cheryl Wolf, and the editors for ensuring that I was facilitated with everything I needed during the publication process.

I express my gratitude to the people who mentored and facilitated me during the time I was doing my PhD work, because of whom today it has been possible for me to write this book. I thank Dr. Sushil Kumar, my PhD guide, for facilitating me in every way possible during my research work. My sincere thanks to Dr. M. Gopal Rao, whom I consider my chemistry mentor, without whose guidance the identification of the chemical compounds in chamomile would not have been possible. I thank Dr. S. Rajeswari, who helped me to develop descriptors for chamomile. I thank my friends Dr. Ajai Prakash Gupta, Dr. Suphala Bajpai, Dr. Manish Kulshreshtha, Dr. Parul Singhal, Dr. Ritika Gupta, B. Sandhya Rani, and Baby Yohannan for their friendship, support, and encouragement during my research.

I gratefully acknowledge my other friends and well-wishers, who have motivated me and helped me directly or indirectly while I was writing this book, whose names I am not able to mention here because of space constraints.

I cannot thank enough my family members for their unwavering encouragement and support. My parents, sisters Pushpita and Sushmita, and brothers-in-law S. Kalyanaraman and K. Rajesh have always been there when I needed them.

Author

Moumita Das earned a PhD in botany from the Central Institute of Medicinal and Aromatic Plants, Lucknow, India. She has published several research papers and articles in the areas of medicinal plants, intellectual property rights, and sustainable development in journals of national and international repute. She has also published a book. Dr. Das is presently working as assistant director at the National Centre for Innovations in Distance Education, Indira Gandhi National Open University, New Delhi, India.

1 Introduction to Chamomile

1.1 INTRODUCTION

Chamomile (*Matricaria recutita* L.), commonly known as German chamomile, is an important medicinal and aromatic plant. The plant belongs to the daisy (Asteraceae) family and the flowers have a characteristic herbaceous fragrance. The flowers are actually not individual flowers but inflorescences. Throughout this book, the word *flowers* would be used to denote the *inflorescences* or *capitula*.

The name *Chamomile* is derived from two Greek words: *Khamai* meaning "on the ground" and *melon* meaning "apple." Pliny the Elder mentioned that the plant has an apple-like smell (Franke 2005), and the name is attributed to the Roman chamomile, the flowers of which have an apple-like aroma (Hanrahan and Frey 2005; The Columbia Encyclopedia 2012). The Roman chamomile (*Chamaemelum nobile* [L.], earlier known as *Anthemis nobilis* [L.]), also belongs to the family Asteraceae and looks similar to the German chamomile. However, there are morphological differences between the flowers of the Roman and German chamomile. Further, the essential oil and chemical constituents of German chamomile and Roman chamomile are markedly different (Mann and Staba 1986). Consequently, their properties and uses are quite different. Therefore, the knowledge of the differences between Roman and German chamomile is useful. However, the discussion of Roman chamomile is beyond the scope of this book as it deals exclusively with the German chamomile and henceforth, the name "chamomile" will be used to denote German chamomile throughout the book.

Chamomile is commonly known by different names all over the world, such as chamomile, flos chamomillae, German chamomile, Hungarian chamomile, *Matricaria* flowers, pinheads, sweet false chamomile, true chamomile, wild chamomile, and Babuna (WHO 1999). Carl Linnaeus made the earliest attempt to systematically classify chamomile and give it the botanical name—*Matricaria*. The name *Matricaria* was chosen by Linnaeus perhaps due to its wide use in treating gynecological diseases, or "diseases of the womb (matrix)" (Franke 2005). The species name attributed by Linnaeus in 1753 (Linnæi 1753) came into controversy and since then, several taxonomists have been working on the correct nomenclature of chamomile.

Chamomile originated in Europe and West Asia (Bisset 1994) and since ancient times, it has been highly valued by the Egyptians, Romans, and Greeks for its medicinal properties. The Egyptians considered the plant sacred and believed it was a gift from the God of the Sun (Salamon 1993). The Saxons considered chamomile as one of the nine sacred herbs and the Egyptians dedicated the plant to the sun god Ra

(Hanrahan and Frey 2005). It is revered so highly in Slovakia that there is a saying that one must bow to the chamomile plant if one comes across it (Salamon 2004, 2007).

It has been used since the time of Hippocrates, the father of medicine, in 500 BCE. The ancient Greeks, Egyptians, and Romans regularly used the chamomile flowers to treat erythema and xerosis caused because of dry weather (Baumann 2007) and as a calming beverage in the form of tea or tisane (Lissandrello 2008). Several eminent scientists of ancient times, such as Hippocrates, Pliny, Dioscorides, Galen, and Asclepiades studied the plant and passed on their knowledge to the subsequent generations through their writings (Salamon 1993). Hippocrates described chamomile as a medicinal plant and chamomile tea was recommended by Galen and Asclepiades (Carle and Gomma 1991/92). During the same period, Mathiolus/Peter Ondej Mathioli described chamomile in his Latin herbarium (Salamon 1993), where he listed the essential oil of chamomile as a remedy against spasms (Carle and Gomaa 1991/92).

Chamomile came into widespread use during the medieval age. It was extensively prescribed by the doctors of the sixteenth and seventeenth centuries for intermittent fevers (Antonielli 1928, as cited in Singh et al. 2011). In 1593, Bock described that chamomile flowers were used in all kinds of medicines and in 1664, Tabernaemontanus reported that chamomile was used in medicine in the form of plasters, ointments, pouches, and medicinal baths. In 1488, Saladin von Asculum described the blue oil of chamomile for the first time. In 1500, Heironimus Brunschwig described the distillation of chamomile oil (Franke 2005).

The chamomile flowers, on hydrodistillation, yield a blue oil that finds extensive use in medicines, cosmetics, and foodstuff. An extract of flowers is also prepared using water, alcohol, and various other solvents. The essential oil and flower extracts contain about more than 120 secondary metabolites, such as chamazulene, (−)-α-bisabolol, apigenin, and luteolin, and many of these are pharmacologically active. Some of these compounds in the essential oil and the extracts are also used in perfumery and flavoring. The flower extract and essential oil possess anti-inflammatory (Al-Hindawi et al. 1989; Carle 1990; Carle and Gomma 1991/92) spasmolytic, carminative, antiseptic (Chetvernya 1986), sedative (Fundario and Cassone 1980), and ulcer protecting (Zaidi et al. 2012) properties. The flower extracts and the essential oil are therefore the ingredients of several folk and traditional herbal remedies, and other complementary and alternative systems of medicine (CAM), such as Homeopathy and Unani. Today, the ethnobotanical knowledge and use of chamomile is extensive and worldwide. It is evident through many studies carried out to explore its use as folk remedies.

Smitherman et al. (2005) studied the use of chamomile as folk remedies among urban African-American community in Michigan. They found that the knowledge of folk remedies existed in the community and was used extensively by the caregivers. They also found that chamomile tea for treating colic was practiced by the caregivers. Zucchi et al. (2013) carried out an ethnobotanical survey on the use of chamomile in Ipameri, Brazil. They found that chamomile was one of the most used plants for medicinal purposes. Šavikin et al. (2013) found in an ethnobotanical study that in southwestern Serbia, chamomile was one of the most commonly used plants for medicinal purposes. Raal et al. (2013) reported that in Estonia, chamomile flowers were frequently used as a self-medication to treat cold and flu. Malik et al. (2013) reported the increasing use of chamomile in homeopathy in Pakistan.

Traditionally, chamomile flowers have been handpicked and used. Gradually, the plants were cultivated for use. Chamomile has been said to be brought into cultivation during the Neolithic period approximately from 9000 to 7000 BC (Salamon 1993). However, the yield of these cultivated plants was low in terms of flowers, oil content, and oil quality. To develop high-yielding varieties, the first attempts in breeding were made a little over 50 years ago (Franke and Schilcher 2007). Several high-yielding varieties were developed, such as Manzana, Degumille, Bodegold, Zloty Lan, Bona, and Goral. Today, chamomile cultivation and use has spread to almost all parts of the world, and the chief producers of chamomile are Argentina, Egypt, Germany, France, Italy, Turkey, Greece, Bulgaria, Yugoslavia, Hungary, Slovakia, and Australia. It was estimated that in 2003, about 50,000 acres were under chamomile cultivation worldwide (Brester et al. 2003).

The worldwide demand and consumption of chamomile is high. The annual consumption of chamomile flowers in Germany alone in 1992 was reported to be about 5000 t (Franz 1992). In Australia, the demand is estimated to be above 50 t (Purbrick and Blessing 2007). In Italy, the demand for chamomile amounts to 1000–1200 t annually, worth €3 million (CBI Ministry of Foreign Affairs 2011). With the increasing use of chamomile formulations in herbal medicines and homeopathy and their increased availability as the over-the-counter drugs, the demand for chamomile is increasing in the South American countries such as Brazil (Freitas et al. 2012), Colombia (Gómez-Estrada et al. 2011), Chile (Burgos and Morales 2010), and Peru (Huamantupa et al. 2011). As the demand for chamomile-based products is increasing, countries such as India and Tasmania (Falzari et al. 2007) are planning to take up chamomile cultivation in a big way.

Considering the over-the-counter availability of chamomile-based drugs and increasing self-medication, the quality, safety, and efficacy of these drugs have achieved paramount importance. Needless to say, there are several issues in the quality, safety, and efficacy of the drugs. These issues are discussed in Sections 1.5.1.7, 1.5.2.4, and 1.5.3.4.

Chamomile flower extracts and oil are being used not only in medicinal formulations and aromatherapy (Buchbauer 1996), but also as foodstuffs for flavoring, in cosmetics (Mann and Staba 1986), cosmeceuticals (Thornfeldt 2005; Baumann 2007; Padma Preetha and Karthika 2009), and dyeing (Çaliş and Yücel 2009). Further, new uses of chamomile have been discovered in diverse areas, such as insect and pest repellence, veterinary, and reclamation of polluted soils. With such evidence, chamomile has proved to be a very important plant for the betterment of our society.

1.2 CHAMOMILE AS FOOD

The chamomile flowers are taken as tea, tisane, and herb beer (Chamomile 2006; McKay and Blumberg 2006). In fact, chamomile tea is one of the most popular herbal teas of the world and almost a million cups are consumed every day

(Srivastava et al. 2010). In addition to tea, the fresh flowers of chamomile can be taken as salads and drinks. The dried flowers can be used as an ingredient in soups and salads to improve the flavor and enhance the nutritional value. Drinks made from chamomile flowers such as chamomile lemonade could be a refreshing drink in summer (Lissandrello 2008).

Not only the whole flowers, but also the essential oil and the extract of the flowers are used in food. The aqueous ethanol extract of chamomile flowers can contain up to 20% of the original dry matter, including 20%–25% of the mineral components and 5% of the free amino acids. It can be added to all kinds of foodstuffs and drinks. Chamomile flower extracts and oil are used to render the characteristic flavor (Furia and Bellanca 1975; Gasič et al. 1989) to the items of food, confectionery, alcoholic and nonalcoholic beverages, and tobacco. Chamomile oil is used for coloring foods as well (Mann and Staba 1986; Emongor et al. 1990).

Because the essential oil of chamomile is antimicrobial, it can used as a coating on various food products (Aliheidari et al. 2013).

1.3 CHAMOMILE AS A COSMETIC INGREDIENT

Various preparations of chamomile based on the whole plant or flower extracts, oil of flower, and compounds isolated from chamomile find uses in cosmetic industries. A variety of skin and hair care preparations contain chamomile flower extracts or oil. The natural antioxidant and antimicrobial effects of chamomile oil renders it a safe cosmetic without any side effects (Khaki et al. 2012). Mouthwashes and toothpastes are also made using chamomile (Laksmi et al. 2011; Nimbekar et al. 2012). Chamomile oil is blended with other essential oils in certain perfumes. The kinds of cosmetics that contain chamomile are creams, ointments, shampoos, soaps, detergents, and perfumes (Mann and Staba 1986).

1.4 OTHER USES OF CHAMOMILE

Chamomile extracts have also found several other uses, such as for mosquito repellence (Thorsell et al. 1970; Thorsell 1988), biological control of pests (Barakat et al. 1985; Mishra 1990), dyeing (Çališ and Yücel 2009), improvement of dairy (Abou Ayana and Gamal El Deen 2011), poultry (Poráčová et al. 2007), and veterinary medicine (Tilford 2004).

The chamomile plant is known for reclamation of sodic soils (Singh 1970) and bioremediation for metals such as cadmium (Chand et al. 2012).

1.5 MEDICINAL USES OF CHAMOMILE

Chamomile is used for medicinal purposes in the traditional system of medicine, homeopathy, and Unani system of medicine. These three systems of medicine are referred to as CAM. These systems use chamomile in specific disease conditions. These systems of medicine have their own unique way of preparing the distinct drug formulations and dispensing them. Each system markets different kinds of drugs.

1.5.1 USE OF CHAMOMILE IN THE TRADITIONAL SYSTEM OF MEDICINE

The use of chamomile for medicinal purposes in the traditional system of medicine is mostly guided by the pharmacopoeias. The pharmacopoeias are authoritative texts that specify how to identify the correct plant, which plant part is to be used, what physical and chemical qualities of the drug are required, and how the drug formulations should be prepared and administered to the patient. These specifications ensure that the drug is effective. In addition to the pharmacopoeias, there are other authoritative texts such as compendiums and monographs that provide guidance on the effective formulations and use of chamomile.

The chamomile flowers used as whole flowers or extracts in alcohol or water are made from the dried flowers. The essential oil of the flower is also used. The whole flowers are used as teas, tinctures, tablets, and compresses. The guidelines in many of the pharmacopoeias specify that the flowers used for medicinal purposes should have a minimum of 0.4% volatile oil content (Schilcher 2005a). The hydroalcoholic extracts of the flowers and the hydrodistilled essential oils are used in creams, lotions, and aromatherapy. The teas, tinctures, and extracts are prepared according to prescribed specifications.

1.5.1.1 Preparation of Tea

Chamomile tea, as described by the German Commission E monograph, is prepared by pouring 150 mL of boiling water to one heaped teaspoon (3 g) of chamomile flowers. It is kept covered for 5–10 minutes and passed through a strainer (Bisset 1994). It has been found that the tea contains the spasm-reducing (spasmolytic) compounds—the flavonoids. The tea, however, cannot reduce internal inflammation because it has extremely low amounts of the anti-inflammatory compounds, which are mainly found in the essential oil. The tea contains only 1%–3% of the essential oil. However, if the tea is externally applied, it reduces inflammation (Schilcher 2005b).

1.5.1.2 Preparation of Tincture and Extract

To prepare tincture or extract, the dried chamomile flowers are homogenized at room temperature in ethanol–water and the liquid is evaporated. For tinctures, the ratio of ethanol to water is kept at 1:5 (Schilcher 2005b). The tincture can be taken in place of tea and is more effective (Bisset 1994). About 10–15 drops of tincture can be added to a glass of tepid water and used for gargle (Salamon 1993).

For extracts, the ratio of ethanol to water is kept at 1:1. The extracts are further dried and concentrated into viscous extracts and added to gels, ointments, and creams. Dry extracts are used to prepare tablets, capsules, and coated pills (Schilcher 2005b).

A high-quality extract should have one of the standard characteristics mentioned in Table 1.1.

1.5.1.3 Essential Oil

The essential oil of chamomile is present in the whole plant. However, the essential oil content is higher in the flowers than in other parts of the plant, and also has higher levels of useful compounds. Therefore, the essential oil of the flowers is mostly used for medicinal and aromatic purposes. The essential oil of chamomile is obtained by the

TABLE 1.1
Standard Characteristics of Chamomile Extract

S. No.	Ethanol–Water Extract (g)	Blue Essential Oil (mg)	(–)-α-Bisabolol (mg)	Chamazulene (mg)	Apigenin 7-Glucoside (mg)
1.	100	150–300	50	3	150–300
2.	100	170	50	—	10–40
3.	100	200	—	—	150

Source: Adapted from Schilcher, H., In R. Franke and H. Schilcher (eds.), *Chamomile: Industrial Profiles,* CRC Press, Boca Raton, FL, 2005b. With permission.

process of steam distillation or hydrodistillation, in which the flowers are subjected to high pressure, temperature, and steam to separate out the essential oil from them. The oil is deep blue or ink blue in color and has a characteristic sweet, grassy smell. It may turn green and then dark brown on oxidation and lose its therapeutic value (Shutes 2012). At the time of distillation, it is extremely concentrated. The quality of the essential oil may differ from one variety of chamomile plant to another, but the various pharmacopoeias clearly mention that to be used for medicinal purpose, the oil content should be 0.4% in the flowers. The oil is prone to vaporization and decomposition, and so it has to be stored carefully in dark bottles under the prescribed temperature.

Essential oils are absorbed into the body on inhalation and through the skin. The compounds penetrate the skin and enter the bloodstream and act as medicines. The essential oil of chamomile is extensively used in aromatherapy, massage, and baths.

1.5.1.3.1 Aromatherapy

Aromatherapy is a technique of healing where the patient is made to inhale the vapors of the essential oil. A few drops of chamomile oil are applied on a piece of cloth or handkerchief or tissue and slowly inhaled. Sometimes a few drops of oil are added to hot water and the steam is inhaled. Chamomile oil vapor is used extensively in aromatherapy to calm a person and reduce pain and anxiety (Ford-Martin and Odle 2005).

1.5.1.3.2 Massage

A massage with oil enables the medicinal compounds to penetrate the skin and enter the bloodstream. For the purposes of massages, the chamomile oil used is diluted with other oils such as olive oil, sunflower oil, or lavender oil. The oil is gently rubbed or massaged on to the inflicted part.

1.5.1.3.3 Bath

In many disease conditions, a hot bath or a cold bath is given to a patient. In hot baths, warm to hot water is used and in cold baths, cold water or ice is used. A few drops of chamomile essential oil are added for healing purposes. Sometimes whole chamomile flowers are put in a small bag and kept in the bath.

Introduction to Chamomile 7

1.5.1.3.4 Compress
A compress is made by steeping a cloth or a towel in a bowl of hot (hot compress) or cold (cold compress) water. A few drops of chamomile oil are added in the water before steeping the cloth or towel. This compress is then applied to the affected part.

1.5.1.4 Disease Conditions in Which Chamomile Is Used
Chamomile is used in the traditional medicines in the treatment of several conditions such as appetite enhancement, asthma, bladder problems, bleeding, blood purifying, bronchitis, callouses, children's diseases, colds, colic, colitis, corns, cramps, dandruff, digestion problems, dizziness, drug withdrawal, earache, eye problems, gas, headache, hemorrhage, hemorrhoids, inflammation, insomnia, jaundice, kidney problems, menstrual problems, migraine, nervous disorders, pain, parasites, spleen disorders, swelling, toothache, worms, and wounds (Salamon 1993). The *Gale Encyclopedia of Alternative Medicine* (2005) lists about 50 disease conditions in humans where chamomile can be used. The homeopathic system (Iyer 1994) and the Unani system (Rashid and Ahmad 1994) also use chamomile to treat a variety of disease conditions. A list of the diseases and the effect of chamomile administration is compiled and provided in Table 1.2.

TABLE 1.2
Disease Conditions and Use of Chamomile

S. No.	Disease/Ailment	Effect of Chamomile
1.	Alcohol withdrawal	Calming and restorative, nerve tonifying
2.	Anorexia nervosa	Reduces anxiety and depression by stimulating appetite, relax the body. The essential oil is used, which is inhaled, massaged, or put in bath water
3.	Anxiety	Reduces phobias and panic disorder by promoting general relaxation of the nervous system. Chamomile acts as an adaptogen by promoting adaptability to stress
4.	Asthma	Promotes free breathing. Essential oil is inhaled to reduce obstruction in the airways
5.	Athlete's foot	Reduces symptoms. Essential oil is applied directly to the toes. The oil can be added to bath water as well
6.	ADHD	Calms the person. Chamomile extract is used for the treatment
7.	Binge eating disorder	Reduces stress, thereby reducing the disorder
8.	Boils	Fights infection. Chamomile is to be applied topically
9.	Bruxism	Prevents grinding of teeth. Chamomile acts as an antispasmodic and central nervous system relaxant. Chamomile is prescribed before going to bed
10.	Bunions	Relieves pain. Promotes wound healing. Used chamomile tea bag is applied to the bunion, which may prove helpful. Massaging with essential oil of chamomile or a cream containing chamomile may provide relief

(Continued)

TABLE 1.2 (*Continued*)
Disease Conditions and Use of Chamomile

S. No.	Disease/Ailment	Effect of Chamomile
11.	Burns	Reduces anxiety. Chamomile tea is recommended
12.	Canker sores	Existing sores are treated with tea. Compresses soaked in the tea are recommended to be applied directly to the mouth. The tea can also be swished around in the mouth for several minutes
13.	Chicken pox	Aids in sleep or promotes sleep
14.	Chills	Prevents chills and cold intolerance. Chamomile tea is recommended
15.	Colic	Reduces bowel inflammation and gas. Chamomile tea is recommended
16.	Conjunctivitis	Prevents discomfort of the eye. An eyewash is made of two to three teaspoons of chamomile flowers added to boiling water to make tea. The tea is cooled. A cool compress is made and put over the eye. Damp tea bags of chamomile may also be used
17.	Constipation	Stimulates movement of digestive and excretory systems
18.	Corns	Thickened skin is dissolved, providing relief. One teaspoon of lemon juice, one teaspoon of dried chamomile flowers, and one crushed garlic clove can be directly applied to dissolve thickened skin
19.	Cradle cap	Oil production of the skin is slowed down. Tannins in the chamomile tea can slow down this process. Chamomile tea can be rubbed onto the skin with a cloth several times per day
20.	Cuts and scratches	Repairs skin damage and encourages new cell growth. Chamomile oil can be sprayed onto the affected area
21.	Dermatitis	Provides relief. Chamomile ointment may be applied to the affected area
22.	Diarrhea	Provides relief. Chamomile infusion to be provided throughout the day
23.	Diverticulitis	Reduces inflammation in cases of uncomplicated diverticulitis
24.	Dry mouth	Stimulates salivary flow. Chamomile tea is recommended
25.	Eczema	Reduces inflammation. Chamomile ointment is used
26.	Epilepsy	Chamomile creates soothing mood. Essential oil is inhaled
27.	Fibromyalgia	Soothes muscle and joint pain. Chamomile tub soak (bath) or compress is recommended
28.	Fractures	Calming effect. Chamomile tea is recommended
29.	Fungal infections	Antifungal. Chamomile tea is used. The used tea bag can be put on the area of infection
30.	Gas	Relieves gas. Chamomile tea is recommended

Introduction to Chamomile

TABLE 1.2 (*Continued*)
Disease Conditions and Use of Chamomile

S. No.	Disease/Ailment	Effect of Chamomile
31.	Gastritis	Counteracts free radicals and inhibits *Helicobacter pylori*. Chamomile tea is used
32.	Heartburn	Provides relief. Chamomile tea is used
33.	Holistic dentistry	Sedative effect and promotes relaxation. Tea or infusion is used
34.	Hypertension	Relieves stress. Essential oil is used
35.	Indigestion	Provides relief. Chamomile tea is used
36.	Inflammatory bowel disease	Reduces inflammation, reduces spasms, antibacterial action. Flowers of chamomile are soaked in water for 10–14 minutes and the tea is taken 3–4 times daily
37.	Insomnia	Promotes sleep. Chamomile tea is used. Putting chamomile flowers inside the pillow is also recommended
38.	Juvenile rheumatoid arthritis	Detoxification of body to reduce symptoms. Chamomile oil massage is recommended
39.	Knee pain	Reduces spasms and swellings. Chamomile tea and oil massage is recommended
40.	Low back pain	Reduces spasms. Chamomile tea and oil massage is recommended
41.	Measles	Reduces restlessness. Chamomile tea is used
42.	Ménière's disease	Promotes relaxation. Chamomile tea and chamomile oil massage is recommended
43.	Menstrual problems	Relieves mood swings, tension, and cramps. Chamomile tea is recommended. Essential oil massage is also recommended
44.	Nausea	Relieves symptoms of nausea and vomiting. Chamomile tea is used
45.	Osteoarthritis	Relieves symptoms. Massage with chamomile oil is recommended
46.	Ear (otitis media)	Reduces the congestion of upper respiratory tract infections
47.	Ovarian cysts	Stimulates blood circulation and healing in ovaries. Compress, made of towels soaked in chamomile oil, wrapped around a hot water bottle is applied to the lower abdomen
48.	Psoriasis	Relief of symptoms. Warm water bath with chamomile flowers is recommended
49.	Radiation injury	Reduces skin inflammation following radiation therapy. Chamomile cream is used
50.	Rashes	Relieves symptoms. Chamomile tea is recommended
51.	Rheumatic fever	Relieves pain. Massage with chamomile oil is recommended

(*Continued*)

TABLE 1.2 (*Continued*)
Disease Conditions and Use of Chamomile

S. No.	Disease/Ailment	Effect of Chamomile
52.	Rosacea	Soothes irritated skin. Cold compress of chamomile tea is recommended
53.	Scarlet fever	Promotes relaxation. Bath with tepid infusion of chamomile is recommended
54.	Stomach ache	Relieves upset stomach, gas, and stomach spasms. Chamomile tea is recommended
55.	Teething problems	Relieves pain. Cloth dampened with chamomile tea is placed in the freezer and used in place of a freezable toy

Source: Salamon, I., *The Modern Phytotherapist*, 13–16, 1993; Ford-Martin, P. and Odle, T. G. in Longe, J.L., ed., *Gale Encyclopedia of Alternative Medicine*, Second Edition, Volume I (A–C). Farmington Hills, MI: Thomson Gale, 2005, p. 123; Cooper, A. in Longe, J.L., ed., *Gale Encyclopedia of Alternative Medicine*, Second Edition, 2005, Encyclopedia.com. http://www.encyclopedia.com/doc/1G2-3435100699.html. Accessed March 8, 2014; Rowland, B. and Odle, T. in Longe, J.L., ed., *Gale Encyclopedia of Alternative Medicine*. 2005. Encyclopedia.com. http://www.encyclopedia.com/doc/1G2-3435100748.html. Accessed March 8, 2014; Turner, J. in J.L. Longe, ed., *Gale Encyclopedia of Alternative Medicine*. 2005. Encyclopedia.com. http://www.encyclopedia.com/doc/1G2-3435100768.html. Accessed March 8, 2014.

1.5.1.5 Herbal Formulations of Chamomile

There are several herbal medicinal formulations of chamomile available in the market in different countries. These are in the form of tablets, capsules, oils, creams, lotions, ointments, soaps, and shampoos. The U.S. Natural Medicine Comprehensive database lists 1154 products containing chamomile (Therapeutic Research Faculty 2012). In Germany, around 150 medicinal preparations are made from chamomile extracts in combination with other plant extracts and at least 18 preparations have the chamomile compounds as the sole ingredient (Schmidt and Vogel 1992). Chamomile drug is mostly sold all over the world with its generic name (Chamomile generic 2012) as well as branded products. There are some branded products as well and some well-known branded formulations of chamomile (World Standard Drug Database 2012). Some of these formulations are listed in Table 1.3.

1.5.1.6 Pharmacopoeias and Monographs

Chamomile is included in the pharmacopoeia of many countries, such as the European, British, French, German, and others (Schilcher 2005a). In addition to the pharmacopoeia, chamomile is also mentioned in the monographs of the German Commission E, European Scientific Cooperative on Phytotherapy (ESCOP), and World Health Organization (WHO).

1.5.1.6.1 European Pharmacopoeia

The *European Pharmacopoeia* is extensively used all over the world as an authoritative reference for manufacturing quality chamomile drug. It provides a description

TABLE 1.3
List of Some Branded Chamomile Formulations

S. No.	Name of the Formulation	Brand	Website
1.	Chamomile ointment	Kamillosan	http://www.medapharma.de/otc/kamillosan/
2.	Carminative tea	Djehuty	http://www.cherryfones.com/carminative-tea.html
3.	Massage balm	Weleda	http://www.weleda.co.uk/aches-+amp-pains/massage-balm-with-calendula-50ml/invt/204004/
4.	Chamomile liquid	HealthAid	http://www.healthaid.co.uk/shopexd.aspx?id = 381
5.	Herbal tea	Jan de Vries	http://www.jandevrieshealth.co.uk/store_main.asp?int_catalog_id = 1&int_category_id = 12&int_subcategory_id = 0&prod = 140
6.	Chamomile flower capsules	Bio-health	http://www.baldwins.co.uk/herbs/capsules/bio-health-chamomile-flowers-250mg-60-vegetarian-capsules
7.	St John's Wort herbal complex	Vega nutritionals	http://www.vegavitamins.co.uk/st-john-s-wort-herbal-complex-prd-256.html
8.	Daily gum and toothpaste	Corsodyl	http://www.corsodyl.co.uk/maintenance/toothpaste.shtml
9.	Sinose spray	Salcura	https://www.salcuraskincare.com/product/sinose/
10.	Elena's trinity soap and shampoo	Elena's Nature Collection	http://elenasnaturecollection.co.uk/our-skin-care-products/#ecwid:category = 1923096&mode = product&product = 8734354

of the morphological and cellular characteristics of the chamomile flowers to enable easy identification using microscopy. It also provides the guidelines to identify the essential oil, methods to detect any adulteration, and guidelines for proper storage, and labeling to ensure the safety of the finished products. An excerpt from the pharmacopoeias is provided below (Schilcher 2005a).

1. Description
    a. Morphological characteristics
        i. Bracts: Involucres in one to three rows, ovate to lanceolate, brownish gray, scarious margin
        ii. Receptacle: Elongated, conical, hollow, without pale
        iii. Ligulate florets: Twelve to ten, white, marginal, elongated ligule, corolla has brownish yellow tube at base
        iv. Tubular florets: Several dozen, yellow, five-toothed corolla tube
        v. Androecium: Five stamens, syngenecious, epipetalous
        vi. Gynoecium: Same in ligulate and tubular florets; ovary inferior, ovoid to spherical; style long; stigma bifid

b. Cellular characteristics
      i. Bracts: Cells thin walled, elongated sclereids in the central region, stomata
      ii. Ligulate florets: Epidermis has thin-walled, polygonal, slightly papillose cells; outer epidermis has sinuous and striated cells
      iii. Tubular florets: Epidermis has longitudinal elongated cells, small groups of papillae present near apex of lobes
      iv. Glandular trichomes: Present on the outer surface of the bract of the corollas of both ligulate and tubular florets, short stalk and head on 2–3 tiers of cells
      v. Androecium: Anther lobes contain calcium oxalate crystals; pollen 30 μm, spherical to triangular with three pores, spiny exine
      vi. Gynoecium: Sclerous ring at the base of the ovary, ovary wall has longitudinally elongated cells with numerous glandular trichomes, alternating with fusiform, radially elongated cells containing mucilage, inner tissues containing calcium oxalate crystals; stigma cells form rounded papilla
2. Essential oil
   a. Essential oil and extract: 4 mL/kg in dried flowers, clear intensely blue viscous liquid, intense, characteristic odor, apigenin-7-glucoside (0.25%), azulenes, levomenol, bornyl acetate, en-yn-dicycloether. Detected by thin layer chromatography (TLC) and gas chromatography–mass spectrometry (GCMS)
3. Storage
   a. The container should be well filled and airtight. The container should be protected from light and stored at a temperature not exceeding 26°C.
4. Labeling
   a. The type of oil should be indicated in the label, such as "rich in bisabolol oxides" or "rich in levomenol."

1.5.1.6.2 *German Commission E Monograph*

The German Commission E monograph provides the botanical name of the chamomile plant. No detailed botanical description of the flowers is provided. It specifies the essential oil content in the flowers used in medicines. It also provides guidelines for the therapeutic uses, dosage, and the mode of administration (Carle and Gomma 1991/92; Schilcher 2005a).

1. Description of the drug
   a. Fresh or dried heads of *M. recutita* L. (synonym: Chamomilla recutita (L.) Rauschert) and their preparation of effective doses.
2. Essential oil
   a. Essential oil and extracts: The flowers should contain not less than 0.4% (v/w) of volatile oil. The main constituents are (−)-α-bisabolol or bisabolol oxides A and B, matricarin, apigenin, and apigenin-7-glucoside

Introduction to Chamomile

3. Therapeutic uses
   a. External Uses: Inflammation of skin and mucous membrane, bacterial diseases of skin, oral cavity and gums, respiratory tract (inhalation of vapors), inflammation of the anogenital region
   b. Internal Uses: Spasms and inflammation of the gastrointestinal region
4. Contraindications
   a. Not known
5. Dosage
   a. For adults
      i. Tea: Boil 3 g of chamomile flowers in 150 mL water. Keep it covered for 5 –10 minutes. Strain with a tea strainer. For gastrointestinal disorders drink three to four times a day between meals. Use as a gargle in case of inflammation of mucous membranes of the mouth or throat
      ii. Poultices: 3%–10% of infusions
      iii. Bath: 50 g of flower in 10 L
      iv. Semisolid preparation: 3%–10% of the preparation should contain chamomile
6. Mode of administration
   a. Liquid and solid preparation for external and internal application

1.5.1.6.3 European Scientific Cooperative on Phytotherapy Monograph

The *ESCOP Monograph* essentially provides the same guidelines as the *European Pharmacopoeia* as far as chamomile is concerned. In addition, it specifies the quality of the essential oil and extract, the therapeutic uses, doses, methods of administration, and contraindications (Schilcher 2005a).

1. Description
   a. Complies with *European Pharmacopoeia*
2. Essential oil
   a. Essential oil and extract
      i. Should contain no less than 4 mg/kg of blue essential oil (0.5%–1.5%)
      ii. Apigenin-7-glucoside (0.5%)
      iii. Sesquiterpenes (bisabolol A, bisabolol oxides A and B, bisabolone oxide A) (50%)
      iv. *cis-* and *trans*-en-yn-dicycloethers (25%)
      v. Matricin (converted to chamazulene) (15%)
      vi. Coumarins (herniarin and umbelliferone), phenolic acids, and polysaccharides (up to 10%)
3. Therapeutic uses
   a. Internal Uses: Gastrointestinal spasms, flatulence
      b. External Uses: Minor inflammations and initiations of skin and mucosa of oral cavity and respiratory tract (inhalation of vapors), and anal and genital region (bath and ointments)

4. Dosage
   a. Internal Use
      i. For adults
         A. Tea: 3 g of the drug is added to 150 mL of hot water, 3–4 times a day
         B. Fluid extract: One part drug to two parts solvent (50% ethanol preferred), 3–6 mL daily
         C. Dry extract: 5–300 mg three times daily
      ii. For elderly: Same as adults
      iii. For children: Proportion of adult dose according to age or body weight
   b. External Use
      i. Compress: 3–10 m/v of infusion, 1% v/v fluid extract, or 5% v/v tincture
      ii. Rinses, Gargle, and Bath: 5 g drug or 0.8 g of extract per liter of water
      iii. Solid/Semisolid Preparation: 3%–10% m/v of extract in the preparation
      iv. Vapor: 10–20 mL alcoholic extract per liter of hot water
5. Method of administration
   a. Oral, local application, and inhalation
   b. No restriction on duration
6. Contraindications
   a. Sensitivity to *Matricaria* or other *Compositae*

1.5.1.6.4 WHO Monograph

The WHO monograph provides a list of common names of chamomile, and botanical and microscopic description of the flowers. It also provides specifications for the essential oil and the purity tests such as those for microbes, pesticide residues, and heavy metals in the drug. It indicates the uses of the drug supported by clinical data (WHO 1999).

1. Definition
   a. The drug is the dry flowering heads of *C. recutita* (L.) Rauschert (synonyms: *M. chamomilla*, *M. recutita* L., *M. suaveolens*). The monograph also provides selected vernacular names such as Babuna, German chamomile, Chamomile, Kamille, and Manzanilla
2. Description of flower heads
   a. Morphological characteristics
      i. Peduncle: Short, up to 2.5 cm long, weak brown to dusky greenish yellow, longitudinally furrowed, more or less twisted
      ii. Receptacle: Conical, narrow, hollow
      iii. Involucre: 20–30 imbricate, oblanceolate and pubescent scales
      iv. Ligulate florets: A few, white, pistillate, three-toothed, four-veined
      v. Tubular florets: Numerous, yellowish orange to pale yellow, perfect, without pappus
      vi. Achenes: Obovoid, faintly, three to five ribbed, no pappus, slightly membranous crown
   b. Cellular characteristics
      i. Receptacles
      ii. Bracteoles: Schizogenous secretory ducts present; phloem fibers present; vessels spiral, annular, reticulate, pitted

Introduction to Chamomile 15

     iii. Androecium: Pollen spherical or triangular, numerous spines
     iv. Gynoecium: Ovaries do not have lignified scale at the base; ovary walls have longitudinal bands of small mucilaginous cells; stigma has elongated papilla at the apex
     v. Glandular hairs: All parts bear glandular hairs with short biseriate stalk and elongated head, several tiered, and each made of two cells

3. Essential Oil
   a. Blue color, not less than 0.4% v/w of essential oil. Total volatile oil is to be determined. For volatile oil, TLC and GLC are to be used. For detection of flavonoids, high-performance liquid chromatography is to be used.

1.5.1.6.5 Compendium of Monographs, Canada

The compendium of monographs of Canada has been developed by Natural Health Products Directorate (NHPD) in 2009 and contains a monograph on chamomile. The monograph provides different names of chamomile and specifies administration, dosage, and risk information. It specifies that the formulations should be made according to the *British Pharmacopoeia*, *European Pharmacopoeia*, and *United States Pharmacopeia* (Health Canada 2009).

1. Proper names
   a. *M chamomilla* L. (Asteraceae) (synonyms: *M. recutita* L.; *C. recutita* L. Rauschert)
2. Common names
   a. German chamomile, Chamomile
3. Source material
   a. Flower
4. Route(s) of administration
   a. Oral, Topical, Buccal (rinse or gargle)
5. Dosage
   a. Oral: Children (2–4 years) and adolescents (5–9 years) should be given 0.3–0.6 g dried flowers per day, and adolescents (10–14 years) and adults ($\geq$14 years) 0.8–24 g dried flowers per day
   b. Topical and buccal: Preparations containing the equivalent of 3%–10% dried flower; 1% v/v fluid extract, 5% v/v tincture
6. Risk information
   a. Caution(s) and warning(s): Consult a health-care practitioner if symptoms persist or worsen
   b. Known adverse reaction(s): Hypersensitivity, such as allergy, has been known to occur in which case, discontinue use
7. Specifications
   a. The finished product must comply with the minimum specifications outlined in the current NHPD Compendium of Monographs. The medicinal ingredient may comply with the specifications outlined in the pharmacopoeial monographs, the *British Pharmacopoeia*, *European Pharmacopoeia*, and *United States Pharmacopeia*.

1.5.1.7 Quality, Safety, and Efficacy Issues of Chamomile Herbal Formulations

The basic raw material of chamomile drugs is the plant, which is mostly cultivated. The issue of quality begins at the cultivation stage. The drugs may be contaminated with pesticides, heavy metals, harmful microbes, and radioactive materials (Salamon and Plačková 2007; Alwakeel 2008). More often than not the drugs and essential oil are found to be adulterated (Martins et al. 2001; Nascimento et al. 2005). Often the packaging is unhygienic, which encourages microbial growth. These issues of quality, safety, and efficacy are described briefly in Sections 1.5.1.7.1 through 1.5.1.7.10.

1.5.1.7.1 Adulteration of Flowers

Chamomile formulations available in the market have been occasionally found to be adulterated with the flowers of other species such as *A. arvensis*, *A. cotula*, *A. montana*, *A. tinctoria*, *C. suaveolens*, *Chrysanthemum leucanthemum*, *M. perforata* (Mann and Staba 1986), *A. nobilis* (Uzma and Khan 1998), *M. aurea*, and *Inula vestita* (Ahmad et al. 2009). Considering the enormous levels of adulteration, the correct botanical identification of the drug assumes great importance. The drug in its powdered form can be microscopically and analytically examined and matched with the description of the drugs as provided in the various pharmacopoeias. Several macroscopic, microscopic, and analytical examination of the powdered chamomile drug have been carried out and further detailed descriptions have been provided (Rashid and Ahmad 1994; Uzma and Khan 1998) to detect adulterations.

1.5.1.7.2 Adulteration of the Essential Oil

The essential oil has been found to be adulterated with the essential oil of the tree *Vanillosmopsis erythropappa*. The oil from this tree is rich in (−)-α-bisabolol and is cheaper than chamomile oil (Leung 1980). Researchers have found that the oil of *M. matricarioides* is also chemically similar to chamomile (Loomis et al. 2004). However, the isotope ratio mass spectrometry (IRMS) study on the oil of *V. erythropappa* shows that the oil has a different chemical composition than chamomile oil. These findings help in detecting and checking adulteration (Verpoorte et al. 2005).

1.5.1.7.3 Pesticide Residues

Chamomile drug samples have been found to contain pesticide residues, which could pose as a health hazard, such as endocrine disruption or even cancer. A study in Egypt found that the chamomile samples contained the pesticide chlorpyrifos at 0.01 mg/kg (Farag et al. 2011). The limit of chlorpyrifos has been set at 0.2 mg/kg by the *European Pharmacopoeia*. The limits have been set at 5.0 mg/kg for seeds, 1.0 mg/kg for fruits, and 1.0 mg/kg for roots by the Codex Alimentarius Commission of WHO (Kosalec et al. 2009). High levels of chlorpyrifos could cause neurological damage and attention deficit hyperactivity disorder (ADHD) in infants (Rauh et al. 2006). A study by Lozano et al. (2012) found that a large number of chamomile teas sold in the European Union contained banned pesticides and also pesticide levels beyond the prescribed limit.

1.5.1.7.4 Heavy Metals

High levels of heavy metals, such as cadmium and lead, are harmful to humans in many ways. These cause a variety of ailments, and are toxic and carcinogenic to humans. Heavy metals have been detected on chamomile samples in a study (Alwakeel 2008). The study found that the chamomile samples contained lead, 0.13 ppm; mercury, 0.08 ppm; aluminum, 1.7 ppm; cadmium, 0.6 ppm; copper, 0.3 ppm; iron, 0.6 ppm; zinc, 943 ppm; and potassium, 228 ppm. It is to be noted that the permissible limits of heavy metals in herbal drugs are lead, 2.0 ppm; mercury, 0.5 ppm; aluminum, 0.2 ppm; cadmium, 0.2 ppm; iron, 15 ppm; and zinc, 5 ppm. The levels of the heavy metals were not alarming, but the levels of aluminum were much higher than the permissible limits. Aluminum is known to cause Alzheimer's disease (Alwakeel 2008). Testing for heavy materials is therefore important for maintaining the quality of the drug.

1.5.1.7.5 Microbial Contamination

Chamomile drug samples have been found to be contaminated with microbes, some of which cause severe diseases and some of these are potentially fatal. Many microbes produce mycotoxins that are teratogenic and carcinogenic. The microbes identified in chamomile samples are *Bacillus cereus*, *Clostridium perfringens*, *Salmonella*, *Escherichia coli*, *Fusarium* spp., *Penicillium* spp., *Absidia* spp., *Cladosporium* spp., *Paecilomyces* spp., *Cryptococcus albidus*, *C. laurentii*, *Rhodotorula glutinis*, *R. mucilaginosa*, *Aspergillus* spp., yeast, mold, *Rhizopus* spp., *Ulocladium* spp., *and Mycelia sterilia* (Mimica-Dukic et al. 1993; Martins et al. 2001; Foote 2002; Carvalho et al. 2009). A study on decontamination methods of microbes advocated stringent measures to guarantee the quality and safety of chamomile drugs (Maximino et al. 2011).

1.5.1.7.6 Other Impurities

The other impurities include foreign matters such as insects, animal matter, and sand. In several studies, it was found that commercial chamomile samples contained significant amounts of foreign matter (Nascimento et al. 2005; Falkowski et al. 2009). Insects such as *Lasioderma serricorne* were found in the drug (Pons et al. 2010). Such contaminants are, needless to say, undesirable in the drug.

1.5.1.7.7 Packaging

The tea bags and containers that hold the drug are important. This is because the hygiene, and the physical and chemical stability of the drug depend on the packaging. A study has recommended the use of CO_2 as an alternative to fumigation with synthetic chemicals for insect control in containers during shipment (Pons et al. 2010). The container of the drug, such as the tea bags or bottles, should be sterilized so that the shelf life of the drug is improved. The stability of the contents should be checked through TLC, GCMS, and so on, which help in detecting the presence or absence of the relevant compounds.

1.5.1.7.8 Labeling

Labeling is to ensure that the quality of information given to the consumer is good (Kunle et al. 2012). The label of the drug container should contain the relevant information about the origin, quality, safety, and dosage of the drug. The label should contain the warnings or risk factors of the drug.

1.5.1.7.9 Safety

The toxicity studies of the chamomile drug are important to ensure that it does not prove harmful or fatal and is safe for use. Several toxicity studies have been carried out with the chamomile drug. Chamomile drug has tolerable effects (Roder 1982) and there appear to be no reports of any gross toxicity and allergenicity caused either by the crude preparations or by the individual compounds of chamomile (Habersang et al. 1979). For testing the toxicity, physicians recommend a patch test on the skin. The label of the container of the chamomile drug should carry a warning about the toxicity effects.

1.5.1.7.10 Efficacy

The efficacy of the drug is to be indicated on the label of the container. To establish the efficacy of the chamomile drug, several pharmacological studies have been carried out on humans, animal models, microbes, and so on. The individual constituents or the compounds of the flower extractable in the essential oil or through the use of solvents determine the specific activities of the drug. The activities of the compounds have been identified on the basis of their effect on the experimental models. The tests have revealed that chamazulene possesses antioxidant and anti-inflammatory properties and provides protection against liver damage. Bisabolol protects against ulcers. Apigenin has anti-inflammatory and spasmolytic properties. The polysaccharides are immunostimulating.

1.5.1.8 Methods to Ensure Quality, Safety, and Efficacy of Chamomile Drugs

Several methods are employed to assess and ensure the quality, safety, and efficacy of chamomile drugs. The quality issues mentioned previously, such as adulterants, pesticides, heavy metals, microbes, and other material, can be easily detected macroscopically, microscopically, and through various analytical tools such as scanning electron microscopy, TLC, GCMS, ultraviolet spectrophotometry, IRMS, and microbial culture methods. Many chemical tests are also carried out to detect the traces of pesticides and heavy metals. Some of the methods are described in Sections 1.5.1.8.1 through 1.5.1.8.6 (WHO 1998, 2011).

1.5.1.8.1 Detection of Adulterants of Chamomile Flowers

The chamomile drug comprising dry flowers can be subject to macroscopic and microscopic observation and compared to the description of the actual drug available in the literature. The macroscopic examination includes visual examination of the drug, which determines the size of the chamomile flowers, the color of flowers, and odor and taste. Chamomile flowers possess a characteristic odor and can be identified by its organoleptic properties.

Introduction to Chamomile

Several pharmacopoeias and research papers have provided a detailed description of the features of the drug observed under the microscope. These are listed under Section 1.5.1.6.

1.5.1.8.2 Detection of Adulteration of the Essential Oil

The authenticity of the essential oil of chamomile can be described by its visual appearance, odor, the specific compounds in it, and also the specific percentage of those compounds of the oil that it possesses. The chamomile essential oil is known to contain chamazulene, (−)-α-bisabolol, bisabolol oxides A and B, bisabolone oxide A, and farnesene, among others. A complete list of the chemical compounds is available in the literature and is also provided in Chapter 4. For testing the oil, the flowers are subjected to steam distillation. The distilled oil is collected and studied for its properties such as color and odor, and then tested through the methods of TLC and GCMS.

1.5.1.8.3 Detection of Total Volatile Oil Content

The chamomile oil is volatile, which means it has the ability to vaporize at room temperature. One of the methods to determine the suitability of the chamomile drug is the amount of essential oil or volatile oil content in it. To determine the total volatile component in the chamomile flowers, the flowers are steam distilled in a Clevenger-type apparatus. The volume of the oil distilled per 100 g flowers is expressed in percentage. This percentage indicates the total volatile oil content.

1.5.1.8.4 Detection of Pesticide Residues

The pesticide residues are found in chamomile drug due to agricultural practices during cultivation. The pesticide residues usually have compounds such as chlorinated hydrocarbons or sulfur-containing dithiocarbamate or organophosphorus. These can be detected by column chromatography (CC) and GC. A purified chamomile extract is specifically prepared for the detection of the pesticide residues in it by GC.

Abdel-Gawad et al. (2011) suggested that the insecticide ^{14}C-ethion could be satisfactorily eliminated from the essential oil of chamomile if adsorbents such as calcium oxide and sawdust were added during the distillation process. Such methods that could efficiently eliminate pesticide residues need to be developed.

1.5.1.8.5 Detection of Heavy Metals

For detecting the levels of heavy metals such as lead and cadmium, several methods of analysis are available to choose from, such as inverse voltammetry or atomic absorption spectrometry.

1.5.1.8.6 Detection of Microorganisms

To detect the microorganisms present in chamomile, a sample of the drug is cultured in a liquid broth or semisolid culture medium in agar. Within a day or 2, the microorganisms grow in the broth or agar medium. These are identified according to their growth or no growth in a specific medium, the type of colony formation, or their morphology.

1.5.1.9 Good Manufacturing Practices for Herbal Formulations per World Health Organization Guidelines

The guidelines for Good Manufacturing Practices (GMP) have been developed by the WHO for herbal medicines. These guidelines are meant to be adopted by the member countries to ensure the quality, safety, and efficacy of the herbal medicines made in their own countries (WHO 2007). The GMP guidelines describe the complete process of manufacture of the herbal formulation right from the collection or harvesting of the plant to the packaging and marketing of the finished product. The guidelines also specify the infrastructural, sanitary, equipment, and management standards that need to be followed by the manufacturers.

1.5.1.9.1 Quality Assurance in the Manufacture of Herbal Medicines

Quality assurance includes all those matters that influence the quality of the herbal drug. Quality of the herbal drugs can be assured through analytical techniques such as high-performance thin-layer chromatography, GC, atomic absorption, and capillary electrophoresis. The WHO guidelines also specify that the manufacturer must assume responsibility for the quality of the drug or formulation.

1.5.1.9.2 Good Manufacturing Practice for Herbal Medicines

All manufacturing practices should be clearly defined and reviewed for consistency of the process. Operators who manufacture the drugs should be properly trained. Any complaints about the marketed drug should be attended to.

1.5.1.9.3 Sanitation and Hygiene

At all levels of manufacturing process, sanitation and hygiene should be practiced. This includes the sanitation and hygiene of the premises and the personnel employed.

1.5.1.9.4 Qualification and Validation

The infrastructure, premises, all equipment, processes and tests, and the documentation used for manufacture should be qualified and validated.

1.5.1.9.5 Complaints

A complaint in the manufactured drug should be reviewed and action should be taken. If necessary, the product should be recalled.

1.5.1.9.6 Product Recalls

A process should be enforced and person should be authorized to recall the defective products from the market. The process should be monitored and the recalled products should be stored in a place separate from the site of manufacture.

1.5.1.9.7 Contract Production and Analysis

Contracts given out to manufacture drugs should be correctly defined and controlled. All manufacturing and marketing processes should be as per the regulations and GMP.

Introduction to Chamomile

1.5.1.9.8 Self-Inspection

The manufacturer should set up a self-inspection system and a team to inspect the quality aspects from time to time. The inspection item list should include the premises, personnel, equipment, validation of the processes and documents, and audit.

1.5.1.9.9 Personnel

The number of personnel should be adequate, trained, and aware of the GMP. Their job responsibilities should be well defined.

1.5.1.9.10 Training

The manufacture should provide training to the personnel in the GMP through approved training programs. In addition, the new recruits should be trained in the jobs they are responsible for.

1.5.1.9.11 Personal Hygiene

All employees should undergo health examinations and trained in the practices of personal hygiene. Direct contact should be avoided between the personnel and the raw materials, or the intermediate product or the finished product.

1.5.1.9.12 Premises

The premises should be located in an area that minimizes contamination. The premises should be constructed well to provide adequate light and ventilation, maintained well, cleaned, and disinfected.

1.5.1.9.13 Equipment

The equipment should be installed properly to avoid any kind of contamination. These should be calibrated from time to time. Cleaning should be done as per validated process.

1.5.1.9.14 Materials

All the materials used in the manufacturing process, such as the raw materials, chemicals, and packaging materials, should be suitable, tested for contamination, and stored carefully.

1.5.1.9.15 Documentation

The documents used for the manufacture and marketing of drugs, such as reference materials, should be authentic and properly recorded. The information on the label and packaging should be based on the correct documentation.

1.5.1.10 National Regulations for Herbal Formulations

Several countries have laws in place to regulate the manufacture and marketing of herbal medicines. A study by the WHO in 2005 revealed that 53 countries had regulations related to the traditional medicine (WHO 2005). Table 1.4 lists some of the countries that have legally binding national regulations in place, which also recommend the use of pharmacopoeia and monographs.

TABLE 1.4
List of Some Countries with National Regulations on the Use of Chamomile Drugs

S. No.	Country	Legally Binding Regulations
1.	Argentina	Resolution 144/98, Farmacopea nacional Argentina, United States Pharmacopoeia, European Pharmacopoeia, British Pharmacopoeia, European Scientific Cooperative on Phytotherapy (ESCOP) Monographs, and the WHO Monographs. Good manufacturing practices (GMP)
2.	Australia	The Therapeutic Goods Act, 1989, British Pharmacopoeia
3.	Brazil	RDC 48/2004, Farmacopéia Brasileira, GMP
4.	Canada	The Food and Drugs Act, 2003, Compendium of pharmaceuticals and specialties, Canadian drug reference for health professionals, Compendium of Nonprescription Products, United States Pharmacopoeia, Herbal medicines, Expanded Commission E Monographs, ESCOP Monographs, WHO Monographs, Pharmacopoeia of the People's Republic of China, Physicians' Desk Reference for Herbal Medicines, British Herbal Compendium, and British Herbal Pharmacopoeia, GMP
5.	Colombia	Decree 677 of 1995 and Decree 337 of 1998, United States Pharmacopoeia, Codex francés, and British Herbal Pharmacopoeia, GMP
6.	Croatia	GMP
7.	Egypt	Pharmacy Law no. 127 of 1955, Egyptian Pharmacopoeia, GMP
8.	Germany	The German Drug Law of 1976, Deutsches Arzneibuch (German pharmacopoeia, DAB), and the European Pharmacopoeia, GMP
9.	Hungary	"Healing products or paramedicine" 1987, Hungarian Pharmacopoeia, GMP
10.	India	The Drugs and Cosmetics Act, 1940, Unani Pharmacopoeia of India, GMP
11.	Israel	No national regulation, British Pharmacopoeia, French Pharmacopoeia, and United States Pharmacopoeia, GMP
12.	Iran	Regulation in 1996, British Pharmacopoeia (not legally binding), Pharmacopoeia of the People's Republic of China (not legally binding), National formulary of Iran (not legally binding), GMP
13.	Kenya	No national regulation, no monographs
14.	Pakistan	The Drugs Act of 1962, Tibbi Pharmacopoeia (not legally binding), GMP
15.	Poland	GMP
16.	Serbia	Law on use of herbal medication of 1993, European Pharmacopoeia, GMP
17.	Slovakia	Law of 1997, Pharmacopoeia Slovaca, Codex Pharmaceutical Slovacus, GMP
18.	Tasmania	GMP

TABLE 1.4 (*Continued*)
List of Some Countries with National Regulations on the Use of Chamomile Drugs

S. No.	Country	Legally Binding Regulations
19.	United Kingdom	Medicines Act, 1968 (2001/83/EC also applies), British Pharmacopoeia, GMP
20.	United States	GMP

Source: WHO, *National Policy on Traditional Medicine and Regulation of Herbal Medicines: Report of a WHO Global Survey,* World Health Organization, Geneva, 2005.

1.5.2 Homeopathy

Homeopathy is a system of therapy, which is based on symptom similarity. It works on the combined principles of two laws: (1) a group of symptoms present in a disease and (2) a group of symptoms caused by the effect of a drug on a healthy human. It means that if a drug in its crude form causes the symptoms of a particular disease in a healthy human, that drug in a heavily diluted form is used to treat the same symptoms in a diseased person. Samuel Christian Hahnemann, the founder of homeopathy, expressed it as *Similia similbus curenteur*, meaning "let likes be cured by likes" (Iyer 1994; WHO 2009; Homeopathy-Chamomilla 2012).

The *Materia Medica* used in homeopathic practice emphasizes that the name of the disease is not the leading feature, but the general character and nature of the disease combined with the constitutional peculiarities of the patient, and the general characteristics of the remedy are important for treatment. The general characteristics of chamomile, as described in the Homeopathic Materia Medica, are as follows (Iyer 1994).

The chief guiding symptoms belong to the mental and emotional group, which lead to this remedy in many forms of disease, especially in diseases of children, where peevishness, restlessness, and colic give the needful indications. Chamomilla is sensitive, irritable, thirsty, hot, and numb. Oversensitiveness from abuse of coffee and narcotics. Pains unendurable, associated with numbness. Night sweats. Child fretful, wants to be carried, wants things, and then does not want them, snappish. One cheek red and the other pale. Diarrhea and colic. Green stools, like rotten eggs, during period of dentition. Sleeplessness of children. Rheumatic pains that drive patient from bed. Excessive restlessness and tossing. Hot and thirsty. Wind colic. Skin moist and hot. Worse by heat, anger, during evening, before midnight, open air, in the wind, eructations. Better in children from being carried, warm, wet weather.

Complimentary: Follows Belladonna in diseases of children and useful in cases spoiled by the use of opium or morphine in complaints of children; Mag C.

1. Antidotes of Chamomilla: Camphor, *Pulsatilla*, nux vomica
2. Compare: Belladonna, *Bryonia*, coffee, *Pulsatilla*, sulfur

1.5.2.1 Disease Conditions

Chamomile is recommended in several disease conditions of the different body parts, such as those of the head and brain, eyes, ears, throat, teeth, and stomach. It is also used to treat pregnancy-related ailments. Specific conditions of women and infants are also treated using chamomile. Table 1.5 lists some of the disease conditions and the use of chamomile in homeopathy.

TABLE 1.5
Use of Chamomile in Some Disease Conditions in Homeopathy

S. No.	Category	Disease Conditions
1.	Head and brain	Rheumatic headache accompanied by tearing pains on one side of the head and earache
2.	Head and brain	Nervous headache caused by cold and sore throat
3.	Head and brain	Hysteria
4.	Head and brain	Epilepsy with headache before and after the spasm, and spasms of the facial muscles
5.	Eyes	Sore eyes of young infants. If eyes are glued together in the morning, chamomile solution in warm water may be applied
6.	Ears	Earache due to inflammation due to cold
7.	Ears	Hardness of hearing connected to cold and sore throat
8.	Throat	Bronchitis in the first inflammatory stage, especially for children
9.	Throat	Sore throat due to cold. Useful especially for children
10.	Teeth	Toothache in children and females, especially when the person has earache, faceache, before menstruation, or the person is of nervous or hysterical nature
11.	Stomach and abdomen	Sour stomach in nursing infants
12.	Stomach and abdomen	Colic or gripping pain in bowels. Chamomile is especially suitable for women and children. It is recommended for colic and pain in pregnant women
13.	Stomach and abdomen	Dyspepsia or indigestion. Effective where gastric derangements are brought about by fits of passion, sour eructations, and regurgitation of food
14.	Stomach and abdomen	Diarrhea, especially for infants. Recommended for condition of diarrhea after a cold
15.	Stomach and abdomen	Mild cases of jaundice, especially if the disease is caused by fits of passion. Useful for infants and it is recommended that one dose of one or two drops of the solution every 4 hours should be given
16.	Stomach and abdomen	Rheumatic fevers accompanied by tearing pains in body parts with a sensation of numbness in the parts
17.	Stomach and abdomen	Gastric and bilious fever accompanied by diarrhea or frequent stools, colicky pains, sleeplessness, or excitement
18.	Diseases of women	Menorrhagia with profuse pains in the abdomen and back

Introduction to Chamomile

TABLE 1.5 (*Continued*)
Use of Chamomile in Some Disease Conditions in Homeopathy

S. No.	Category	Disease Conditions
19.	Diseases of women	Dysmenorrhea with pains like labor pains, violent abdominal cramps. Dark discharges of coagulated clots
20.	Pregnancy	Diarrhea during pregnancy
21.	Pregnancy	Hysterical fits or fainting during pregnancy, due to fits of anger or excitement
22.	Pregnancy	Palpitation of heart during pregnancy for nervous persons
23.	Pregnancy	Toothache during pregnancy in carious teeth and violent pains in teeth
24.	Pregnancy	Neuralgia during pregnancy and increased irritability
25.	Pregnancy	Miscarriage with excessive restlessness, severe pain in the back, pains resembling labor pains and each pain followed by discharge of dark colored blood
26.	Pregnancy	During labor pains if there is great mental excitement
27.	After childbirth	After delivery if there is lot of nervousness, restlessness, and excitement
28.	After childbirth	Lochia or discharges after childbirth. Chamomile is recommended if lochia is suppressed due to cold followed by diarrhea and colic
29.	After childbirth	Milk fever with nervous excitement, tenderness of the breasts
30.	Infants	Snuffles or obstruction of the nose in infants accompanied by runny nose
31.	Infants	Crying of infants if there is reason to think that it is crying due to some pain
32.	Infants	Colic in infants
33.	Infants	Restlessness and wakefulness in infants accompanied by flatulence and feverishness
34.	Infants	Teething (dentition)
35.	Infants	Prickly heat with fever and restlessness
36.	Infants	Discharge from ears with pain
37.	Infants	Rickets with restlessness and irritability, colic and diarrhea
38.	General	Sciatica when pains are worse at night
39.	General	Sleeplessness in children due to severe pain and in nervous women
40.	General	Toothache immediately on drinking coffee

Source: Iyer, T.S., *Beginners Guide to Homeopathy*, B. Jain Publishers, New Delhi, 1994.

1.5.2.2 Formulations in Homeopathy

Chamomile is used in the liquid form as a tincture in homeopathy. This tincture may be used by adding to water or to tiny sugar pellets. The process of making tincture is described in this section.

The whole plant of chamomile is used to prepare the tincture. The plant is harvested when it is at the flowering stage. This stage is considered to have optimum healing properties. The whole plant is cut into small pieces and macerated with equal parts of 35% alcohol and left for some time. This is followed by filtration to obtain a liquid. The liquid is in a very crude form and is called the mother tincture. The mother tincture is diluted repeatedly and used for therapy (Homeopathy-chamomilla 2012). It is believed that the higher the dilutions, the greater the effectiveness of the drug or potency. To make the dilutions, one part of mother tincture is taken and diluted with 99 parts of distilled water. It is shaken vigorously and the process is called *succussion*. This dilution is called 1c and is considered weak. To make stronger potencies, one part of 1c is mixed with 99 parts of distilled water to make 2c. Further dilutions repeated four times yield 6c, which is administered. Other potencies used are 30c and 200c (Olsen 2012).

1.5.2.3 Homeopathic Pharmacopoeia

The practitioners of homeopathic system of medicine follow *The United States Homeopathic Pharmacopoeia* and the *British Homeopathic Pharmacopoeia*, in addition to the *Homeopathic Materia Medica*.

1.5.2.3.1 The United States Homeopathic Pharmacopoeia

The Homeopathic Pharmacopoeia of the United States (HPUS) was published in 1878 and provides a description of the chamomile plant, the active principle, and the plant part used. It provides specification for the form of use; guidelines for the collection; methods to reduce contamination, preparation, and dilution; and dispensing of the drug (*The United States Homeopathic Pharmacopoeia* 1878).

1. Description of the plant.
   a. Chamomilla. *M. chamomilla*. Feverfew.
   b. This annual is a native of Europe, but is occasionally seen in the flower gardens of this country (United States). It prefers a gravelly soil and grows in both cultivated and uncultivated lands. It is about 2 ft. high, has a branching stem, and bears a profusion of flowers composed of white petals and a yellow disc. Those used officially are imported from Germany.
2. Active principle.
   a. Oleum anthemidis, quercitron, crystallizable principle.
3. Part used.
   a. The whole plant when in flower.

The general practices recommended in the pharmacopoeia for quality, safety, and efficacy are as follows:

1. Collection of plant.
   a. Every plant should be gathered only from those localities to which it is indigenous and their surrounding environmental conditions should be taken into account. If the whole plant is to be used, the most favorable time for gathering is when this is partly in flower and partly in seed.

Introduction to Chamomile

Collection is prohibited in the rainy season because the oils, resins, volatile principles, and so on, are not secreted during this period.
2. Treatment to reduce contamination.
   a. The plants should be procured whole. Granular and powdered forms should not be used at the initial stage of making the drug as they are prone to adulterants and contaminants. As soon as the plant material is procured, it should be converted into tincture. If a delay is unavoidable, then proper attention should be paid toward wrapping, boxing, and placing the plant in a cool, dry place.
3. Form of preparation.
   a. Tincture, made by macerating one part in five of dilute alcohol for 1 week and by filtering; greenish-brown color; contains the taste and odor of the plant.
4. Preparation of the tincture.
   a. Chamomile tincture is prepared by the method of expression. The plant is cut into very small pieces, it is bruised to pulp in a mortar, then enclosed in a loose muslin bag, and subjected to great pressure as in a screw press. The expressed juice is mixed with an equal part, by weight, of alcohol and allowed to stand in a cool place for 8 days, at the end of which time it is filtered and is ready for use.
5. Dilution of the tincture.
   a. One part of the tincture or solution is mixed with nine parts of the vehicle, which could be water, dilute alcohol, or alcohol. This is strongly shaken or succussed from 100 to 200 times in a vial. The solvents used in dilution such as water should be distilled, and the alcohol should be made nearly anhydrous.
6. Dispensing.
   a. Medicine prepared for homeopathic uses is dispensed in three forms: liquids, powders, and pellets or globules.

1.5.2.3.2 British Homeopathic Pharmacopoeia

The *British Homeopathic Pharmacopoeia* provides a detailed list of common names in addition to the botanical names, a description of the flower, the parts of the plant used, the time for collection, preparation, and forms the drug. In addition, it also specifies measures to ensure the quality of the drugs (*British Homoeopathic Pharmacopoeia* 1876).

1. Description.
   a. Name: Chamomilla.
   b. *M. chamomilla*. Nat. ord., Compositae.
   c. Synonyms: *Chamaemelum vulgare, C. nostras, Leucanthemum*.
   d. Common names: Wild chamomile, bitter chamomile, corn feverfew.
   e. Foreign names: German, Feld-Kamille, Mutter-Kraut; French, Camomille commun; Italian, Matricaria; Spanish, Matricaria.
   f. Grows in most parts of Europe, in corn fields, waste grounds, and roadsides. Flowers from May to August.

2. Botanical characters.
   a. Receptacle naked, almost perfectly cylindrical, hollow. Very similar to the well-known fetid chamomile (*A. cotula*), but distinguished from it by having no scales on the receptacle.
3. Parts employed.
   a. Whole plant.
4. Time for collecting.
   a. When in flower.
5. Preparation.
   a. Tincture, corresponding in alcoholic strength with 20 OP spirit.
6. Proper forms for dispensing.
   a. Tincture, pilules, or globules.

The *British Homeopathic Pharmacopoeia* also mentions the following measures to maintain the quality, efficacy, and safety of the drug.

1. Water.
   a. The water used for dilution should be the purest. It should not possess color, taste, or smell. Evaporated in a clean glass capsule, it should leave no visible residue.
2. Alcohol.
   a. The alcohol used for dilution should be pure. It should be colorless, transparent, very mobile and inflammable, of a peculiar pleasant odor, and has a strong spirituous burning taste. It should burn with a blue flame, without smoke. With a specific gravity of 0.8298, it should remain clear when diluted with distilled water. Its odor and taste should be purely alcoholic.
3. Plant material.
   a. It should be collected fresh intact, never in the form of powder. When the whole plant is used, it should be gathered when it is partly in flower and partly in seed. In the case of biennials, the collection should be done on the spring of the second year. After the fresh materials are collected, they should be prepared as soon as possible to avoid deterioration. In case of an unwanted delay, the plants should be packed carefully in tin cases (ordinary botanical boxes) and kept as cool as possible.
4. Tincture.
   a. In every instance, the dry crude substance is taken as the starting point from whence to calculate the strength and, with very few exceptions, the mother tinctures contain all the soluble matter of 1 oz. of the dry plant in 10 fl. oz. of the tincture.

1.5.2.4 Quality Issues of Chamomile Homeopathic Formulations

Because the basic starting material of the drug is the chamomile plant, the quality, safety, and efficacy issues of homeopathic formulations are similar to the herbal formulations. These issues are described briefly in Sections 1.5.2.4.1 through 1.5.2.4.4.

Introduction to Chamomile

1.5.2.4.1 Adulterants

The drugs may contain traces of unsafe starting material, such as the original plant itself may be an adulterant, which can prove toxic (WHO 2009).

1.5.2.4.2 Heavy Metals and Pesticide Residues

There could be contaminants arising from unsafe manufacturing practices as well. Trace amounts of heavy metals and pesticides may be present because of such malpractice.

1.5.2.4.3 Microbial Contamination

Presence of microbes, such as harmful bacteria or fungi, that produce harmful and lethal mycotoxins cannot be ruled out in the drugs.

1.5.2.4.4 Issues of Efficacy

Adverse effects in homeopathy are not expected by homeopaths because of the negligible quantities of active substances in a remedy (Stub et al. 2012). There is a lot of debate on the clinical efficacy of homeopathic medicines (Freckelton 2012). Nonetheless, many consumers, pharmacists, physicians, and other health-care providers continue to use or practice homeopathic medicine and advocate its safety and efficacy (Johnson and Boon 2007). There is a significant body of clinical research including randomized clinical trials and meta-analyses of such trials, which suggest that homeopathy has actions that are not placebo effects (Fisher 2012).

1.5.2.5 Methods to Improve Quality, Safety, and Efficacy of Homeopathic Drugs

The methods to improve the quality of homeopathic medicines involve a combination of the following pharmacopoeial standards and also detection methods to identify adulterants and contaminants. In addition, the packaging of the drugs should strictly follow standard norms. Most of the detection methods are the same as those described for herbal medicines.

1.5.2.5.1 Detection of Adulterants

Because the whole plant of chamomile is used for the preparation of the mother tincture, the original plant material of chamomile should be identified correctly according to the pharmacopoeia or monograph that has the description of the plant. Plants from other adulterant species, if found, should be removed before extracting the juices from the plants.

1.5.2.5.2 Detection of Pesticide Residues

Pesticide residues could be present in the mother tincture of chamomile because of the agricultural practices involved in the cultivation process. These pesticides are potentially harmful and pose safety issues. These pesticides can be detected using the techniques of CC or GC.

1.5.2.5.3 Detection of Heavy Metals

Heavy metals, such as lead or cadmium, and also other metals, such as aluminum, may be present in the mother tincture. These should be detected through techniques such as inverse voltammetry or atomic absorption spectrometry.

1.5.2.5.4 Detection of Microbes

The potentially harmful microbes and mycotoxins that might be present in the mother tincture should be tested using the microbial culture methods and the recommended tests for mycotoxins.

1.5.2.5.5 Detection of Foreign Material

Foreign material that may be present when the plant is being processed to prepare the mother tincture should be carefully removed. These foreign matter could be insects, animal excreta, any other undesirable and toxic matter, and sand. These could be detected visually through the use of microscopes and removed.

1.5.2.5.6 Packaging

The packaging should be done carefully in sterilized containers to ascertain not only the safety of the drug but also the stability and long shelf life of both the package and the drug.

1.5.2.6 Guidelines for Quality, Safety, and Efficacy of Homeopathic Drugs

In 2009, the WHO prepared a technical document on the safety issues of homeopathic drugs with an aim to support the national regulatory authorities—and the manufacturers of homeopathic medicines—in ensuring the safety and quality of homeopathic medicines. The guidelines detailed in the technical document are briefly presented in Sections 1.5.2.6.1 through 1.5.2.6.7 (WHO 2009):

1.5.2.6.1 Identification of Source Material

The source material is the most important aspect of homeopathy. Its correct identification is a must. For this, the WHO specifies the following requirements:

1. Scientific name of the plant
2. Stage of growth
3. Part of plant used
4. Whether cultivated or collected from the wild, and the place
5. Comparison of the specimen with an illustrated description of an authentic specimen for macroscopic and microscopic characteristics
6. Analytical determination of marker substances or standard substances

1.5.2.6.2 Complementary Tests

The complementary tests should be performed on the raw plant material for the following:

1. Foreign matter
2. Total ash

Introduction to Chamomile

3. Water content
4. Bitterness value
5. Loss on drying
6. Radioactive contamination

1.5.2.6.3 Limit Tests

1. Limit tests should be performed for pesticides, heavy metals, microbes, mycotoxins, and any other relevant matter.
2. Limit tests should be done at the unprocessed or raw stage.
3. Limit tests and ranges should comply with pharmacopoeia standards.

1.5.2.6.4 Mother Tincture

The following data for the mother tincture should be presented, or their absence needs to be justified as follows:

- Method of preparation according to the pharmacopoeia
- Appearance and description
- Identity tests
- Purity tests
- Stability tests

1.5.2.6.5 Finished Product

Homeopathic final products should be tested to determine the following:

- Identity and content
- Quality of dosage
- Residual solvents, reagents, or incidental contamination
- Stability

1.5.2.6.6 Diluents and Excipients

The diluents used in homeopathy, such as the distilled water or alcohol, should be of the pharmacopoeial standards.

The manufacturer should ensure the following:

- All excipients and diluents included in the final product are listed in the documentation and label.
- If new excipients and diluents are included, sufficient data on their safety and quality are provided to national health authorities.

1.5.2.6.7 Labeling

The labeling requirements (among others) are listed as follows:

- Name and address of manufacturer, packager, or distributor
- Manufacturer's batch number
- Content of the product in the container
- Statement that identifies the product as homeopathic

- Scientific name of the active substance, the degree of dilution/potency, and a reference to the pharmacopoeia that was used for the method of preparation
- Indications
- Directions for use and dosage requirements
- Storage conditions
- Warning that advises the user to consult a doctor or qualified health-care professional if the symptoms persist or worsen

1.5.2.7 National Regulations on Homeopathy System of Medicine

Several countries have national laws and regulation in place for the practice, manufacture, and marketing of homeopathic drugs. Some of these are described in Sections 1.5.2.7.1 through 1.5.2.7.3.

1.5.2.7.1 Europe

Homeopathy as a distinct therapeutic system is recognized by law in Belgium (1999), Bulgaria (2005), Germany (1998), Hungary (1997), Latvia (1997), Portugal (2003), Romania (1981), Slovenia (2007), and the United Kingdom (1950) (Camdoc Alliance 2010). In the European Union, the law governing the manufacture and marketing of homeopathic drugs came into force in 2001 under the Directive 2001/83/EC of the European Parliament and of the Council of November 6, 2001 on the Community code relating to medicinal products for human use (Directive 2001).

According to the Article 14 of the Directive, only those homeopathic medicines will be registered for sale, which will have 1 part per 10,000 mother tincture or 1/100th of the smallest dose used in allopathy are administered orally or externally, and no specific therapeutic indication appears on the labeling. There have been several modifications to this Directive. In 2003, the safety issues were brought in the homeopathic medicines in the Directive 2003-63.EC. In the Annexure of this Directive of 2003, in Part III, it is regulated that the quality requirements should be incorporated in the starting material and all intermediate steps of the manufacturing process of the homeopathic drugs. The finished product should be subject to controlled tests and stability tests, and any missing information should be justified (Directive 2003).

In the third amendment of the Directive in 2004, it was stated that if new scientific evidence so warrants, the Commission may amend the third indent of the first subparagraph by the procedure referred to in Article 121(2) (Directive 2004/27). The third indent deals with the dilution requirement of 1/10,000, which is considered unscientific by some researchers and practitioners of medicine.

1.5.2.7.2 United States

The Federal Food, Drug, and Cosmetic Act recognizes the homeopathic drugs and its supplements as official drugs and standards in the Sections 201 (g)(1) and 501 (b), respectively (FDA 1995).The HPUS is also recognized as official under Section 201(j) of the Act, and it is required that the method of preparation of the homeopathic drugs should be according to the HPUS. In addition to the HPUS, there is a compendium to the HPUS, which contains specifications and standards of preparation, content, and dosage of the homeopathic drugs.

According to the Act, in order for a drug to be recognized as homeopathic drug:

1. It should be listed in the HPUS, an addendum to it, or its supplements.
2. The potencies of homeopathic drugs are specified in terms of dilution, that is, 1× (1/10 dilution), 2× (1/100 dilution), and so on.
3. Homeopathic drug products must contain diluents commonly used in homeopathic pharmaceutics.

To maintain the quality of homeopathic drug, labeling containing all the specifications is mandatory. The labeling is categorized for prescription drugs and over-the-counter drugs.

The General Labeling Provisions should include the following:

1. Name and place of business of the manufacturer, packer, or distributor.
2. Directions for use (not for prescription drugs).
3. Statement of the quantity and amount of ingredient(s) expressed in homeopathic terms, for example, 1× and 2×.
4. Documentation must be provided to support those products or ingredients that are not recognized officially in the HPUS.
5. Established name that may include both English and Latin names.

1.5.2.7.2.1 Prescription Drugs
The products must comply with the general labeling provisions mentioned previously. In addition, the label should have a drug legend that says "Caution: Federal law prohibits dispensing without prescription," a statement of identity and a declaration of net quantity of contents and statement of dosage.

1.5.2.7.2.2 Over-the-Counter Drugs
Product labeling must comply with the general labeling provisions mentioned previously. In addition, it should have a principal display panel, statement of identity, declaration of net quantity of contents, indications for use likely to be understood by lay persons, directions for use, and warnings.

1.5.2.7.3 India
The manufacture and marketing of the homeopathic drugs in India are regulated by the Drugs and Cosmetics Act, 1940.

Section 2 (dd) of the Act specifies what drug is considered homoeopathic medicines. According to the Act, a drug is considered as homeopathic medicine if it meets the following criteria:

1. It is recorded in *Homoeopathic Pharmacopoeia.*
2. Its therapeutic efficacy has been established through long clinical experience as recorded in authoritative homoeopathic literature of India and abroad.
3. It is prepared according to the techniques of homoeopathic pharmacy.
4. Covers combination of ingredients of such homoeopathic medicines but does not include a medicine that is administered by parenteral route.

In 1973, The Homeopathy Central Council Act, 1973 (India) was enacted to provide for the constitution of a Central Council of Homoeopathy and maintenance of a Central Register of Homoeopathy and for matters connected therewith. In 2003, the Department of Ayurveda, Yoga and Naturopathy, Unani, Siddha, and Homoeopathy (AYUSH) was established, which deals with the education and practice of homeopathy, the manufacture and formulation of homeopathic drugs, and their marketing.

Subsequent amendments in the Drugs and Cosmetics Act, 1940 incorporated the issues of the quality, safety, and efficacy of the homeopathic drugs. The Section 3(7) of the Act specifies that testing of the homeopathic drugs should be carried out in the designated national laboratories.

There are provisions to ensure the quality, safety, and efficacy of the drugs, which are manufactured in India or those that are imported and sold in the Indian market. This is provided in the Second Schedule of the Act (The Drugs and Cosmetics Act 1940, p. 49) and presented below.

1. Drugs included in the *Homoeopathic Pharmacopoeia of India*. The standards required for the drugs manufactured according to the *Homeopathic Pharmacopoeia of India* are the following:
   a. The label should display the list of ingredients.
   b. Standards of strength, quality, and purity, as may be prescribed.
2. Drugs not included in the *Homoeopathic Pharmacopoeia of India*, but which are included in the *Homoeopathic pharmacopoeia of the United States* or *the United Kingdom*, or the *German Homoeopathic Pharmacopoeia*: For such drugs, the standards of identity, purity, and strength prescribed in such pharmacopoeia and other standards as may be prescribed should be followed.
3. Drugs not included in the *Homoeopathic Pharmacopoeia of India, the United States*, or *the United Kingdom*, or the *German Homoeopathic Pharmacopoeia*: For such drugs, the label should display the formula of list of ingredients and such other standards as may be prescribed by the central government should be followed.

1.5.2.7.3.1 Conditions of License

The marketing of homeopathic drugs in India as per the Section 67 (A) of the Drugs and the Cosmetics Act, 1940 require licenses that are issued by the government after stringent scrutiny. The license is granted after ensuring that the premises in which the drugs are manufactured and stocked are clean and hygienic. The following conditions are specified in the Section 85 H (e) of the Act with relation to the quality, safety, and efficacy of the mother tincture:

1. Crude drug used shall be identified.
2. Alcohol content shall be determined.
3. Containers should be of clean, neutral glass.
4. Hygienic conditions shall be scrupulously observed during the manufacturing process.

The Section 88 of the Act also specifies the labeling of containers. The label of the container should have the following information:

1. Name and address of the manufacturer
2. Scientific name of the substance
3. Purpose for which it has been manufactured

1.5.3 Unani System of Medicine

The Unani system works under the principle that disease is a natural process and the symptoms are the reactions of the body to the disease. The Unani system of medicine originated in Greece and was developed by the Arabs. Buqrat (Hippocrates) is known to be the father of this system of medicine, and the contribution of Jalinus (Galen) is significant. Ibn Sina (Avicenna) was a physician of this system whose work Al-Quanoon (Canon of Medicine) is one of the most important medical books for more than six centuries (NFUM 2006; Unani Formulations 2012). The Unani system of medicine is practiced widely in the Arabian countries as well as in Indian subcontinent under various names such as Greco Arab Medicine, Arabic Medicine, Tibb-e-Sunnati, Traditional Iranian Medicine, Eastern Medicine, and Uighur Medicine.

Over the centuries, the Unani system has imbibed the traditional medicines prevalent in Egypt, Syria, Persia, India, Middle East, Far East, and Central Asian countries (NFUM 2006). Therefore, it follows a combination of several working principles of the body including the hippocratic principles of humors, which are Dam (Blood), Balgham (Phlegm), Safra (Yellow bile), and Sauda (black bile) (Rahman et al. 2008). The drugs used in Unani are of herbal, animal, or mineral origin. The herbal drugs are single-origin drugs or compound formulations. The drugs are taken internally or applied externally. Internally the drugs may be taken as tablets, pills, and powders. For external use, the drug formulations are made as ointments and medicated oils (Kabir 2003).

Chamomile is called Babuna in the Unani system of medicine. It is used as a single drug or as a compound formulation with other components. The temperament (Mijaz) of the drug is hot and dry. The mode of administration is oral or local and as directed by the physician. It is used as a dilator, demulcent, resolvent, diuretic, emmenagogue, abortifacient, relaxant, brain tonic, and expectorant.

1.5.3.1 Disease Conditions

The various disease conditions in which chamomile is recommended are distension, headache, cold, stomatitis, conjunctivitis, scabies, itch, jaundice, real stones, fever, ileus, cystalgia, and herpes cornea. The reference to this drug and its use for treating the disease conditions can be traced back to the Al-qanun Fil-tibb, Vol II by Abu Ali Ibn Sina in the eleventh century. Some other disease conditions in which Babuna is used have been compiled in Table 1.6.

1.5.3.2 Formulations of Chamomile

In the Indian subcontinent, the Unani system of medicine uses chamomile as an ingredient in several of its formulations (Ali 1979). The different forms of chamomile formulations used in Unani medicines are the ointments (majoon, jawarish,

TABLE 1.6
Disease Conditions under Which Chamomile Is Recommended in the Unani System of Medicine

S. No.	Disease Conditions
1.	Indigestion
2.	Polyuria
3.	Anorexia
4.	Dementia
5.	Amnesia
6.	Sexual debility
7.	Dysuria
8.	Rheumatism
9.	Swelling(s)
10.	Visceritis
11.	Lumbago
12.	Earache
13.	Pneumonia
14.	Anterior mesodmitis
15.	Mediastinal pleurisy

Source: NFUM, *National Formulary of Unani Medicine,* Central Council for Research in Unani Medicine, India, 2006; Rashid, M.A. and Ahmad, F., *Hamdard Medicus,* 37, 73–81, 1994.

zimad, and qairooti) and the medicated oil (Raughan) (Rashid and Ahmad 1994; NFUM 2006). Their form and method of preparation are described below:

1. Majoon: It is a semisolid preparation of the dried chamomile flower in an edible base. The base is prepared by adding purified honey, sugar, or jaggery to water and boiling over a slow fire. After it acquires the required consistency, the base is purified by adding lime juice and alum. The powdered drug is mixed with a little clarified butter (ghee) and added to the base to produce a semisolid preparation. The majoon is taken internally.
2. Jawarish: It is a type of majoon but with a few differences. Its base is semisolid but of more liquid consistency than majoon. Another difference is that the powdered drug is coarser than majoon.
3. Zimad: The powered form of the drug is called zimad. The dried flowers are finely powdered and passed through a sieve of 100 mesh size. The powder is added to heated wax and then cooled. The product zimad is in the form an ointment (Marham), which is applied externally.
4. Qairooti: It is a semisolid preparation, just like an ointment. It is used externally. To prepare qairooti, the oils (Raughan-e-badam, Raughan-e-gul, or any other oil mentioned in the text) are heated and then wax or fat is dissolved in the oils. The dried chamomile flowers are mixed and stirred until a semisolid mass is formed.

5. Raughan-e-Babuna: The dried chamomile flowers are steeped in an appropriate oil to prepare raughan (oil). Such a type of preparation is made by the following method: 4 parts (by weight) dried flowers are soaked in 5 parts (by weight) of sesame and kept in a covered glass jar. This jar is exposed to sunlight for 40 days. The material is taken out after 40 days and crushed with hands to obtain a thorough suspension. The suspension is filtered using a fine cloth. The medicated oil obtained is used externally.

Table 1.7 lists the different Unani formulations and their forms, methods of administration, therapeutic uses, and the mechanisms of actions (Unani Formulations 2012).

In addition to the formulations mentioned in Table 1.6, some other formulations available in the market are listed as follows (NFUM 2006):

1. Majoon-e-Hafiz-ul-Ajsad
2. Qairooti Babuna Wali
3. Qairooti-e-Arad-e-Baqla

TABLE 1.7
Different Formulations, Forms, Administration, and Action of Chamomile Drugs in the Unani System of Medicine

S. No.	Formulation	Form of Drug	Form of Administration	Action of Drug
1.	Jawarish Baboonah	Semisolid	Oral	Stomachic
2.	Majoon-e-Falasifa	Semisolid	Oral	Stomachic, digestive, appetizer, sedative
3.	Zimad-e-Muballil-ut-Teeb	Ointment	Topical	Anti-inflammatory
4.	Zimad-e-Sumbul-ul-Teeb	Ointment	Topical	Anti-inflammatory
5.	Raughan-e-Babuna-Sada	Oil	Topical	Analgesic, anti-inflammatory
6.	Raughan-e-Babuna-Qawi	Oil	Topical	Analgesic, anti-inflammatory
7.	Dawa-e-Waja-ul-Ain	Tablets, pills, powder, decoction, paste	Oral	Severe pain of the eyes
8.	Nutool Barai Tahabbuj	Liquid	Poured on affected part	Edema
9.	Majoon-e-foodanaj	Semisolid	Oral	Antipyretic, cold
10.	Inkebab	Vapor	Inhalation	Headache
11.	Raughan Nardin	Oil	Topical	Coldness of stomach, colic
12.	Majoon-e-Atiyatullah	Semisolid	Oral	Piles, hemorrhoids

4. Qairooti-e-Mamool
5. Zimad Kharateen Shingrafi
6. Zimad Muqawwi
7. Zimad Niswan
8. Raughan Muqawwi-e-Asab
9. Raughan Samaat Kusha Jadeed
10. Bekh-e-Babuna
11. Tukhm-e-Babuna

1.5.3.3 Unani Pharmacopoeia

The Unani pharmacopoeia describes chamomile as Babuna. In fact three separate descriptions are provided for Babuna, Gul-e-Babuna, and Tukhm-e-Babuna as follows:

1. Babuna (*The Unani Pharmacopoeia of India* 2009)
    a. Babuna is a crude drug comprising chamomile flowers.
    b. Name: *M. chamomilla* L. of Asteraceae
        i. Morphology
            A. Peduncles: 0.20–0.40 cm in diameter, 1.5–2.0 cm long
            B. Receptacle: Discoid with involucral bracts
            C. Sepals: Pappus with brown margins
            D. Petals: Ligulate, white, elongate, tridentate
            E. Androecium: Stamens with short filament, epipetalous, and connate
            F. Gynoecium: Ovary bicarpellary, syncarpous, unilocular
            G. Seed: Anatropous, black, single in each ovary on basal placentation, vertically three to five ribbed
        ii. Chemical
            A. Total ash: Not more than 7.50%
            B. Acid-insoluble ash: Not more than 1.55%
            C. Alcohol-soluble matter: Not less than 12.00%
            D. Water-soluble matter: Not less than 20.00%
2. Gul-e-Babuna
    a. Gul-e-Babuna consists of the floral shoots of chamomile
    b. Name: *M. chamomilla* L. of Asteraceae
        i. Cellular
            A. Petal: The transverse section has uniseriate, adaxial, and abaxial epidermal layers containing unicellular covering hair; sandwiching homogenous parenchymatous mesophyll, few cells containing cuboid or rhomboid calcium oxalate crystals
            B. Anthers: Dithecous, tetralocular anther lobes, obtuse, entire pollen grains globular, tectum smooth, 5–6 µm in diameter
            C. Ovule (seed): Unitegmic, albuminous
3. Tukhm-e-Babuna
    a. Description provided is the same as Gul-e-Babuna.

1.5.3.3.1 Dosage

1. The doses, unless otherwise stated are regarded suitable for adults when administered orally two to three—times in 24 hours. The frequency and the amount of the therapeutic agent will be the responsibility of the medical practitioner.
2. If in case of administration of the drug by a route other than oral, the single dose for such administration is mentioned.

The Unani pharmacopoeia provides not only the description of the plant but also several single and compound formulations of chamomile, such as Raughan-e-Babuna (single drug), and Majoon-e-Falasifa (compound formulation). But most importantly, it provides the methods for testing the samples of the formulations in Appendices 1–5 of the pharmacopoeia. Appendix 1 specifies the apparatus to be used for the tests of the samples. The use of weights and measures, volumetric glassware, sieves, and so on are specified. Appendix 2 specifies the methods to determine microbial contamination and also the determination of quantitative data, such as foreign matter, total ash, volatile oil content, TLC, and alkaloid estimation. The Appendix also specifies limit tests for arsenic, heavy metal and pesticide contamination, and GC. Appendix 3 specifies physical tests such as those for determining the refractive index. Appendix 4 specifies the reagents and solutions. Appendix 5 specifies estimations of tannins and determination of elements such as aluminum or mercury.

1.5.3.4 Quality Issues of Unani Formulations

The quality issues of the Unani medicines are similar to the quality issues of the herbal formulations of traditional medicine since the starting material is a plant. The flowers could be adulterated with flowers from other plant species (Joharchi and Amiri 2012) or contaminated with pesticides, heavy metals, or microbes during cultivation or postharvest handling. Quality issues are present regarding the efficacy of the drug formulations (Rahman et al. 2008). Quality issues also arise during the manufacturing and storage of the drugs. Further, the issues of quality arise when the drug is marketed.

1.5.3.5 Quality Control of Unani Formulations

The National Formulary for Unani medicine and the Unani pharmacopoeia has laid down specifications for standardized formulation of Unani drugs. The Unani pharmacopoeia has extensively described the methods that are to be followed to ensure the manufacture of quality Unani formulations.

1.5.3.5.1 Formulations

The process of preparation of the formulations has been standardized by the National Formulary of Unani Medicine, Ministry of Health and Family Welfare, Government of India, based on authentic Unani literature (NFUM 2006).

1.5.3.5.2 Purity

For drugs originating from plants, they should be free from the following (NFUM 2006, pp. xxix–xxx):

1. Insects, foreign matter, animal excreta, fungus growth, mold, or other evidence of deterioration (toxic, injurious, or harmful) and to show no abnormal substances, odor, color, or sliminess.
2. Any unnatural and unusual impurity for which the rational considerations require that it be absent and it should not be putrefied or decomposed form.

The National Formulary specifies that the foreign impurities in drug should be cleaned by sieving or washing.

1.5.3.5.3 General Process of Preparation

1.5.3.5.3.1 Grinding

The general process of drug preparation involves making a powder of the chamomile drug. The particle has to be of a specific mesh size. The process of making the powder involves grinding in a mortar and pestle, made of stone, iron, wood, porcelain, or glass. Sometimes they are rubbed on a flat grinding stone.

1.5.3.5.3.2 Washing

In some preparations, the chamomile flowers are not powdered but directly used. These flowers are washed for a few hours before the formulation is prepared.

1.5.3.5.4 Specified Precautions to Be Observed during Preparation

1. Pills and tablets
   a. The pills or tablets are made by taking a small mass and mixed with a water-soluble adhesive. A weighed amount of this mass is taken and rolled between the fingers. Specific oils are used during rolling the mass to avoid sticking of the mass to the fingers. It is specified that the pills and tablets should neither be too hard nor too soft. The tablets are to be preserved carefully in clean, well-dried jars and stored in a cool and dry place to avoid contamination.
2. Ointment (Majoon)
   a. The majoon is prepared by mixing one component with another. During its preparation, care should be taken to stir continuously to allow proper mixing. The mixture should not come in contact with moisture under any condition. The majoon is recommended to be preserved in dried and clean glass, china clay, or tin-coated special containers. During preservation, if the majoon gets dry, it can be brought to normal consistency by adding purified honey or a thick syrup made of sugar.
3. Medicated oil (Raughan)
   a. The process of extraction should be strictly according to the "General Methods of Preparation." The oil should always be of the required consistency, flavor, color, and tests as given in the Unani texts. These oils

should be preserved in clean and dry glass jar containers under hygienic conditions in a cool and dry place.

1.5.3.5.5 Storage
1. Container and its cover must not interact physically or chemically with the substance that it holds so as to alter the strength, quality, or purity of the substance.
2. Container should be tight and well closed. It should protect the contents from contamination, moisture or extraneous solid, efflorescence, deliquescence or evaporation, and loss of substance under ordinary or customary conditions of handling, shipment, storage, or sale.

1.5.3.6 National Legislation in India

The Unani formulations are manufactured and marketed as per the regulations of the Drugs and Cosmetics Act, 1940, as amended in 1964, and the Drugs and Cosmetics Rules, 1945 (The Drugs and Cosmetics Act 1940). The Act has defined a Unani drug and laid the regulations for marketing quality, safe, and effective Unani drugs. In this Act, a Unani drug is defined as a drug, which includes all medicines intended for internal or external use in the diagnosis, treatment, mitigation, or prevention of diseases in accordance with the formula described in the authoritative books of Unani (Tibb) systems of medicine. The Act also specifies misbranded, spurious, and adulterated drugs. According to the Act, a Unani drug shall be deemed to be adulterated if

- It consists of any filthy, putrid, or decomposed substance.
- It has been prepared, packed, or stored under insanitary conditions.
- Its container is composed of any poisonous or deleterious substance.
- It bears or contains a color other than one that is prescribed.
- It contains any harmful or toxic substance.
- Any substance has been mixed therewith so as to reduce its quality or strength.
- It has been substituted wholly or in part by any other drug or substance.

The Drugs and Cosmetic Rules of 1945 provides proper labeling of the Unani drugs.

1.5.3.6.1 Labeling
The label of the drug should contain the following information, among others:

1. Name of the drug
2. Reference to the method of preparation thereof as detailed in the authoritative books
3. Correct statement of the net content in terms of weight, measure, or number
4. Name and address of the manufacturer
5. Number of the license under which the drug is manufactured
6. Batch number

7. Date of manufacture
8. Words "Unani medicine."
9. Words "FOR EXTERNAL USE ONLY" if the medicine is for external application

To fulfill the objectives of the Drug and Cosmetics Act, 1940, the Government of India set up the pharmacopoeia committee for Unani medicine in 1964. In 1970, a Pharmacopoeial Laboratory for Indian Medicine was established to work for evolving standards for Unani drugs. In 1981, as a result of extensive deliberations by the Unani Pharmacopoeia Committee, *National Formulary of Unani Medicine* was compiled. Following this, in 2009, *The Unani Pharmacopoeia of India* was compiled, which comprises hitherto unstudied and unreported standards for single drugs of plant origin included in the *National Formulary of Unani Medicine*.

REFERENCES

Abdel-Gawad, H., Abdel Hameed, R. M., Elmesalamy, A. M., and Hegazi, B. 2011. Distribution and elimination of ^{14}C-Ethion insecticide in chamomile flowers and oil. *Phosphorus Sulfur and Silicon and the Related Elements* 186(10): 2122–2134.

Abou Ayana, I. A. A. and Gamal El Deen, A. A. 2011. Improvement of the properties of goat's milk labneh using some aromatic and vegetable oils. *International Journal of Dairy Science* 6(2): 112–123.

Ahmad, M. A., Zafar, M., Hasan, A., Sultana, S., Shah, G. M., and Tareen, R. B. 2009. Chemotaxonomic authentication of herbal drug chamomile. *Asian Journal of Chemistry* 21(5): 3395–3410.

Al-Hindawi, M. K., A1-Deen, I. H. S., Nab, M. H. A., and Ismail, M. A. 1989. Anti-inflammatory activity of some Iraqi plants using intact rats. *Journal of Ethnopharmacology* 26(2): 163–168.

Ali, S. S. 1979. *Unani Adviya Mufridah*. New Delhi: Bureau for Promotion of Urdu.

Aliheidari, N., Fazaeli, M., Ahmadi, R., Ghasemlou, M., and Emam-Djomeh, Z. 2013. Comparative evaluation on fatty acid and *Matricaria recutita* essential oil incorporated into casein-based film. *International Journal of Biological Macromolecules* 56: 69–75.

Alwakeel, S. S. 2008. Microbial and heavy metals contamination of herbal medicines. *Research Journal of Microbiology* 3(12): 683–691.

Antonielli, G. 1928. Matricaria chamomilla L. and Anthemis nobilis L. in intermittent fevers in medicines. *Biological Abstract*. 1928; 6:21145.

Barakat, A. A., Fahmy, H. S. M., Kandil, M. A., and Ebrahim, N. M. M. 1985. Toxicity of the extracts of black pepper, cumin, fennel, chamomile and lupine against *Drosophila melanogaster*, *Ceratitis capitata* and *Spodoptera littoralis*. *Indian Journal of Agricultural Sciences* 55(2): 116–120.

Baumann, L. S. 2007. Less-known botanical cosmeceuticals. *Dermatologic Therapy* 20: 330–342.

Bisset, N. G. (ed.). 1994. Matricaria flos. *Herbal Drugs and Phytopharmaceuticals. A Handbook for Practice on a Scientific Basis*. Stuttgart, Germany: Medpharm Scientific Publishers; Boca Raton, FL: CRC Press.

Brester, G., Swanser, K., and Watts, T. 2003. Market opportunities and strategic directions for specialty herbs and essential oil crops in Montana. A Report Prepared for Montana Department of Agriculture, U.S. Department of Agriculture Federal-State Marketing Improvement Program. http://www.ams.usda.gov/AMSv1.0/getfile?dDocName=STELPRD3247954 (Accessed September 25, 2012).

British Homoeopathic Pharmacopoeia. 1876. The British homoeopathic society. http://chestofbooks.com/health/materia-medica-drugs/British-Homoeopathic-Pharmacopoeia/Chamomilla.html (Accessed October 04, 2012).

Buchbauer, G. 1996. Methods in aromatherapy research. *Perfumer and Flavorist* 21(3): 31–36.

Burgos, A. N. and Morales, M. A. 2010. Qualitative study of use medicinal plants in a complementary or alternative way with the use of among of rural population of the Bulnes City, Bío-Bío Region, Chile. *Boletín Latinoamericano y del Caribe de Plantas Medicinales y Aromáticas* 9(5): 377–387.

Çaliş, A. and Yücel, D. A. 2009. Antimicrobial activity of some natural textile dyes. *International Journal of Natural and Engineering Sciences* 3(2): 58–60.

Camdoc Alliance. 2010. The regulatory status of complementary and alternative medicine for medical doctors in Europe. http://www.efpam.eu/status.pdf (Accessed October 04, 2012).

Carle, R. 1990. Anti-inflammatory and spasmolytic botanical drugs. *British Journal of Phytotherapy* 1(1): 33–39.

Carle, R. and Gomma, K. 1991/92. Medicinal uses of Matricariae Flos. *British Journal of Phytotherapy* 2(4): 147–153.

Carvalho, S., Stuart, R. M., Pimentel, I. C., Dalzoto, P. do R., Gabardo, J., and Zawadneak, M. A. C. 2009. Fungi contamination in the chamomile, anis and mate teas. *Revista do Instituto Adolfo Lutz (Impr.)* 68: 91–95.

CBI Ministry of Foreign Affairs. 2011. MAPs for cosmetics in Italy. www.cbi.eu/?pag=85&doc=6214&typ=mid_document (Accessed September 25, 2012).

Chamomile Generic. 2012. http://www.igenericdrugs.com/?s = Chamomile&showfull = 1 (Accessed September 27, 2012).

Chamomile. 2006. Benders' Dictionary of Nutrition and Food Technology, s.v. "chamomile". http://www.credoreference.com/entry/whdictnutr/chamomile (Accessed September 4, 2012).

Chand, S., Pandey, A., and Patra, D. D. 2012. Influence of vermicompost on dry matter yield and uptake of Ni and Cd by chamomile (*Matricaria chamomilla*) in Ni- and Cd-polluted soil. *Water Air and Soil Pollution* 223: 2257–2262.

Chetvernya, S. A. 1986. A comparative study of phenols in inflorescence of two species of *Matricaria chamomilla* L. *Restitution and Resurrection* 22(2): 373–377.

The Columbia Encyclopedia. 2012. Chamomile. Sixth Edition. Encyclopedia.com. http://www.encyclopedia.com/doc/1E1-chamomil.html (Accessed September 25, 2012).

Cooper, A. Scarlet Fever. In J.L. Longe, ed., *Gale Encyclopedia of Alternative Medicine*. 2005. (Encyclopedia.com.) http://www.encyclopedia.com/doc/1G2-3435100699.html (Accessed March 8, 2014).

Directive 2001. 2001. Directive 2001/83/EC of the European Parliament and of the Council of 6 November 2001 on the Community code relating to medicinal products for human use. *Official Journal of the European Communities* L 311: 67–128. http://www.homeopathyeurope.org/regulatory-status/eu-regulations/homeopathic-medicines-1/EC.pdf (Accessed February 11, 2013).

Directive 2003-63.EC. 2003. Commission Directive 2003/63/EC of 25 June 2003 amending Directive 2001/83/EC of the European Parliament and of the Council on the Community code relating to medicinal products for human use. *Official Journal of the European Union* L 159: 46–94. http://www.homeopathyeurope.org/regulatory-status/eu-regulations/homeopathic-medicines-1/Directive%202003-63-EC.pdf (Accessed February 11, 2013).

Directive 2004. 2004. Directive 2004/27/EC of the European Parliament and of the Council of 31 March 2004 amending Directive 2001/83/EC on the Community code relating to medicinal products for human use. *Official Journal of the European Union* L 136: 34–57. http://www.homeopathyeurope.org/european-union/eu-regulations/homeopathic-medicines-1/Directive%202004-27-EC.pdf/view (Accessed February 11, 2013).

Directive 2004/27. 2004. Current Directive 2004/27 of the European Parliament and of the Council of 31 March 2004 Amending Directive 2001/83/EC. pp. 1–22. http://www.echamp.eu/fileadmin/user_upload/Positions/Resolutions_and_Communiques/Synopsis_of_the_Past_and_Current_EU_Legislation.pdf (Accessed February 11, 2013).

The Drugs and Cosmetics Act. 1940. The Drugs and Cosmetics Act, 1940 as Amended by the Drugs (Amendment) Act, 1955, the Drugs (Amendment) Act, 1960, the Drugs (Amendment) Act, 1962, the Drugs and Cosmetics (Amendments) Act, 1964, the Drugs and Cosmetics (Amendments) Act, 1972, the Drugs and Cosmetics (Amendments) Act, 1982, the Drugs and Cosmetics (Amendments) Act, 1986 and the Drugs and Cosmetics (Amendments) Act, 1995 and the Drugs And Cosmetics Rules, 1945 as Corrected up to the 30th April, 2003, Government of India Ministry of Health and Family Welfare (Department of Health). http://www.cdsco.nic.in/html/Copy%20of%201.%20D&CAct121.pdf (Accessed February 13, 2013).

Emongor, V. E., Chweya, J. A., Keyo, S. O., and Munavu, R. M. 1990. Effect of nitrogen and phosphorus on the essential oil yield and quality of chamomile (*Matricaria chamomilla* L.) flowers. *Traditional Medicinal Plants*. pp. 33–37. http://www.greenstone.org/greenstone3/nzdl?a=d&d=HASHa9287526d39203650f9874.7.6.np&c=cdl&sib=1&dt=&ec=&et=&p.a=b&p.s=ClassifierBrowse&p.sa= (Accessed September 30, 2012).

Falkowski, G. J. S., Jacomassi, E., and Takemura, O. S. 2009. Quality and authenticity of samples of chamomile tea (*Matricaria recutita* L.—Asteraceae). *Revista do Instituto Adolfo Lutz* 68(1): 64–72.

Falzari, L., Menary, R. C., and Dragar, V. 2007. Feasibility of a chamomile oil and dried flower industry in Tasmania. *ISHS Acta Horticulturae* 749: 71–80.

Farag, R. S., Abdel Latif, M. S., Abd El-Gawad, A. E., and Dogheim, S. M. 2011. Monitoring of pesticide residues in some Egyptian herbs, fruits and vegetables. *International Food Research Journal* 18: 659–665.

FDA. 1995. CPG Sec. 400.400 conditions under which homeopathic drugs may be marketed. Inspections, compliance, enforcement, and criminal investigations. http://www.fda.gov/ICECI/ComplianceManuals/CompliancePolicyGuidanceManual/ucm074360.htm (Accessed February 11, 2013).

Fisher, P. 2012. What is homeopathy? An introduction. *Frontiers in Bioscience (Elite Edition)* 1(4): 1669–1682.

Foote, J. C. 2002. The microbiological evaluation of chamomile. A PhD dissertation, Texas Tech University. pp. 1–44. http://repositories.tdl.org/ttu-ir/bitstream/handle/2346/12512/31295017083568.pdf?sequence=1 (Accessed September 29, 2012).

Ford-Martin, P. and Odle, T. G. 2005. Aromatherapy. In J.L. Longe, ed. *Gale Encyclopedia of Alternative Medicine*. Second Edition, Volume I (A–C). Farmington Hills, MI: Thomson Gale, p. 123.

Franke, R. 2005. Plant sources. In: R. Franke and H. Schilcher (eds.) *Chamomile: Industrial Profiles*. Boco Raton FL: CRC Press, p. 44.

Franke, R. and Schilcher, H. 2007. Relevance and use of chamomile (*Matricaria recutita* L.). *ISHS Acta Horticulturae* 749: 29–43. http://www.actahort.org/books/749/749_2.htm (Accessed September 25, 2012).

Franz, C. 1992. Genetica biochimica e coltivazione della camomilla (*Chamomilla recutita* [L.] Rausch.). *Agricoltura Ricerca* 131: 87–96.

Freckelton, I. 2012. Death by homeopathy: Issues for civil, criminal and coronial law and for health service policy. *Journal of Law and Medicine* 19(3): 454–478.

Freitas, A. V. L. de., Coelho, M. de. F. B., Azevedo, R. A. B. de., and Maia, S. S. S. 2012. The herbalists and the marketing of medicinal plants in Sao Miguel, Rio Grande do Norte, Brazil. *Revista Brasileira de Biociências* 10(2): 147–156.

Fundario, A. and Cassone, M. C. 1980. The effect of chamomile, cinnamon, absithium, mace and origanum essential oils on rat operant conditioning behaviour. *Bollettino Della Società Italiana di Biologia Sperimentale* 56(22): 2375–2380.
Furia, T. E. and Bellanca, N. 1975. *Fenarolis' Handbook for Flavour Ingredients*. Second Edition, Volume I. Cleveland, OH: CRC Press, p. 771.
Gasič, O., Lukič, V., Adamovič, R., and Durkovic, R. 1989. Variability of content and composition of essential oil in various chamomile cultivars. *Herba Hung* 28: 21–28.
Gómez-Estrada, H., Díaz-Castillo, F., Franco-Ospina, L., Mercado-Camargo, J., Guzmán-Ledezma, J., Domingo, M. J., and Gaitán-Ibarra, R. 2011. Folk medicine in the northern coast of Colombia: An overview. *Journal of Ethnobiology and Ethnomedicine* 7: 27. http://www.springerlink.com/content/l271176071x65657/fulltext.pdf (Accessed September 25, 2012).
Habersang, S., Leuschner, F., Isaac, O., and Thiemer, K. 1979. Pharmacological studies on toxicity of (-)-alpha-bisabolol. *Planta Medica* 37: 115–123.
Hanrahan, C. and Frey, R. J. 2005. Chamomile. *Gale Encyclopedia of Alternative Medicine*. Second Edition, Volume I (A–C). Farmington Hills, MI: Thomson Gale, pp. 409–411.
Health Canada. 2009. German chamomile. Compendium of monographs. http://www.hc-sc .gc.ca/dhp-mps/alt_formats/pdf/prodnatur/applications/licen-prod/monograph/mono _germ_chamom_allem-eng.pdf (Accessed September 27, 2012).
Homeopathy—Chamomilla. 2012. German chamomile *Chamomilla recutita* syn *Matricaria chamomilla*. http://www.herbs2000.com/homeopathy/chamomilla.htm (Accessed October 03, 2012).
Huamantupa, I., Cuba, M., Urrunaga, R., Paz, E., Ananya, N., Callalli, M., Pallqui, N., and Coasaca, H. 2011. Richness, use and origin of expended medicinal plants in the markets of the Cusco City. *Revista Peruana de Biología* 18(3): 283–291.
Iyer, T. S. 1994. *Beginners Guide to Homeopathy*. New Delhi: B. Jain Publishers.
Joharchi, M. R. and Amiri, M. S. 2012. Taxonomic evaluation of misidentification of crude herbal drugs marketed in Iran. *Avicenna Journal of Phytomedicine* 2(2): 105–112.
Johnson, T. and Boon, H. A. 2007. Where does homeopathy fit in pharmacy practice? *Journal of Pharmaceutical Education* 71(1): 7.
Kabir, H. 2003. *Samsher's Morakkabat (Unani Formulations)*. Aligarh: Samsher Publishers and Distributors, p. 12.
Khaki, M., Sahari, M. A., and Barzegar, M. 2012. Evaluation of antioxidant and antimicrobial effects of chamomile (*Matricaria chamomilla* L.) essential oil on cake shelf life. *Journal of Medicinal Plants* 11(43): 9–18.
Kosalec, I., Cvek, J., and Tomić, S. 2009. Contaminants of medicinal herbs and herbal products. *Arhiv za higijenu rada i toksikologiju* 60: 485–501.
Kunle, O. F., Egharevba, H. O., and Ahmadu, P. O. 2012. Standardization of herbal medicines: A review. *International Journal of Biodiversity and Conservation* 4(3): 101–112.
Laksmi, T., Geetha, R. V., Ramamurthy, J. G., Rummila Anand, V.A., Roy, A., Vishnu priya, V., and Ananthi, P. 2011. Unfolding gift of nature-herbs for the management of periodontal disease: A comprehensive review. *Journal of Pharmacy Research* 4(8): 2576–2580.
Leung, A. Y. 1980. Encyclopedia of common natural ingredients: Used in food, drugs and cosmetics. New York: John Wiley, p. 296.
Linnæi, C. 1753. Species Plantarum Exhibentes Plantas Rite Cognitas ad Genera Relatas, Cum Differentiis Specificis, Nominibus Trivialibus, Synonymis Selectis, Locis Natalibus, Secundum Systema Sexuale Digestas. Volume II, pp. 890–891. http://biodiversitylibrary.org /page/358911; http://biodiversitylibrary.org/page/358912 (Accessed September 25, 2012).
Lissandrello, M. 2008. Healing Foods: Chamomile. http://www.vegetariantimes.com/article /healing-foods-chamomile/ (Accessed September 25, 2012).

Loomis, T. F., Ma, C., and Daneshtalab, M. 2004. Medicinal plants and herbs of Newfoundland. Part 1. Chemical constituents of the aerial part of pineapple weed (*Matricaria matricarioides*). *DARU* 12(4): 131–135.

Lozano, A., Rajski, Ł., Belmonte-Valles, N., Uclés, A., Uclés, S., Mezcua, M., and Fernández-Alba, A. R. 2012. Pesticide analysis in teas and chamomile by liquid chromatography and gas chromatography tandem mass spectrometry using a modified QuEChERS method: Validation and pilot survey in real samples. *Journal of Chromatography A* 8: 109–122.

Malik, F., Hussain, S., Ashfaq, K. M., Tabassam, S., Ahmad, A., Rashid Mahmood, R., and Mahmood, S. 2013. Assessment of frequently accessible homeopathic mother tinctures for their pharmacopoeal specifications in Pakistan. *African Journal of Pharmacy and Pharmacology* 7(21): 1374–1381.

Mann, C. and Staba, E. J. 1986. The chemistry, pharmacognosy and chemical formulations of chamomile. *Herbs Spices and Medicinal Plants* 1: 236–280.

Martins, M. H., Martins, M. L., Dias, M. I., and Bernardo, F. 2001. Evaluation of microbiological quality of medicinal plants used in natural infusions. *International Journal of Food Microbiology* 68(1–2): 149–153.

Maximino, F. L., Barbosa, L. M. Z., Andrade, M. S., Camilo, S. B., and Furlan, M. R. 2011. Evaluation of fungal decontamination of chamomile (*Chamomilla recutita* [L.] Rauschert) through different home procedures at two temperatures. *Revista Brasileira de Plantas Medicinais* 13(4): 396–400.

McKay, D. L. and Blumberg, J. B. 2006. A review of the bioactivity and potential health benefits of chamomile tea (*Matricaria recutita* L.). *Phytotherapy Research* 20(7): 519–530.

Mimica-Dukic, N., Pavkov, R., Lukic, V., and Gasic, O. 1993. Study of the chemical composition and microbial contamination of chamomile tea. *ISHS Acta Horticulturae* 333: 137–142.

Mishra, P. N. 1990. Studies on the ovicidal action of some natural products on the eggs of brinjal leaf beetle, *Henosepilachna viginioctopunctata Fabrication, Science and Culture* 56(I): 50–52.

Nascimento, V. T., Lacerda, E. U., Melo, J. G., Lima, C. S. A., Amorim, E. L. C., and Albuquerque, U. P. 2005. Quality control of medicinal plant products commercialized in the city of Recife (Pernambuco, Brazil): Erva-doce (*Pimpinella anisum* L.), quebra-pedra.

(*Phyllanthus* spp.), espinheira santa (*Maytenus ilicifolia* Mart.), and chamomile (*Matricaria recutita* L.). *Revista Brasileira de Plantas Medicinais* 7(3): 56–64.

NFUM. 2006. National Formulary of Unani Medicine. First Edition. Central Council for Research in Unani Medicine, Department of Ayush, Ministry for Health and Family Welfare, Government of India. http://www.ccrum.net/wp-content/uploads/2012/07/data/National_Formulary_of_Unani_Medicine_Part_I.pdf (Accessed October 10, 2012).

Nimbekar, T., Wanjari, B., and Bais, Y. 2012. Herbosomes—Herbal medicinal system for the management of periodontal disease. *International Journal of Biomedical and Advance Research* 3(6): 468–472.

Olsen, S. 2012. How are homeopathic medicines made. http://be-well-now.org/how-are-homeopathic-medicines-made/ (Accessed October 03, 2013).

Padma Preetha, J. and Karthika, K. 2009. Cosmeceuticals—An evolution. *International Journal of ChemTech Research* 4: 1217–1223.

Pons, M. J., Cámara, A. G., Guri, S., and Riudavets, J. 2010. The use of carbon dioxide in big bags and containers for the control of pest in food products. *Julius-Kühn-Archiv* 425: 414–418.

Poráčová, J., Blasčáková, M., Zahatňanská, M., Taylorová, B., Sutiaková, I., and Sály, J. 2007. Effect of chamomile essential oil application on the weight of eggs in laying hens Hisex Braun. *ISHS Acta Horticulturae* 749: 203–206. http://www.actahort.org/books/749/749_23.htm (Accessed September 25, 2012).

Purbrick, P. and Blessing, P. 2007. Chamomile demand, cultivation & use in Australia. *ISHS Acta Horticulturae* 749: 65–70. http://www.actahort.org/books/749/749_4.htm (Accessed September 25, 2012).

Raal, A., Volmer, D., Sõukand, R., Hratkevitš, S., and Kalle, R. 2013. Complementary treatment of the common cold and flu with medicinal plants—Results from two samples of pharmacy customers in Estonia. *PLoS ONE* 8(3): e58642.

Rahman, S. Z., Khan, R. A., and Latif, A. 2008. Importance of pharmacovigilance in Unani system. *Indian Journal of Pharmacology* 40(Suppl 1): S17–S20.

Rashid, M. A. and Ahmad, F. 1994. Pharmacognostical studies on the flowers of *Matricaria chamomilla* L. *Hamdard Medicus* 37(2): 73–81.

Rauh, V. A., Garfinkel, R., Perera, F. P., Andrews, H. F., Hoepner, L., Barr, D. B., Whitehead, R., Tang, D., and Whyatt, R. W. 2006. Impact of prenatal chlorpyrifos exposure on neurodevelopment in the first 3 years of life among inner-city children. *Pediatrics* 118(6): e1845–e1859. http://pediatrics.aappublications.org/content/118/6/e1845.full.html (Accessed September 29, 2012).

Roder, E. 1982. Secondary effects of medicinal herbs. *Deutsche Apotheker-Zeitung* 122(14): 2081–2092.

Rowland, B. and Odle, T. Stomachaches. In J.L. Longe, ed., *Gale Encyclopedia of Alternative Medicine*. 2005. (Encyclopedia.com.) http://www.encyclopedia.com/doc/1G2-3435100748.html (Accessed March 8, 2014).

Salamon, I. 1993. Chamomile. *The Modern Phytotherapist* :13–16.

Salamon, I. 2004. The Slovak gene pool of German chamomile (*Matricaria recutita* L.) and comparison in its parameters. *Horticultural Science* 31(2): 70–75.

Salamon, I. 2007. Large-scale production of chamomile in Streda Nad Bodrogom (Slovakia). *ISHS Acta Horticulturae* 749: 121–126. http://www.actahort.org/books/749/749_12.htm (Accessed September 25, 2012).

Salamon, I. and Plačková, A. 2007. Environmental risks associated with the production and collection of chamomile flowers. *ISHS Acta Horticulturae* 749: 211–261.

Šavikin, K., Zdunić, G., Menković, N., Zivkovic, J., Ćujić, N., Tereščenko, M., and Bigović, D. 2013. Ethnobotanical study on traditional use of medicinal plants in South-Western Serbia, Zlatibor district. *Journal of Ethnopharmacology* 146(3): 803–810.

Schilcher, H. 2005a. The legal situation of German chamomile: Monographs. In: R. Franke and H. Schilcher (eds.). *Chamomile: Industrial Profiles*. Boca Raton, FL: CRC Press, pp. 7–38.

Schilcher, H. 2005b. Traditional use and therapeutic indications. In: R. Franke and H. Schilcher (eds.). *Chamomile: Industrial Profiles*. Boca Raton, FL: CRC Press, pp. 265–274.

Schmidt, P.C. and Vogel, K. 1992. Chamomile—Evaluation of the stabilities of chamomile preparation. *Apotheker-Zeitung* 132(10): 462–468.

Shutes, J. 2012. German chamomile (*Matricaria recutita*), essential oil monographs, East West School for Herbal and Aromatic Studies. http://theida.com/ew/wp-content/uploads/2012/01/German-chamomile7.pdf (Accessed September 25, 2012).

Singh, L. B. 1970. Utilisation of saline-alkali soils for agro-industry without prior reclamation. *Economic Botany* 24(4): 439–442. http://www.jstor.org/stable/4253178 (Accessed April 04, 2012).

Singh, O., Khanam, Z., Misra, N., and Srivastava, M. K. 2011. Chamomile (*Matricaria chamomilla* L.): An overview. *Pharmacognosy Reviews* 5(9): 82–95.

Smitherman, L. C., Janisse, J., and Mathur, A. 2005. The use of folk remedies among children in an urban black community: Remedies for fever, colic, and teething. *Pediatrics* 115: e297–e304.

Srivastava, J. K., Shankar, E., and Gupta, S. 2010. Chamomile: A herbal medicine of the past with bright future. *Molecular Medicine Reports* 3(6): 895–901.

Stub, T., Alræk, T., and Salamonsen, A. 2012. The Red flag! Risk assessment among medical homeopaths in Norway: A qualitative study. *BMC Complementary and Alternative Medicine* 12(1): 150.
Therapeutic Research Faculty. 2012. Natural Medicines Comprehensive Database. http://naturaldatabase.therapeuticresearch.com/home.aspx?cs=&s=ND (Accessed January 03, 2013).
Thornfeldt, C. 2005. Cosmeceuticals containing herbs: Fact, fiction, and future. *Dermatologic Surgery* 31: 873–880.
Thorsell, W. 1988. Introductory studies of plant extracts with mosquito repelling properties. *Fauna Flora (Stock H)* 83(5): 202–207.
Thorsell, W., Mikiver, A., Mikiver, M., and Malm, E. 1970. Plant extracts as protectants against disease-causing insects. *Entomologisk Tidskrift* 100(3/4): 138–141.
Tilford, G. 2004. The calming herb chamomile. *The Whole Dog Journal.* http://www.whole-dog-journal.com/issues/7_2/features/Calming-Herb-Chamomile_5607-1.html (Accessed September 26, 2012).
Turner, J. Teething Problems. In J.L. Longe, ed., *Gale Encyclopedia of Alternative Medicine.* 2005. (Encyclopedia.com.) http://www.encyclopedia.com/doc/1G2-3435100768.html (Accessed March 8, 2014).
Unani Formulations. 2012. The Traditional Knowledge Digital Library. http://www.tkdl.res.in/tkdl/langdefault/Unani/Una_Unani-Glance.asp?GL=Eng (Accessed October 10, 2012).
UPI. *The Unani Pharmacopoeia of India.* 2009. First Edition, Part II, Volume I. pp. 221, 232, 257.
USHP. *The United States Homeopathic Pharmacopoeia.* 1878. First Edition. Chicago: Duncan Brothers Publishers, pp. 24–31, 95. http://ia600709.us.archive.org/9/items/unitedstateshomo00chic/unitedstateshomo00chic.pdf (Accessed October 04, 2012).
Uzma, N. and Khan, M. A. 1998. Palynological studies of *Matricaria chamomilla* L. (Babuna) and its related genera. *Hamdard Medicus* 41(4): 94–97.
Verpoorte, R., Choi, Y. H., and Kim, H. K. 2005. Ethnopharmacology and systems biology: A perfect holistic match. *Journal of Ethnopharmacology* 100: 53–56.
WHO. 1998. *Quality Control Methods for Medicinal Plant Materials.* Geneva: World Health Organisation.
WHO. 1999. Flos Chamomillae. WHO Monographs on Selected Medicinal Plants. Volume I. pp. 86–92. www.who.int/medicinedocs/en/d/Js2200e/ (Accessed September 28, 2012).
WHO. 2005. *National Policy on Traditional Medicine and Regulation of Herbal Medicines: Report of a WHO Global Survey.* Geneva: World Health Organization.
WHO. 2007. WHO Guidelines on Good Manufacturing Practices (GMP) for Herbal Medicines. pp. 1–15. http://apps.who.int/medicinedocs/documents/s14215e/s14215e.pdf (Accessed September 12, 2012).
WHO. 2009. Safety Issues in the Preparation of Homeopathic Medicines. World Health Organization. p. 4. www.who.int/medicines/areas/traditional/Homeopathy.pdf (Accessed October 03, 2012).
WHO. 2011. Quality Control Methods for Herbal Materials. Geneva: World Health Organisation. whqlibdoc.who.int/publications/2011/9789241500739_eng.pdf (Accessed January 30, 2013).
World Standard Drug Database. 2012. http://216.122.144.54/cgi-bin/drugcgic/INGR?117977281+0 (Accessed September 27, 2012).
Zaidi, S. F., Muhammad, J. S., Shahryar, S., Khan, U., Gilani, A. H., Jafri, W., and Sugiyama, T. 2012. Anti-inflammatory and cytoprotective effects of selected Pakistani medicinal plants in *Helicobacter pylori*-infected gastric epithelial cells. *Journal of Ethnopharmacology* 141(1): 403–410.
Zucchi, M. R., Oliviera Júnior, V. F., Gussoni, M. A., Silva, F. C., and Marques, N. E. 2013. Ethnobotanical survey of medicinal plants in Ipameri City—Goiás State. *Revista Brasileira Plantas Medicinais* 15(2): 273–279.

2 Chamomile
Botany

2.1 INTRODUCTION

Chamomile originated in the regions of southern Europe, North Africa, and West Asia. The plant has numerous traditional names not only in these countries but also all over the world where it has spread. In this context, it is interesting to note that the WHO monograph on chamomile lists no less than 34 selected vernacular names (WHO 1999) and Franke listed 68 vernacular names of chamomile (Franke 2005). A study says that in Spain there are about 62 different species called chamomile or manzanilla of which only 30 are considered as chamomile (de Santayana and Morales 2010). Considering the variety of names, the chances of adulteration are high and the correct identification of chamomile sometimes may pose a problem. This problem is compounded by the confusing botanical nomenclature of chamomile. However, as more and more taxonomic and metabolomic data are accumulating, it is expected that the correct taxonomic nomenclature of chamomile would be established. However, it would be interesting to have a look at the origins of the classification and botanical nomenclature of chamomile, and where it stands now.

2.2 COMMON NAMES

Chamomile is known locally by several names in different countries. Table 2.1 lists some of the commonly used names of chamomile (Salamon 1993; Rashid and Ahmad 1994; Salamon and Honcariv 1994; Franke 1995; WHO 1999; Skaria et al. 2007; de Santayana and Morales 2010).

2.3 OCCURRENCE

The herbaceous annual chamomile plant is adapted to the temperate climate. It is found growing in the wild in the areas where it originated. These areas extend from the south and east European countries to the Near East countries. The plant is an annual in many countries but biannual in many other countries. Chamomile grows in its wild form and where it is also cultivated chiefly in the countries mentioned in Table 2.2.

TABLE 2.1
Various Names of Chamomile in Different Countries

Language	Name
English	German chamomile, Hungarian chamomile, wild chamomile, common chamomile, pinheads, sweet false chamomile, feverfew
German	Kamillenbluten, Kleinekimille, Feld kamille, Echte kamille, Mutterkraut
French	Fleur de camomile, petite camomille, chamomille, matricaire
Russian	Romaska aptecnaja, romaska obodrannaja
Italian	Chamomilla vulgare, capomilla
Spanish	Manzanilla, camomila borda
Hungarian	Szekfu kamilla
Bulgarian	Lajka
Turkish	Papatya
Slovak	Rumanek
Arabic	Gul-e-Baboonah
Indian	Babuna, babunphul, seemaseventhi pu, sima jevanthi pushpam, shime-shavantige, sinacha mauli pushapamu
Japanese	Kamitsure
Danish	Kamille
Polish	Rumaneck pospolity
Portuguese	Camomila
Czech	Hermánek pravý

2.4 BOTANICAL CLASSIFICATION AND NOMENCLATURE OF CHAMOMILE

The classification of any plant starts with its description. A correct description ensures its correct classification. Chamomile was described as early as Hippocrates (460–377 BC) and Dioscorides (first century AD). Subsequently, it was described by the German botanist Tabernaemontanus (1522–1590) in 1564; Linnaeus in 1753, 1755, and 1763; and the Spanish botanist José Query Martínez (1695–1764) in 1762–1764 (Woo 1989; de Santayana and Morales 2010).

Dioscorides, in his *Materia Medica*, described three types of chamomile: *Leucanthemum*, *Chrysanthemum,* and *Eranthemon*. It is now believed that *Leucanthemum* could be "our common chamomile flower"; *Matricaria recutita* and *Chrysanthemum* could be the yellow chamomile; probably *Anthemis tinctoria* and *Eranthemon* could be the red chamomile; Eranthemon could probably be *Adonis aestevis* L. or *Adonis flammea* Jacq as suggested by Franke. However, de Santayana and Morales argue that Dioscorides said all chamomiles have a yellow button (or flower heads) and as *Adonis aestevis* or *Adonis flammea* belongs to the Ranunculaceae family, and they do not have a flower head, it is unlikely that *Eranthemon* could be *Adonis aestevis* L., or *Adonis flammea* Jacq (de Santayana and Morales 2010).

TABLE 2.2
Cultivation of Chamomile in Different Regions of the World

S. No.	Region	Country	Reference
1.	South Europe	Spain	Franke (2005, p. 79)
2.		Italy	Dellacecca (1996), Leto et al. (1994a–c)
3.	Eastern Europe	Bulgaria	Franke (2005, p. 79)
4.		Serbia	Stevanovič et al. (2007)
5.		Romania	Franke (2005, p. 79)
6.		Russia	Franke (2005)
7.	Central Europe	Hungary	Galambosi et al. (1988)
8.		Slovakia	Salamon (1992, 1996)
9.		Czech Republic	Holubář (2005)
10.		Poland	Seidler-Lozykowska (2007)
11.		Germany	Franke and Schilcher (2007)
12.		Croatia	Šiljkovic and Rimanić (2005)
13.		Yugoslavia	Franke (2005, p. 79)
14.	Northern Europe	Norway	Dragland (1996), Dragland et al. (1996)
15.		Finland	Galambosi et al. (1988, 1991)
16.	Africa	Egypt	Aly and Hussein, (2007)
17.		Kenya	Cinc et al. (1984)
18.	Middle East	Turkey	Franke (2005, p. 79)
19.		Iran	Azizi et al. (2007)
20.	Australian continent	Australia	Purbrick and Blessing (2007)
21.		New Zealand	Franke (2005, p. 79)
22.	Asia	Japan	Ohe et al. (1995)
23.		Korea	Young et al. (1992)
24.		China	Franke and Schilcher (2007)
25.		India	Chandra et al. (1979), Kapoor (1982), Ghosh (1989)
26.		Pakistan	Ahmad et al. (2011)
27.		Afghanistan	Franke (2005, p. 79)
28.	South America	Brazil	Correa (1995), Donalisio (1986)
29.		Cuba	Acosta et al. (1986)
30.		Mexico	Franke (2005, p. 80)
31.		Argentina	Franke (2005, p. 80)
32.		Chile	Franke (2005, p. 82)

Tabernaemontanus described six types of chamomile: Our common chamomile, yellow chamomile, red chamomile, Roman chamomile, filled Roman chamomile, and filled Roman chamomile of another genus. Franke is of the opinion that Roman chamomile and filled Roman chamomile refer to *Chamaemelum nobile*. However, the botanical name of filled Roman chamomile of another genus is not clearly determined (Franke 2005).

Quer described five types of chamomile: Our common chamomile, Roman chamomile, delicate chamomile, stinking chamomile, and scentless chamomile.

Delicate chamomile refers to *Matricaria aurea*, stinking chamomile probably refers to *Anthemis arvensis* or *Anthemis cotula*, and scentless chamomile could refer to *Anacyclus clavatus* or *Tripleurospermum inodorum* (syn. *Matricaria inodora*, *Matricaria maritima*) (de Santayana and Morales 2010).

In the subsequent years, many botanists described chamomile in different European countries. Among them were the French botanists Joseph Pitton de Tournefort and Sebastien Vaillant, the Dutch botanist Thomas Francois Dalibard, and the Swiss botanist Gaspard Bauhin. Their work was duly included by Linnaeus who added the chamomile plants described by them as synonyms of chamomile.

Linnaeus classified chamomile and grouped it under the genus *Matricaria*. The name *Matricaria* was coined by him. In 1753, he described five species of *Matricaria*: *Matricaria parthenium*, *Matricaria maritima*, *Matricaria chamomilla*, *Matricaria argentea*, and *Matricaria recutita*. However, later he made some changes in the species names *chamomilla* and *recutita*, and renamed the species *Matricaria recutita* as *Matricaria suaveolens*. He dropped the name recutita altogether from his subsequent works. These changes gave rise to a lot of confusion in the nomenclature of chamomile, which continues till today. A brief account of his work and the nomenclatural problems of chamomile are provided in Section 2.4.1.1.

Linnaeus described the genus *Matricaria* as follows (Linnaei 1754; Linnaeus 1775):

> The cup is hemispherical, calyx scales linear, imbricated and nearly equal. The compound flower has rays, the tubular flowers are hermaphrodite, numerous, and the ray florets are feminine in most cases. The tubular flowers are funnel shaped and the tip segmented into five, which are free. The female ray floret is oblong and tridentated. The stamens are hairy and very short, with cylindrical capillary-like (tubulose) anthers. The germen in both ray and tubular florets is oblong and naked. The style is like a thread and along the length of the stamina, the stigma in the hermaphrodite flowers is bifid and free. The seed is single, oblong and without any down.

Linnaeus described the species *chamomilla* and *suaveolens* as follows (Linnaeus 1775):

> *M. chamomilla. Corn Feverfew* with conical receptacles, patent rays, naked seeds, and scales up on the cup equal.
> *M. suaveolens. Sweet-scented Feverfew* with conical receptacles, deflex rays, naked seeds, and the scales of the cup equal.

Even after the work of Linnaeus, botanists in Europe continued to work independently on the classification of chamomile and used several other botanical names for the genus and the species (Franke 2005). These are listed in Table 2.3. The current status of these plant names has been provided in *The Plant List*, which is a working list prepared by the Royal Botanic Gardens, Kew and Missouri Botanical Garden (*The Plant List* 2012).

2.4.1 LINNAEUS ON CHAMOMILE

Linnaeus worked extensively from 1735 at the hothouses of George Clifford at Amsterdam and used the many living and dried specimens in Clifford's possession. He published his work *Hortus Cliffortianus* in 1737. In the subsequent years,

TABLE 2.3
Various Names Given to Chamomile by the Botanists after the Work of Linnaeus in 1753. Also Presented Is the Current Status of the Names as per *The Plant List* (2012)

S. No.	Year	Name	Name of the Botanist	Status of the Name (as in 2012)
1.	1761	*Matricaria chamaemilla*	John Hill	It is an unresolved name
2.	1779	*Leucanthemum chamaemelum*	Jean Baptiste Antoine Pierre de Monnet de Lamarck	It is an unresolved name
3.	1782	*Matricaria patens*	Jean-Emmanuel Gilibert	It is an unresolved name
4.	1782	*Chamomilla patens*	Jean-Emmanuel Gilibert	*Chamomilla patens* Gilib. is a synonym of *Matricaria chamomilla* L
5.	1796	*Matricaria tenuifolia*	Richard Anthony Salisbury	It is an unresolved name
6.	1800	*Chrysanthemum chamomilla*	Johann Jakob Bernhardi	It is a synonym of *Matricaria chamomilla* L
7.	1809	*Matricaria pusilla*	Carl Ludwig von Willdenow	It is a synonym of *Matricaria chamomilla* L
8.	1821	*Chamomilla vulgaris*	Samuel Frederick Gray	It is considered a synonym of *Matricaria chamomilla* L
9.	1837	*Matricaria pyrethroides*	Augustin Pyramus de Candolle	It is a synonym of *Matricaria chamomilla* var. *coronata* J.Gay ex Boiss
10.	1841	*Lepidotheca*	Thomas Nuttal	*Lepidotheca suaveolens* (Pursh) Nutt. is a synonym of *Matricaria matricarioides* (Less.) Porter
11.	1843	*Chamomilla meridionalis*	Karl (Carl) Heinrich Emil (Ludwig) Koch	It is a synonym of *Matricaria chamomilla* var. *coronata* J. Gay ex Boiss
12.	1843	*Matricaria coronata*	Wilhelm Daniel Joseph Koch	It is a synonym of *Matricaria chamomilla* var. *coronata* J.Gay ex Boiss
13.	1844	*Matricaria kochianum*	Carl (Karl) Heinrich 'Bipontinus' Schultz	The status of the name is unknown
14.	1844	*Tripleurospermum disciformis*	Carl (Karl) Heinrich 'Bipontinus' Schultz	It is an accepted name of a species in the genus *Tripleurospermum*
15.	1844	*Sphaeroclinium nigellifolium*	Carl (Karl) Heinrich 'Bipontinus' Schultz	It is a synonym of *Cotula nigellifolia* var. *nigellifolia* (DC.) K. Bremer & Humphries
16.	1844	*Courrantia chamomilloides*	Carl (Karl) Heinrich 'Bipontinus' Schultz	It is a synonym of *Matricaria chamomilla* L

(*Continued*)

TABLE 2.3 (*Continued*)
Various Names Given to Chamomile by the Botanists after the Work of Linnaeus in 1753. Also Presented Is the Current Status of the Names as per *The Plant List* (2012)

S. No.	Year	Name	Name of the Botanist	Status of the Name (as in 2012)
17.	1844	*Gastrostylum praecox*	Carl(Karl) Heinrich 'Bipontinus' Schultz	It is a synonym of *Tripleurospermum parviflorum* (Willd.) Pobed
18.	1844	*Rhytidospermum inodorum*	Carl(Karl) Heinrich 'Bipontinus' Schultz	It is a synonym of *Tripleurospermum maritimum* subsp. *inodorum* (L.)
19.	1846	*Dibothrospermum pusillum*	Joseph (Josef) Friedrich Knaf	It is a synonym of *Tripleurospermum maritimum* subsp. *inodorum* (L.)
20.	1847	*Chamaemelum inodorum*	Roberto de Visiani	It is a synonym of *Tripleurospermum maritimum* subsp. *inodorum* (L.)
21.	1849	*Trallesia matricarioides*	Antonio Maurizio Zumaglini	It is an unresolved name
22.	1850	*Akylopsis suaveolens*	Johann Georg Christian Lehmann	It is a synonym of *Matricaria matricarioides* (Less.) Porter
23.	1851	*Chamomilla courrantiana*	Augustin Pyramus de Candolle	It is a synonym of *Matricaria chamomilla* L
24.	1862	*Matricaria bayeri*	August (Agoston, Agost) Kanitz	It is a synonym of *Matricaria. chamomilla* L
25.	1864	*Otospermum glabrum*	Heinrich Moritz Willkomm	It is an accepted name of a species in the genus *Otospermum*
26.	1867	*Matricaria obliqua*	Joseph Dulac	The name is unresolved
27.	1874	*Heteromera fuscata*	Auguste Nicolas Pomel	It is an accepted name of a species in the genus *Heteromera*
28.	1916	*Chamomilla suaveolens*	Per Axel Rydberg	It is a synonym of *Matricaria matricarioides* (Less.) Porter

he worked on the classification and assigned a short phrase name in Latin to a species, which served to describe the species and set it apart from other species. He also included the synonyms of earlier authors (Anonymous 2012). After 16 years of further work, he published *Species Plantarum* (first edition in 1753 and second edition in 1763) and *Flora Suecica* (first edition in 1745 and second edition in 1755). He referred *Hortus Cliffortianus* for his subsequent works.

2.4.1.1 *Hortus Cliffortianus*

In *Hortus Cliffortianus*, Linnaeus described three types of *Matricaria* (Linnaeus and Ehret 1737, pp. 415–416). He did not attempt to designate any species to *Matricaria* in this work, rather he listed the prevalent synonyms of *Matricaria* at that time. The three types of *Matricaria* described by him are provided as follows:

Chamomile: Botany

1. *Matricaria* "foliis supra-decompositis setaceis, pedunculis folitariis": This Latin description finds mention in his subsequent works in describing *Matricaria chamomilla*. Linnaeus listed nine synonyms in this category. Among one of the names was *Leucanthemum dioscoridis*, which he possibly meant to be the same as that originally described by Dioscoroides.
2. Matricaria "foliis duplicato-pinnatis, petiolis solitariis": This Latin description finds mention in his subsequent works in describing *Matricaria argentea*. He listed two synonyms in this category.
3. *Matricaria* "foliis compositis planis, folios ovatis incises, pedunculis ramosis": This Latin description finds mention in his subsequent works in describing *Matricaria parthenium*. He listed 18 synonyms in this category.

In addition to the genus *Matricaria*, he had also listed four categories of the genus *Anthemis* in *Hortus Cliffortianus* (p. 145). The first category of *Anthemis* was included in *Matricaria* in his later works.

2.4.1.2 *Flora Suecica* (First Edition)

In 1745, in the first edition of *Flora Suecica*, Linnaeus described two categories of *Matricaria* (item numbers 701 and 702) (Linnaei 1745). These are provided as follows:

1. 701: *Matricaria* "receptaculis conicis, radiis deflexis, seminibus nudis, squamis calcycinis margine aequalibus."
   In addition to this description, he added as synonym *Anthemis* "foliis pinnato-decomposistis, laciniis setaceis" from category 1 of the genus *Anthemis* listed in *Hortus Cliffortianus* (p. 415). He also added two more synonyms from *Anthemis (Chamaemelum nobile* and *Chamaemelum officinarum)*. He described the pharmaceutical uses of the plant as Chamomillae Roman herba, Flores, and oleum.
2. 702: *Matricaria* "receptaculis hemisphaericis, radiis patentibus, seminibus coronato marginis, squamis calcycinis margine exsoletis."
   To this category, in addition to the description above, he added the description of "foliis supra-decompositis setaceis, pedunculis folitariis," which he took from the category 1 of *Matricaria* listed in *Hortus Cliffortianus*. Further, he retained only four synonyms of the nine listed in *Hortus Cliffortianus*. He also described that the plant and the flowers were used in pharmacy in the forms of Chamomillae Nostratis, syrups and oils.

2.4.1.3 *Species Plantarum* (First Edition)

Subsequently, in 1753, Linnaeus, in the first edition of *Species Plantarum* described five species of *Matricaria*. These species were *Matricaria parthenium, Matricaria maritima, Matricaria chamomilla, Matricaria argentea,* and *Matricaria recutita* (Linnaei 1753). Much of the descriptions were derived from his earlier works in *Hortus Cliffortianus* and *Flora Suecica* (first edition). Our concern lies with the two species of *Matricaria chamomilla* and *Matricaria recutita* only, and so, only these two are described here.

2.4.1.3.1 Matricaria chamomilla

In the first edition of *Species Plantarum*, Linnaeus described *Matricaria chamomilla* as "receptaculis hemisphaericis, radiis patentibus, seminibus coronato marginis, squamis calcycinis margine obsoletis."

As we can see, this description was provided in *Flora Suecica* (first edition), item number 702. In addition, he also added here the description of *Matricaria* "foliis supra-decompositis setaceis, pedunculis folitariis" from category 1 in *Hortus Cliffortianus*. He retained only three synonyms (compared to four in item number 702) from category 1 in *Hortus Cliffortianus*. He provided a reference of both *Flora Suecica* (first edition) and *Hortus Cliffortianus* (p. 415) in describing this species.

2.4.1.3.2 Matricaria recutita

Linnaeus described *Matricaria recutita* as "receptaculis conicis, radiis deflexis, seminibus nudis, squamis calcycinis margine aequalibus." In describing *Matricaria recutita*, he referred to *Flora Suecica* (first edition), item number 701. It is to be noted that he did not retain the synonym *Anthemis* "foliis pinnato-decomposistis, laciniis setaceis" from category 1 of the genus *Anthemis* listed in *Hortus Cliffortianus* (p. 415) that he had added to 701 of *Flora Suecica*. Further, he did not retain any synonym of item 701. Instead, he provided two synonyms that were totally new. He provided the reference of *Flora Suecica* (first edition), but he did not provide the reference of *Hortus Cliffortianus*.

2.4.1.4 *Flora Suecica* (Second Edition)

In his second edition of *Flora Suecica* in 1755, he described three species of *Matricara*: *Matricaria chamomilla*, *Matricaria inodora*, and *Matricaria suaveolens* (Linnaei 1755). He introduced the new names *inodora* (scentless) and *suaveolens* (sweetly fragrant) for two species, in this edition.

2.4.1.4.1 Matricaria chamomilla *(Item Number 764)*

In the second edition of *Flora Suecica*, Linnaeus described *Matricaria chamomilla* as "receptaculis conicis, radiis patentibus, seminibus nudis, squamis calcycinis margine aequalibus." He provided the reference of item number 702 of *Flora Suecica* (first edition). However, it is important to note that in item number 702 of *Flora Suecica* (first edition) he had described *Matricaria chamomilla* as "receptaculis hemisphaericis, radiis patentibus, seminibus coronato marginis, squamis calcycinis margine exsoletis." Now, in the second edition of *Flora Suecica* he introduced a change in the description of two characters of *Matricaria chamomilla,* such as the receptacle was now described as conical instead of hemispherical and the seeds were without a crown or pappus. He retained all the other details of *Matricaria chamomilla* as described in *Species Plantarum* of 1753, such as the pharmaceutical use of Chamomile Nostratis, syrups and oil. In addition, he mentioned that *Matricaria chamomilla* was administered to consumptive patients in Finland and also used in prediluvia.

2.4.1.4.2 Matricaria inodora *(Item Number 765)*

Linnaeus described *Matricaria inodora* as "receptaculis hemisphaericis, radiis patentibus, seminibus coronato marginis, squamis calcycinis margine exsoletis." He provided the reference of item number 702 of *Flora Suecica* (first edition). This description is the same as that provided in the reference of item number 702 of *Flora Suecica* (first edition). However, from 702, he kept only the description of *Matricara* and did not include anything else, such as the synonyms or pharmaceutical use. Instead, he introduced the synonyms *Chaememelum inodorum* and *Chamaemelum maritimum*.

2.4.1.4.3 Matricaria suaveolens *(Item Number 766)*

Linnaeus described *Matricaria suaveolens* as "receptaculis conicis, radiis deflexis, seminibus nudis, squamis calcycinis margine aequalibus." This description is the same as that provided in the item number 701 of *Flora Suecica* (first edition), and Linnaeus provided the same reference of 701. This description of *Matricaria suaveolens* was provided to *Matricaria recutita* in *Species Plantarum* of 1753. The synonyms of *Species Plantarum* of 1753 were retained in *Matricaria suaveolens* of *Flora Suecica* (first edition). It seems that Linnaeus just changed the name of the species from *recutita* to *suaveolens*, keeping all the descriptions and the synonyms intact. However, he did not retain the pharmacological usage as described in 701, namely chamomillae Roman herba, Flores, and oleum. All this has added to the confusion in the nomenclature of chamomile.

2.4.1.5 *Species Plantarum* (Second Edition)

In his second edition of *Species Plantarum*, Linnaeus described five species of *Matricaria*: *Matricaria parthenium*, *Matricaria maritima*, *Matricaria chamomilla*, *Matricaria argentea*, and *Matricaria suaveolens* (Linnaei 1763). Just as in the second edition of *Flora Suecica*, he removed the species *recutita* and described *suaveolens* in its place. The species *recutita* was never mentioned by him again.

2.4.1.6 Differences in Classification

Many notable differences in the classification of chamomile have been found between the first and the second editions of *Species Plantarum*. In the second edition, he followed the description in the second edition of *Flora Suecica*. He left out the species *recutita* and brought in the description of *suaveolens*. He retained the changed description of *Matricaria chamomilla* (receptaculis conicis, radiis patentibus, seminibus nudis, squamis calcycinis margine aequalibus) and provided a reference to *Flora Suecica* (second edition) about the pharmaceutical uses of the plant. He retained the changes of *Matricaria suaveolens* made in the second edition of *Flora Suecica*, and referred to item 766, which did not list any pharmaceutical uses of the plant.

2.4.1.7 English Translation of Linnaeus's Work

In 1775, an English translation of Linnaeus's work was published on the generic and specific description of British plants (Linnaeus 1775). In this work, Linnaeus described the genus *Matricaria* under the class Syngenesia and called it feverfew. He described five species of *Matricaria* in Britain namely *parthenium* (feverfew), *maritima* (sea feverfew), *chamomilla* (corn feverfew), *inodora* (field feverfew), and *suaveolens* (sweet-scented feverfew).

In this work, Linnaeus also described the genus *Anthemis* and called it "camomile" in the vernacular. He described four species found in Britain namely *cotula* (stinking mayweed), *arvensis* (corn camomile), *maritima* (sea camomile), and *tinctoria* (common ox-eye).

2.4.2 Nomenclature

The International Code of Botanical Nomenclature (ICBN), has conserved the name *Matricaria recutita* L. (1753) for chamomile. The ICBN Vienna Code of 2006 lists *Matricaria recutita* as typus conservandus in its Appendix IV *Nomina Specifica Conservanda et Rejicienda* (International Code of Botanical Nomenclature 2006). However, debate is still going on among taxonomists as to what is the correct name of the medicinal chamomile.

The confusion started when Linnaeus, in 1753, erroneously included the species *chamomilla*, which had cornonate achenes, in the genus *Matricaria* L., which did not agree with the generic description. Further, in 1935, *Matricaria chamomilla* was assigned the type species for the genus *Matricaria* by Hitchcock and Green (1935). This type specimen was accepted as a lectotype by Phillips in 1951 (Jeffery 1979). The error by Linnaeus was pointed out, according to Rauschert, almost 200 years later by Hylander in 1945. He said that Linnaeus, originally in 1753, had named the scentless chamomile as *Matricaria chamomilla*, but in the 1755 edition of *Flora Suecica*, he created a new binomial *Matricaria inodora* to name the scentless chamomile and for *Matricaria chamomilla* he simultaneously created a new diagnostic character. These changes by Linnaeus, according to Hylander and Rauschert, were against the rules of nomenclature. Hylander was of the opinion that the correct name of true chamomile was *Matricaria recutita* 1753, but it was never used. Instead, *Matricaria chamomilla* 1755 continued to be used for true chamomile, which according to Rauschert was incorrect (Rauschert 1974). According to Rauschert, the name *Matricaria chamomilla* 1753 was for the odorless chamomile and so it could not be applied to true chamomile. He argued that *Matricaria chamomilla* had been already selected as a standard type species for the genus *Matricaria* by Hitchcock and Green, so the genus *Matricaria* should be confined to the odorless chamomile. According to him, the correct name of the genus of true chamomile should be *Chamomilla* as described by S. F. Gray in 1821, and the name of chamomile should be *Chamomilla recutita*. This name is still used by *Flora Europea* to describe chamomile.

Jeffery (1979) did not agree with Rauschert. He was of the opinion that the lectotype chosen by Phillips was incorrect and should be rejected. This was because of the reason that the specimen *Matricaria chamomilla* had coronate achenes, which did not agree with the diagnostic character of ecoronate achenes of the genus *Matricaria* as described by Linnaeus in his *Genera Plantarum* in 1754. Jeffery argued that Pobedimova in 1961 chose *Matricaria recutita* as the lectotype, which agreed with the generic description of *Matricaria* provided by Linnaeus in 1754. However, the species *Matricaria chamomilla* described in 1755 in *Flora Suecica* also agreed with the generic description. Therefore, according to Jeffery, there were two candidates for the species type, that is, *Matricaria recutita* L. (1753) and *Matricaria chamomilla* L. (1755), which agreed with the description of having ecoronate achenes. Jeffery advocated that *Matricaria recutita*, as already chosen by Pomedinova, must be accepted as the correct name as per the Article 8 of the ICBN. Article 8 of the ICBN states that the author who first designated a lectotype must be followed, but may be superseded if it can be shown that the choice was made arbitrarily or was based on the misinterpretation of the protologue (Jeffery 1979). Jeffery also advocated that the name *Chamomilla* S. F. Gray (1821) must be considered as a synonym.

Applequist, in 2002, considered this change in name by Rauschert as confusing and undesirable and attempted to reassess the nomenclature of *Matricaria*. According to her, unknown to Rauschert, in 1974, Grierson has designated a specimen of chamomile namely Herb. Clifford: 415 *Matricaria* 1 BM (Herb. Clifford: 415; BM-000647192) as the lectotype of the genus *Matricaria*. Grierson noted that this specimen was of the variety of chamomile where the achenes of ray florets were coronate (ray-coronate) and the achenes of the tubular florets were ecoronate. He proposed the name *Matricaria chamomilla* var. *recutita*. Applequist argued that this specimen actually matched the description of *Matricaria chamomilla* L. 1753 very well and Grierson's typification defined the 1755 *Matricaria chamomilla* L. as a mere emendation of the 1753 *Matricaria chamomilla* L. name. According to Applequist, the correct name of chamomile is *Matricaria chamomilla* L. 1753.

Applequist further pointed out the observations of other authors that there could be plants of *Matricaria chamomilla* that could have either ecoronate achenes or coronate achenes, or both types of achenes. She, therefore, opined that the coronate or ecoronate character of achenes might not be consistent enough for designating varieties. In this view, a plant of *Matricaria chamomilla* with all coronate achenes should be recognized as the variety *Matricaria chamomilla* var. *chamomilla*, and a plant of *Matricaria chamomilla* with all ecoronate achenes should be recognized as *Matricaria chamomilla* var. *recutita* (Applequist 2002).

Another attempt of reassessment of the nomenclature of chamomile was made by Hansen and Christensen in 2009 (Hansen and Christensen 2009). According to them, the original description of *Matricaria chamomilla* by Linnaeus fitted the scentless mayweed, but the specimen Herb. Clifford: 415 is preferable to chamomile.

The current literature uses the names *Matricaria chamomilla* and *Matricaria recutita* as synonyms. The *Flora of Great Britain and Ireland* has also adopted the name *Matricaria recutita* to describe chamomile in its 2006 edition.

2.4.3 FLORA EUROPEA

The *Flora Europea* classifies chamomile as *Chamomilla recutita* S. F. Gray (Tutin et al. 2010). The *Flora* has, however, retained *Matricaria recutita* and *Matricaria chamomilla* as the synonyms.

> Annuals; leaves alternate, irregularly 2 to 3 pinnatisect with numerous linear segments; capitula small to medium pedunculate or subsessile; involucral bracts in two or more rows with a scarious margin; receptacle conical, hollow; scales absent; ligulate florets when present, female, white; inner florets hermaphrodite, tubular, 4 to 5 lobed; achenes mucilaginous when wet, slightly compressed, obliquely truncate above, dorsal face convex, ventral face with 3–5 narrow whitish longitudinal ribs; pappus absent, or a small corona or auricle.
>
> Lingules usually present; tubular florets 5 lobed; stems 2, 10–60 cm; capitula 10–25 mm diameter including ligules; plant glabrous (recutita).
>
> **Chamomilla S. F. Gray**

2.4.4 FLORA OF GREAT BRITAIN AND IRELAND

The *Flora of Great Britain and Ireland* describes chamomile as *Matricaria recutita* L. and includes *M. chamomilla* and *Chamomilla recutita* (L.) Rauschert as synonyms (Sell and Murrell 2006). The description of the chamomile plant (genus *Matricaria recutita* L.) in the *Flora of Great Britain and Ireland* as defined by C. Jeffery (1979) is provided below.

> Annual herb with fibrous roots, sweetly scented when fresh.
> Stems (2-) 10–60 cm, pale yellowish-green, erect, ridged, rather slender, flexuous, glabrous, leafy, branched above, sometimes nearly to the base.
> Leaves alternate, 1–7 × 1.5–2.5 cm, medium green, oblong or oblong-oblanceolate in outline, rounded at apex, finely bipinnatisect, all lobes narrowly linear or almost filiform and subacute or mucronate at the apex, glabrous sessile or with shirt petioles, swollen at base, and semiamplexicaul.
> Capitula (1-) 8–120 (300), solitary and terminal on stems and branches and forming a loose corymb, 10–25 mm in diameter, peduncles 30–100 mm, with pale ridges and dark green channels, glabrous.
> Involucral bracts in 2-(-3) rows, 1.5–3.0 × 1.0–1.5 mm, green, with pale brown scarious margins, oblong, obtuse at apex.
> Flowers of 2 kinds, the outer 6-8 mm, ligulate and female, the ligules white, shortly 3 lobed and often markedly reflexed soon after flowers open, the inner about 1.5 mm, tubular and bisexual, the tube expanded rather suddenly in the upper half, yellow, thinly glandular externally, with 5, deltoid-acute lobes at apex.
> Receptacle conical, regularly pitted, hollow, without scales.
> Achenes about 1.0 × 0.4 mm, brownish, oblong-cylindrical, slightly curved, thinly glandular, with 4–5, pale, longitudinal ribs on the adaxial face, truncate at apex; pappus absent or represented by a membranous rim.
> Flowers 6–8. Visited by flies and small bees, $2n = 18$.
> The reflexed ligules and protruding receptacle, imitating a circumcised penis (i.e., *recutita*), and the sweet scent help to recognize this species, but are not diagnostic. To be certain of the identification the mature achenes should be examined.

2.4.5 TAXONOMIC KEY

The taxonomic key of chamomile is as follows (Claphman et al. 1962; Tutin et al. 1964):

Ovules completely enclosed in an ovary which has a style and a stigma; Pollen grains adhering to the stigma and fertilization effected by pollen tube; Xylem containing vessels.

Angiospermae
Embryo with 2 cotyledons; vascular bundles of the stem arranged in single ring, cambium present; leaves reticulate veined; flowers typically 4–5 merous.

Dicotyledones
Petals united into a longer or shorter tube, very rarely free or absent.

Metachlamydeae
Herbs, shrubs or trees; leaves exstipulate; flowers hermaphrodite or unisexual, epigynous, actinomorphic to zygomorphic, arranged in heads, surrounded by one or more

Chamomile: Botany

series of free or connate bracts; calyx small, often apappus; corolla usually 4–5 lobed; stamens 5, anthers syngenecious; ovary 1 loculed, ovule solitary, placentation basal; fruit an achene, seeds without endosperm.

Asterales

Herbaceous; often with oil canals; flowers in capitula surrounded by an involucre; anthers cohering in a tube around the style; ovary inferior, unilocular with one ovule having a single integument. Style single below but branching above into 2 stigmatic arms; pollination entomophilous; fruit an achene, seed dispersal by wind, animals.

Asteraceae

Leaves usually spirally arranged and pinnatifid; involucral bracts with broadly scarious margin; head with tubular hermaphrodite fertile disc florets, ray florets ligulate and female or neutral; anthers blunt or rounds at the base; style arms of disc florets usually truncate the base; pappus absent or rudimentary.

Anthemidae

2.4.6 Classification Based on New Taxonomic Tools

With the new advances in taxonomy, several taxonomic tools have been used to identify and demarcate the diversity in chamomile. These include chemotaxonomic tools, such as metabolic fingerprinting (Daniel et al. 2008) and flavonoids (Sharifi-Tehrani and Ghasemi 2011), molecular tools, such as molecular marker DNAs (Solouki et al. 2008; Okoń and Surmacz-Magdziak 2011), and chromosomal tools, such as karyomorphology (Abd El-Twab et al. 2012).

With the emergence of new taxonomic divisions, chamomile has been classified and new ranks have been assigned to it. The National Center for Biotechnology Information (NCBI 2013) classifies chamomile in more detail. It is shown in Table 2.4.

2.5 ANATOMICAL CHARACTERISTICS OF CHAMOMILE

As the chamomile flowers are of medicinal and economic interest, the anatomical characters of the flower of chamomile have been studied thoroughly. The description of the structure of the flowers has been already provided in Chapter 1 under the Section 1.5.1.6. Here it is described very briefly, with some additional information. There are some studies on the structure of the leaves and the roots as well, which are discussed in Sections 2.5.1 through 2.5.5.

2.5.1 Stem

The anatomy of the stem has been studied by Huțanu-Bashtawi et al. (2009), and their observations are provided here. The cross section of the stem shows irregular shape with ridges. The ridges are more prominent in the cross sections of the upper parts of the stem than the basal part. Epidermis and collenchymas are present. In the upper part of the stems, the collenchyma is found just below the ridges in groups, but in the lower part of the stems, the collenchyma cells spread inwards toward the vascular bundles. The cortex is thin with only 5–6 layers of cells. The vascular bundles form a ring around the pith and are collateral and open. In the upper part of

TABLE 2.4
Classification of Chamomile according to NCBI Taxonomy

S. No.	Rank	Name
1.	Unranked	Cellular organisms
2.	Superkingdom	Eukaryota
3.	Kingdom	Viridiplantae (green plants)
4.	Phylum	Streptophyta
5.	Unranked	Streptophytina (charophyta)
6.	Unranked	Embryophyta (higher plants)
7.	Unranked	Tracheophyta (vascular plants)
8.	Unranked	Euphyllophyta (euphyllophytes)
9.	Unranked	Spermatophyta (seed plants)
10.	Unranked	Magnoliophyta (Angiospermae)
11.	Unranked	Eudicotyledons (Dicotyledoneae)
12.	Unranked	Core eudicotyledons (core eudicots)
13.	Subclass	Asterids (Asteridae)
14.	Unranked	Campanulids (Asteranae)
15.	Order	Asterales (Campanulales)
16.	Family	Asteraceae (Compositae)
17.	Subfamily	Asteroideae
18.	Tribe	Anthemideae (Anthemideae class)
19.	Genus	*Matricaria*
20.	Species	*recutita*

Source: http://names.ubio.org/browser/details.php?namebankID = 2659013; http://www.uniprot.org/taxonomy/35493.
Classification: According to *NCBI Taxonomy*.

the stem, the big vascular bundles alternate with small vascular bundles, but in the lower parts of the stem, the big vascular bundles are predominant and small vascular bundles are very few. The phloem is located external to xylem. There are a very few phloem elements found and these are sheathed in sclerenchymatous cells. There are no phloem isles. The xylem tissue has a few libriform fibers near the cambium. The pith comprises large parenchymatous cells.

Glandular hairs have been found in the stem. The glandular hairs are multicellular and biseriate. They consist of two basal cells, two peduncle cells, and six cells that form the secretory head. The structure is covered by a cuticle and secretion occurs only after the cuticular layer breaks. These glandular hairs secrete essential oil and sesquiterpene lactones. In addition, schizogenous glands are also found in the stem, which secrete essential oil and sesquiterpene lactones (Andreucci et al. 2008).

2.5.2 Leaf

The transverse section of the leaf shows a cuticle. The epidermis is single layered on adaxial and abaxial surfaces. The epidermal cell walls are anticlinal undulate to sinuate on both surfaces. Stomata are present on both surfaces. Mesophyll cells are dorsiventral and the palisade is usually mono- or bilayered and discontinuous above the midrib. Three vascular bundles, one large and two small, are found. Trichomes are present (Inceer and Ozcan 2011). Glandular hair schizogenous glands are also found in the leaf (Andreucci et al. 2008).

2.5.3 Peduncle

The cross section of the peduncle shows that the margin has ridges and furrows. Cuticle is present. The epidermis is single layered comprising rectangular cells. The hypodermis comprises 2–3 layers of cells under the ridges only. Hypodermis is absent in the furrows. The hypodermis is followed by 3–4 rows of chlorenchyma cells. Endodermis is not distinguishable. There are 9–12 vascular bundles present below the ridges and are endarch. The vessels have spiral thickenings. The pith has thin-walled cells that are comparatively larger than the other cells. A lot of parenchyma cells have oil droplets (Rashid and Ahmad 1994). Secretory ducts are present in peduncle (Andreucci et al. 2008).

2.5.4 Inflorescence or Capitulum

The microscopic studies on the flowers of chamomile by Rashid and Ahmad (1994) revealed that the bracts comprise small rectangular parenchyma cells. The longitudinal section of the conical receptacle shows that it is hollow. The wall is thin and comprises elongated parenchyma cells interspersed with thin vascular strands.

The ray or ligulate floret is female and comprises polygonal parenchymatous cells without any chlorophyll. The bracteole has thin-walled rectangular cells. The stigma has fimbriae and two vascular strands run through the length of the style.

The disc or tubular floret is hermaphrodite. The anther filaments form a thick-walled connective. The cells of the pollen sac are thin walled above and thick walled near the connective.

The pollen grains are about 21–23 µm in size, rounded and have a spiny exine (Rashid and Ahmad 1994).

Glandular hairs have been found on the involucral bracts, ray and disc florets, and ovary. Some secretory ducts have been found in the bract and stigma.

Nectariferous glands have been found in the disc or tubular florets. The nectary (93 µm × 163 µm; height × diameter) is shaped like a ring above the ovary. The nectary has a single-layer epidermis and 5–8 layers of specialized nectariferous parenchyma. The gland has branches of vascular bundles. Nectar is released through a modified stomata, which is 15–20 µm wide (Sulborska 2011).

2.5.5 Roots

The root of chamomile is a typical dicot root but it has the presence of quadrangular schizogenous ducts or secretory ducts in between the endodermis and the cortex, which secrete lipids (Andreucci et al. 2008).

2.6 REPRODUCTIVE BIOLOGY

Chamomile plants exhibit high variability. This variability is related to the entomophilous mode of pollination. The inflorescence of the plant is characteristically structured to favor pollination by insects. The receptacle bearing the flowers is erect and provides easy accessibility to the insects. In the receptacle, the whorl of flowers matures centripetally. In each flower, the stigma emerges out of the syngenesious anthers 24 hours before the anthers dehisce and insect pollination occurs with pollen from a different flower. In a study by Kuberappa et al. (2007), it was found that the flowers opened from 06.30 AM to 07.00 AM and anther dehiscence was initiated at 07.00 AM. The dehiscence of the anthers and release of pollen grains take place inside the tube of the already pollinated flower (Claphman et al. 1962). The pollen adheres to the stigma of the flower and is made available to the visiting insect during or after the emergence of the style. The honeybees have been found by Kuberappa et al. (2007) to collect nectar from the base of the corolla tube. Kuberappa et al. (2012), in South India, found mainly three species of honeybees namely *Apis cerana*, *Apis florae*, and *Trigona iridipennis* foraging. They found that *Apis florae* visited a plant 7.31 times in 5 minutes. This species was followed by *Apis cerana* (3.36 visits per plant in 5 minutes), *Trigona iridipennis* (2.65 visits per plant in 5 minutes), and other species of pollinators (2.36 visits per plant in 5 minutes). As a result of insect pollination, inbreeding is not common in chamomile. All these factors ensure outbreeding in the population, thus creating tremendous variability. In addition to creating variability, Kuberappa et al. (2012) found that the modes of pollination indeed increased the productivity of chamomile. They found that open pollination gave more yield of flowers and oil compared with the plants grown in plots caged with *Apis cerana*, *Apis florae*, and *Trigona iridipennis*.

The knowledge about the nature of flowers in the capitula permits designing of self- and cross-pollinations in the plant breeding experiments.

2.6.1 STAGES OF FLOWERING

The chamomile capitulum has been divided into four stages of flower development by the breeders (Franz 1980). The stages are based on the morphology of the flowers.

1. Stage I: Buds
2. Stage II: Ligulate flowers developed
3. Stage III: Tubular flowers opened
4. Stage IV: Decaying flower buds

These stages of flowering are linked with the levels of chemical constituents present in the plant during that stage. Using this classification, the stage of harvesting of the flowers can be easily determined.

2.7 CYTOLOGY AND PLOIDY EFFECTS

The annual herbaceous chamomile plant is a natural diploid. The chromosome number of chamomile has been reported to be $2n = 18$ (Darlington and Wylie 1955; Heywood and Harborne 1977). The chromosomes are small and the largest chromosome is 4.5 μm long. The ratios of long arms to short arms vary from 1.0 to 2.77 (Bara 1979). The basic karyotype consists of six metacentric, one submetacentric, and two subtelocentric chromosomes. The nuclear organizer region is present near the centromere on chromosome 5 and in short arms of the subtelocentric chromosomes (Falistocco et al. 1994). Abd El-Twab et al. (2008) found that the chromosome complements of chamomile measured 34.9 μm, with 14 median, 2 submedian, and 2 subterminal centromeric chromosomes. They also reported one pair of satellite chromosomes.

Madhusoodanan and Arora (1979) had classified the chromosomes based on their centromeric indexes and sizes. Karyotypic analysis of the diploid *Matricaria chamomilla* showed that of 18 chromosomes 2 pairs were with median (V), 5 pairs with submedian (L), and remaining 2 pairs with subterminal centromeres (J). The last was the smallest in the complement and can further be distinguished from each other by the presence of satellite on the short arm in one pair. The submedian chromosome could be distinguished on the basis of long/short arm ratio.

The karyotype of chamomile chromosomes was studied by G, C, and OR banding to obtain a more detailed information. Muravenko et al. (1999) analyzed the G-banding patterns of the chromosomes of chamomile. The G-banding revealed 5–10 bands per chromosome. They could identify all the nine homologous chromosome pairs using this technique. They also reported that the G-banding pattern in different cultivars of chamomile was the same, indicating a species-specific character. When they carried out computer-based imaging and analysis, they found 18 G-positive bands per chromosomes with varying staining intensities. From this data, based on the band size, position, and staining intensity, the chamomile genome ideogram was constructed. This allowed the study of the characteristics of the G-banded chromosome structure. Samatadze et al. (2001) found that the chromosome sizes in chamomile are small, which does not allow easy identification of homologous pairs. They used C-banding and OR-banding techniques to identify homologous chromosomes of chamomile. They classified nine pairs of C- and OR-banded chromosomes based on their banding patterns, sizes, and karyotypes. It was found that C-banding revealed some intra- and intervariety C-band. The OR-chromosome banding patterns in karyotypes of all varieties were similar and no OR-band polymorphism was detected. The researchers suggested that the C-banding patterns should be regarded as an important characteristic of plant genomes, and the OR-banding technique is more suitable for the investigation of the chromosome fine structure.

Cytological manipulations have been carried out to generate tetraploids with an aim to improve the quality and quantity of the essential oil and upgraded agronomic traits. Tetraploidization has been carried out using colchicine for 6 hours at 20°C (Franz and Isaac 1984), and 5- to 7-day-old seedlings have been exposed to 0.05% colchicine for 4–6 hours. The tetraploid plants have been screened and further selected for high yields. The tetraploids ($4n = 36$) are fertile and stable.

For easy identification of diploids and tetraploids, a method had been devised where the morphological characters, namely number of pollen germ pores, number of chloroplasts, and number of stomata per leaf is used. It had been found that the diploid plants have 3 germ pores in pollen, a highest number of 22 chloroplasts per pair of guard cells, and a lowest number of 31.2/mm^2 stomata per leaf. In contrast to the diploids, the tetraploid plants have 4 germ pores in pollen, a lowest number of 23 chloroplasts per pair of guard cells, and a highest number of 11.7/mm² stomata per leaf (Letchamo et al. 1994). These differences between diploids and tetraploids may not be repeatable with different varieties of chamomile.

To determine the ploidy level of cells, an efficient method using flow cytometric measurement of DNA content was developed by Carle et al. (1993). Small amount of plant material is homogenized and stained with DAPI (4′,6-diamino-2-phenylindole). The intensity of the fluorescence per nucleus is determined electronically. This intensity gives the exact measurement of the amount of DNA present in each nucleus, thereby allowing the determination of ploidy status of the cell or plant. This method also helps in the detection of aneuploids and chimeras (Carle et al. 1993).

Molecular markers or DNA markers are considered a valuable tool to determine the genetic variability or polymorphism in plant populations to enable selective plant breeding. The genetic diversity of chamomile based on DNA markers has been studied. Solouki et al. (2008) carried out a study on 25 landraces collected from Iran and Europe. Twenty nine primers were used and 369 bands were detected. Of the 369 bands, 314 bands were polymorphic. Pirkhezri et al. (2010) examined the genetic variation in 21 populations in Iran using random amplified polymorphic DNA analysis. They used 18 primers and obtained 220 bands, of which 205 bands were polymorphic.

2.8 PHYSIOLOGY

The yield of flowers and essential oil of chamomile is closely linked to the environmental conditions. The α-bisabolol and chamazulene contents of the essential oil and oil productivity of the chamomile are affected by the environment of cultivation. Lal et al. (1993) found that plants of identical genotypes grown in Egypt produced more oil than those grown in southern Germany. The Egyptian oil was richer in α-bisabolol and the German oil in chamazulene.

Gosztola et al. (2010) studied the effect of weather on the production of essential oil and the levels of the chemical constituents of the oil. They found that a moderately warm and wet weather produced the highest amount of essential oil and α-bisabolol. A moderate variability of the content of chamazulene and low variability of bisabolol oxide A and B were observed with the changes in climate. Thus, the environmental conditions influence the physiological makeup of the plant and the production of secondary metabolites. Environmental conditions such as water deficiency or salinity may cause stress in the plants. During stress conditions, plants produce stress metabolites. Umbelliferone was identified as a stress metabolite of chamomile by Repčák et al. (2001). In an experiment conducted by applying $CuCl_2$ to chamomile plants to induce abiotic stress, it was found that the levels of umbelliferone rose to 10 times the normal levels in the leaves of the plant. The levels of umbelliferone were found to be high, not only in abiotic stress but also in biotic stress caused by the infection of the powdery mildew *Erisyphe chicoracearum* DC. Another stress metabolite is methyl jasmonate.

Several physiological experiments have been carried out on the chamomile plant mainly with the aim to reveal the effects of the environmental conditions of light, temperature (Franz et al. 1975), and photoperiod, along with nutrition and growth

Chamomile: Botany

regulators on the flower and oil yield and quality. Studies on germination of chamomile seeds have also been carried out with an aim to increase the yield of the plants. Experimental studies have been carried out on chamomile plants to determine abiotic stress tolerance to salinity, water deficit, and heavy metal. In addition, the various types of biotic stress of parasite infestation, such as viruses, bacteria, fungi, and insects, have been studied. These are presented in Sections 2.8.1 through 2.8.15.

2.8.1 Germination of Chamomile Seeds

The effect of salinity (0, 25, 50, 75, 100, 200, and 250 mM NaCl), hot water (60°C and 70°C), sulfuric acid (98%), and calcium sulfate (1 and 10 mM) on the germination of chamomile seeds was studied by Zehtab-Salmasi (2008). Hot water, sulfuric acid, and calcium sulfate are used to soften the seed coats so that germination is easier. It was found that germination was about 50% till 100 mM NaCl, but at 200 mM NaCl it fell to 29%. Germination of seeds pretreated with 70°C was significantly higher than that with 60°C. There was no significant difference observed in the rate of germination in the treatments with sulfuric acid and calcium sulfate.

The effect of static magnetic field on germination was investigated by Pourakbar in 2013. It was found that a static magnetic field improved the germination of chamomile seeds. The best results were obtained at a magnetic field of 50 mT for 60 minutes. The enzymatic activities of amylase, dehydrogenase, and protease were found to be significantly increased in the magnetically treated germinating seeds (Pourakbar 2013).

2.8.2 Effect of Light

The chamomile plants were grown in an experimental setup with a light duration of 16 hours in the 24-hour cycle. The intensities of light were 90,000, 70,000, and 32,000 erg/cm^2/s. Temperature was kept constant at 20°C and relative humidity at 72%. The mature flowers were collected and analyzed for flower and oil yield and quality. The reduction in the light intensity from 90,000 to 32,000 erg/cm^2/s resulted in a decrease in the number of flowers, size of the flowers, dry weight of the flowers, essential oil yield, and chamazulene content (Saleh 1973). It is inferred that a high light intensity is favorable for flower and oil yield.

2.8.3 Effect of Temperature

Saleh (1971) exposed chamomile plants to variable day and night temperatures, under a constant 25°C day and 15°C night temperature regime. The plants were illuminated with light of 9.5 erg/cm^2/s × 104 erg/cm^2/s during 16 hours in the 24-hour cycle. The result indicated that optimum chamazulene synthesis in the flowers occurred at 25°C day temperature and 15°C night temperature. Another study by Fatma et al. (1999) revealed that the levels of chamomile and α-bisabolol oxides A and B were found to be slightly increased when the seeds were treated at low temperature. When the seeds were treated with 6°C for 15 days before sowing, the essential oil percentage and yield were high (Fatma et al. 1999).

2.8.4 Effect of Photoperiod and Diurnal Rhythms of Secondary Metabolites

Repčák et al. (1980) studied the effect of photoperiod and diurnal rhythms on the levels of the chemical constituents of chamomile. Two chamomile cultivars were grown under varying photoperiods (light periods) of 8, 13, and 18 hours per 24-hour cycle. Best performance in terms of flower and oil yield in both the cultivars was observed when the photoperiod was of 18 hours. In a separate study, it was found that the sesquiterpenes in the chamomile exhibited a 12-hour cycle in their accumulation of highest concentration. It was found that α-bisabolol showed a 24-hour rhythm, whereas chamazulene and bisabolol oxides A and B showed 12-hour diurnal cycles. The maximum concentration of α-bisabolol was observed to be at night (Repčák et al. 1980).

2.8.5 Effect of Nitrogen, Potassium, and Phosphorus

Krstic-Pavlovic and Dzamic (1984) found out that the highest yield of chamomile flowers as well as essential oil percentage was obtained using NPK 60:30:30 and NPK 40:20:20. Letchamo (1993) found that the plant height, the straw yield, number of productive tillers, primary branches, and volume of the flower heads significantly increased due to the increased levels of N (NH_4NO_3) application. He concluded that nitrogen application (1.2 g/2 kg soil and 1.6 g/2 kg soil) during growing conditions has a positive effect on the yield of the essential oil of chamomile genotypes and favors the content of its active substances, but the response is affected by the genotype. Emongor et al. (1990) found that nitrogen significantly increased the essential oil yield and its constituents. They concluded that nitrogen might have increased the carbohydrate accumulation and the gibberellin and auxin concentrations in chamomile plants, thereby increasing the essential oil yield by the secretory ducts, cavities, or glandular hairs. Nikolova et al. (1999) found that the best yield of chamomile plant was achieved when treated with NPK in the ratio of 40:28:32. They found that nitrogen and potassium increased the yields, whereas phosphorus increased the essential oil content and α-bisabolol level.

The deficiency of nitrogen (nitrate) in chamomile plants was studied (Kováčik et al. 2011a). It was found that, as expected, the nitrate deficiency reduced the nitrogen-based compounds (free amino acids and soluble proteins) and increased the carbon-based compounds such as the reducing sugars, soluble phenols, coumarins, phenolic acids, and flavonoids. The levels of proline, a stress protective amino acid, decreased and the activity of phenylalanine ammonia lyase (PAL) was stimulated in the nitrogen deficient plants. Eliasová et al. (2012) studied the effect of nitrogen deficiency on tetracoumaroyl spermine. They found that tetracoumaroyl spermine was present mainly in tubular flowers and reached maximum levels in the third phase of flowering when the corolla of the tubular flowers starts to open. The researchers found that the levels of tetracoumaroyl spermine were not affected by nitrogen deficiency.

2.8.6 Effect of Calcium, Sulfur, and Magnesium

Nikolova et al. (1999) found that calcium was important for increasing the yield and dry matter, whereas sulfur increased inflorescence weight and essential oil content. Upadhyay and Patra (2011) studied the influence of calcium and magnesium on the growth of chamomile and yield of flowers and found that the combination of equal parts of calcium and magnesium at the rate of 200 mg per pot (10 kg of soil) increased the flower yield and the oil content.

2.8.7 Effect of Indole Acetic Acid, Gibberellin, Kinetin, and Other Growth Regulators

Indole acetic acid, gibberellin, and kinetin are plant growth regulators that control the growth and development of the plants. Indole acetic acid is known to increase growth through cell elongation, gibberellins enhance growth, flowering, and increase in the essential oil constituents, and kinetin increases the essential oil yield of chamomile. Salicylic acid (SA) is a growth regulator, which has been found to influence the growth of plants and regulate the phenolic metabolism. Mostafa et al. (1983) had studied the combined effect of growth regulators (gibberellins, ethephon, cycocel, and B-9) and nitrogen fertilization on growth, flowering, oil yield, and chamazulene percentage of chamomile. They concluded that nitrogen fertilizer at 300 kg/feddan (1 feddan = 0.42 ha) and lower concentration of the growth regulators (gibberellins 10–100 ppm; ethephon 1–10 ppm; cyocel 50–500 ppm; and B-9 50–500 ppm) were most effective in increasing the plant height, flower yield, and oil and chamazulene percentage. Fatma et al. (2010) observed that the vegetative growth improved when the plant leaves were sprayed with 50 mg/L indole acetic acid, 100 mg/L gibberellin (GA_3), and 50 mg/L kinetin. Total carbohydrate, total nitrogen, and crude protein content also increased at the above concentrations of the phytohormones. The number of flowers and their fresh and dry weights increased with 50 mg/L indole acetic acid, 50 and 100 mg/L gibberellin (GA_3), and 50 mg/L kinetin. Spraying with 50 mg/L kinetin increased the essential oil content. The levels of the chemical constituents in the oil also varied with the varying concentrations of the phytohormones. The percentage of α-bisabolol was found to be the highest at 100 mg/L indole acetic acid or kinetin and the percentage of chamazulene was found to be the highest at 100 mg/L gibberellin.

SA influences the growth of chamomile plants and also the metabolism of phenols. At 50 μM, SA promoted plant growth, whereas at 250 μL it inhibited growth. At higher doses, SA increased PAL activity and the levels of soluble phenolic acids (Kováčik et al. 2009).

Sharafzadeh and Bayatpoor (2012) observed that treating chamomile plants with naphthaleneacetic acid and spermidine had an enhancing effect on the levels of the essential oil constituents.

The regulators have an important role in adapting the plant to metal tolerance. Kováčik et al. (2012a) treated aluminum-exposed chamomile plants with four regulators; 2-aminoindane-2-phosphonic acid (AIP), SA, sodium nitroprusside (SNP), and dithiothreitol (DTT). The effect of these regulators on the physiological parameters of tissue water content, soluble proteins, reducing sugars, K^+ content, root lignin content, and free amino acids of the aluminum-exposed plants was studied. SA had the negative impact on the aluminum-induced changes. SA and DTT stimulated the increase in the levels of phenolic acids and flavonols and soluble phenols in the shoot. SNP had the least visible effect, which suggested a protective effect of nitric oxide. The researchers concluded that phenolic metabolites may affect shoot aluminum uptake.

Repčák and Suvák (2013) reported that application of methyl jasmonate increases the levels of (Z)- and (E)-2-β-D glucopyranosyloxy-4-methoxycinnamic acids (GMCA), the precursor of herniarin. When attacked by insects, GMCA levels are increased. This suggests that methyl jasmonate could be used to regulate coumarin accumulation during pest attack.

2.8.8 Water Stress

The effect of water stress (both excess water and water deficit) on the plant was studied by Pirzad et al. (2011). They found that although the parameters of leaf-relative water content, proline levels, and water soluble sugars were unaffected, the levels of total amounts of chlorophyll *a* and *b* were affected. The researchers found that in both the conditions of excess water and water deficit, the levels of chlorophyll *a* and *b* in the leaves were significantly decreased. Bączek-Kwinta et al. (2011), however, reported that there was no significant decrease in the chlorophyll *a/b* during drought, but the levels did increase after rehydration. They also found that the chlorophyll fluorescence parameters during water shortage and rehydration changed. The parameters studied for the dark- and light-adapted leaf were minimal, maximal, and variable fluorescence of the Photosystem II (PSII) reaction centers. They observed that during drought conditions, there was a significant increase in these parameters of the dark-adapted and the light-adapted leaf. There was a reduction in the linear electron transport rate, quantum efficiency of PSII electron transport, and photochemical quenching coefficient. These changes were observed until rehydration. In addition, they found that the carotenoid content increased after rehydration. They concluded that during drought conditions the photosynthetic apparatus of chamomile plants gets impaired.

The possibility of water-deficit stress tolerance of chamomile through the use of hexaconazole was investigated by Hojati et al. (2011). Hexaconazole at 15 mg/L was found to induce better drought tolerance in chamomile. It was deduced that hexaconazole produced this effect through inducing changes in the growth parameters and morphological and biochemical characteristics, such as productivity and the relative water, protein, and proline contents. Hexaconazole was also found to induce changes in the nonenzymatic and enzymatic antioxidant levels and apigenin-7-glucoside content, which suggested that these compounds are related to water deficit stress in chamomile plants.

2.8.9 Salt Stress

Salt stress is one of the most common abiotic stresses. High salt concentrations cause water deficit, ion nutrient imbalance, and oxidative stress. These can cause modification in root morphology and even death of the plant. A study was conducted to determine the effect of salt stress and growth of the plant and the quality and quantity of the essential oil in chamomile. It was found that an increase in salinity caused a reduction in the growth of the plant. High salinity levels decreased plant height, number of branches, number of flowers, and essential oil content. The effect of salinity was highest on the dry flower weight (Dadkhah 2010). The effect of salinity on the growth of the chamomile plant was also studied by Ghanavati and Sengul (2010).

They found that with increasing NaCl doses the dry matter yield was decreased. The concentration of Na increased in the plants with increasing concentration of salt. In a subsequent study, Ghanavati et al. (2011) treated chamomile plants with different concentrations of NaCl at different stages of growth. The effect of salt treatment on the characteristics of plant height, root length, number of leaves, number of nodes, fresh weight of stem, and dry weight of stem was measured. The results indicated that the dry weight decreased with increasing doses of salt.

The impact of salinity combined with nitrogen deficiency on copper uptake was studied. Kováčik et al. (2012b) observed that salinity suppressed the growth, water content, soluble proteins, and reducing sugars more than nitrogen deficiency. The peroxidase activities were pronounced under salinity accompanied by an increase in proline levels. Activities of phenolic enzymes were suppressed and lignin content increased under salinity. The uptake of K, Fe, and Mg were reduced under salinity conditions.

Dagar et al. (2011) concluded in a review that chamomile plants can be successfully cultivated on alkaline/saline soils.

Amino acids have been found to have a positive effect on the growth of chamomile plants in saline soils. Omer et al. (2013) reported that spraying amino acids improved the height, fresh and dry weight of flowers, and increased the level of chlorophyll *a*, proline, and polyphenol.

Efforts have been made to find out ways in which the plants could be made to tolerate salt stress. Plants produce methyl jasmonate, a volatile compound, in response to many biotic and abiotic stresses. It acts as a signaling compound to other plants indicating a stress condition. The effect of methyl jasmonate on chamomile was studied by Salimi et al. (2012). Their aim was to find out if methyl jasmonate improved the salt tolerance of chamomile through bringing in a change in the properties of membrane ion uptake and osmotic adjustment in the cell. The effect of methyl jasmonate on flower weight was also determined. Different concentrations of methyl jasmonate were sprayed at three different growth stages of the plants. In addition, salt stress at different concentrations of salt was applied. They found that methyl jasmonate prevented the entry of salt and allowed preferential absorption of K^+. It was concluded that methyl jasmonate induced an increase in proline content, which led to cell membrane stability and salt tolerance.

In another study in which the chamomile plants were subjected to salt stress and iron deficiency stress (Siahsar et al. 2011), two salinity stress-induced polypeptides were produced, one of 14 kDa and the other of 18 kDa molecular weight. It is yet to be found how these polypeptides help the plant to adapt to salinity stress.

Nasrin et al. (2012) studied the effect of SNP and spermidine on the alleviation of salt stress of chamomile plants. They pretreated the chamomile plants with SNP and spermidine and found that SNP caused enhancement of plant growth, reduction of salt stress-induced malondialdehyde and hydrogen peroxide, and increased ascorbate peroxidase activity. Spermidine also enhanced plant growth, reduction of malondialdehyde and hydrogen peroxide, and increased guaiacol peroxidase activity. The researchers concluded that spermidine probably acts through the nitric oxide pathway in reducing salt stress.

Tai et al. (2013) reported that with the increasing concentrations of NaCl in the soil, the morphological, anatomical, and biochemical changes occur in the chamomile plant. Morphologically, the plant growth is stunted and the yield is low. Anatomically, under salt stress, the number of collenchyma cells and vacuoles in the stem increase and vessels in vascular bundle are more developed. The air cavities in the root cortex and leaf parenchyma increase. Biochemically, the levels of chlorophyll increase, and then decrease. The soluble sugars, proline, and peroxidase activity increase.

2.8.10 Heat Stress

With a view to cultivate chamomile under hot conditions, studies are being conducted. The effect of SA on chamomile plants grown under heat stress conditions was studied by Ghasemi et al. (2013). They cultivated chamomile plants under high temperature conditions and treated the plants with different doses of SA. They found that SA caused heat tolerance in chamomile plants as evident by a higher plant height, flower diameter, and fresh and dry weight. They recommended that SA could be used at a concentration of 100 mg/L to treat heat stress.

2.8.11 Toxic Metals

Cadmium, nickel, and lead are the three most important metals that are toxic not only for plants but for humans as well. Other metals such as zinc and copper are useful to the plant in low doses, but at higher doses they become toxic. Metals such as K, Ca, Mg, Al, Cu, Cr, Hg, and Cd have accumulated by the buds and flowers of chamomile (Kováčik et al. 2012c). Since the flowers are used as drug, the levels of these metals are required to be below toxic levels. Several experimental studies have been carried out to determine the effect of these metals on chamomile plant. The level of accumulation of these toxic metals in the plant has also been studied. Grejtovsky et al. (2001) studied the effect of increasing doses of cadmium (0, 0.3, 3, and 30 mg/kg of soil) on the essential oil percentage and the content of some secondary metabolites of chamomile. The essential oil content was found to be decreased by 10% on an average in dry flowers. The content of chamazulene and α-bisabolol in the oil decreased by 28% and 20%, respectively. Interestingly, the levels of β-farnesene and en-yn-dicycloethers increased with increasing concentrations of cadmium. No visible symptoms were observed in the plants. The yield of flowers was not influenced significantly by cadmium. A study by Pavlovič et al. (2006) on the response of chamomile plants to cadmium treatment indicated that the plant is a cadmium accumulator species and exhibits high tolerance to cadmium. This finding was supported by studies made by Salamon et al. (2007a,b). Kováčik et al. (2006) studied the effect of high (60–120 μL) and low (3 μL) concentrations of cadmium on the metabolism of chamomile. It was found that cadmium only slightly altered metabolism, which indicated the plant's tolerance to cadmium. At high concentrations of cadmium the roots, instead of the shoot, accumulated cadmium. This finding that the roots accumulated more cadmium was also reported by Lesíková et al. (2007). The researchers concluded that chamomile was not a hyperaccumulator of cadmium and, therefore, not suitable for cadmium phytoremediation. This finding was reconfirmed by Kováčik (2013). Kováčik et al. (2007) studied the effect of nitrogen deficiency on the quantitative changes in levels of some free amino acids and coumarins (herniarin and its glucosidic precursors (Z)- and (E)-GMCA; umbelliferone) in the leaf rosettes of chamomile. The biosynthesis of coumarins is connected with aromatic amino acid metabolism and nitrogen uptake. They found that there was a decrease in levels of all detected amino acids, besides histidine. The GMCA increased sharply, herniarin increased slowly, and the content of umbelliferone was low. They concluded that nitrogen deficiency is not an inducing factor for stress accumulation of herniarin and umbelliferone. They also opined that the increase of herniarin glucosidic precursors was apparently due to enhancing PAL activity under nitrogen deficiency, and nitrogen-free carbon skeletons were shunted in to the phenylpropanoid metabolism, including biosynthesis of (Z)- and (E)-GMCA. In another study, Kováčik and Bačkor (2007) found that most benzoic acids (p-hydroxybenzoic, syringic, and vanillic), cinnamic acids (caffeic, chlorogenic, o- and p-coumaric, and ferulic), herniarin, and (Z)- and (E)-GMCA increased with prolonged nitrogen deficiency. The PAL activity increased suggesting it was an important biochemical factor contributing to the observed increase of phenolic compound accumulation by producing nitrogen-free skeletons of t-cinnamate for subsequent pathways of phenylpropanoid metabolism. They also found that the nonaltered extent of lipid peroxidation expressed as malondialdehyde amounts revealed that membrane integrity was not affected by N deficiency.

Kummerova et al. (2010) studied the effect of zinc (12–180 µM) and cadmium (12 µM) on the physiological characteristics of chamomile. They found that with increasing zinc concentrations and without cadmium the roots accumulated higher amounts of zinc in the shoots. When cadmium was added, at higher concentrations of zinc (120–180 µM) there was an accumulation of zinc in the root. The presence of cadmium and zinc (12–120 µM) decreased the accumulation of cadmium in the roots. On the other hand, the amount of cadmium increased in the leaves of the plants treated with cadmium and zinc together. Higher concentrations of zinc (120–180 µM) and a mixture of zinc and cadmium negatively affected the dry mass, carotenoid content F_V/F_F, and root respiration rate. Chlorophyll a was reduced to a higher extent than chlorophyll b.

It was shown that phenols have important role in cadmium and Ni uptake. A study was conducted in vitro to investigate the effects of AIP, a potent PAL inhibitor, on the accumulation of cadmium and nickel in chamomile. It was found that AIP reduced the PAL activity. In plants that were cultured with Ni or cadmium with AIP, the PAL activity was higher. Some increase in proline and tyrosine was observed, whereas a reduction in the levels of coumarin-related compounds, cell-wall-bound phenols, and phenolic acids was observed (Kováčik et al. 2011b).

A study was carried out to examine the effect of vermicompost on Ni and cadmium uptake by chamomile plants. It was found that the application of vermicompost at 2.5 g/kg of soil significantly enhanced the accumulation of Ni and cadmium. The researchers found that although a sizeable amount of Ni and cadmium got accumulated in the flowers, the essential oil did not contain any heavy metal (Chand et al. 2012).

The effect of calcium on cadmium toxicity was studied and it was found that high levels of calcium alleviated the harmfulness of cadmium. Researchers suggested that the calcium might mediate the accumulation of cadmium in chamomile plants (Farzadfar and Zarinkamar 2012). In another study, Farzadfar and Zarinkamar (2013), using a hydroponic experimental setup, concluded that addition of calcium to the growth media decreased cadmium uptake and other cadmium-associated toxicity such as activity of antioxidant enzymes and accumulation of reactive oxygen species. However, there was an increase in the growth of the biomass. They concluded that this reduction in cadmium accumulation could be attributed to Ca-induced reduction of cadmium concentration through reducing cell surface negativity and competing for cadmium ion influx.

The effect of copper chelates on reducing cadmium accumulation was studied by Král'ová et al. (2007). They found that copper accumulated in roots and when applied in the form of a copper chelate, it led to translocation of copper in the shoots. The effect of copper on chamomile diploid and tetraploid cultivars was studied by Kováčik et al. (2011c). They found that the diploid plants had accumulated more copper in their shoots and roots. Further, the potassium content and phenolic metabolism were more reduced in the diploid cultivars. In addition, the root tissue and dry biomass were also more reduced. The soluble proteins were depressed and sulfur-containing amino acids increased in both tetraploids and diploids. In a subsequent study, Kováčik et al. (2012c) found that tetraploid plants had decreased accumulation of toxic metals but increased levels of phenolic metabolites, making them safe and suitable for pharmaceutical purposes.

The effect of lead on chamomile plants was studied by Grejtovský et al. (2008). The chamomile plants were treated with four different concentrations of lead (5, 25, 50, and 75 µmol/L). They found that most of the lead was accumulated in the roots and a small amount was found in the shoots and leaves. The root biomass reduced significantly by 46%. The accumulation of lead (50 µmol/L dose) was investigated in the flowers and it was found that the levels of lead accumulated were three times higher than the normal control plants. However, the level of accumulated lead was much less (2 mg/kg) than the permissible levels of lead (10 mg/kg) by the WHO.

2.8.12 Effect of Inorganic and Organic Fertilizers and Mycorrhiza

An experiment using hydroponic cultures of chamomile was carried out to compare the effect of organic (humic acid) and inorganic (75% Steiner solution) fertilizers on the yield of flowers and essential oil. The researchers found that inorganic fertilizers have a positive influence on the morphological parameters. It was, however, found that the organic fertilizer increased the levels of α-bisabolol in the essential oil (Juárez-Rosete et al. 2012).

The effect of different species of symbiotic mycorrhiza and phosphate solubilizing bacteria (*Bacillus coagulans*) on the qualitative characters, and quantitative traits of chamomile was studied (Farkoosh et al. 2011). It was found that there was an enhancement of the qualitative characters of oil yield and content of chamazulene and α-bisabolol. Chamomile plants have been found to support spores of arbuscular mycorrhizal fungi (Uroviche et al. 2012).

Sharafzadeh et al. (2012) reported that vermicompost was observed to enhance the levels of chamazulene, α-bisabolol, α-bisabolol oxide A, α-bisabolone oxide A, α-bisabolol oxide B, and en-yn-dicycloether in the chamomile essential oil.

2.8.13 Effect of Weedicide

Weedicides are used during cultivation of chamomile crop. The choice of appropriate weedicide that does not damage the plant and leaves little residue is important. Schilcher (1978) reported that several weedicides were used, such as atrazine, prometon, prometryn, linuron, and 2,4-dichlorphenoxyacetic acid. He found that the growth of the plants and the composition of the essential oils were strongly influenced by the more substantial atrazine treatments. He also found that application of linuron after germination damaged the cultivated plants more than the weeds. Mackova (1992) studied the effect of another herbicide, dichlorprop. Dichlorprop is a chlorophenoxy herbicide similar in structure to 2,4-dichlorphenoxyacetic acid. It was found in the study that chamomile plants were found to be less sensitive toward dichlorprop. There was no negative impact of dichlorprop on the content and composition of essential oil and on the fertility of the seeds.

2.8.14 Biotic Stress

The chamomile plants are subjected to many biotic stresses (Petanović et al. 2000; Ale-Agha et al. 2002; Bharat 2003; Fuss et al. 2005; Pérez Bocourt et al. 2005; Plescher 2005; Fránová et al. 2007). The parasites that cause biotic stress are various, such as viruses, bacteria, fungi, and insects. The pathogens are specific to different environments. They are known to occur on chamomile plants in different parts of the world and cause a lot of loss. In fact, it has been reported that in Germany, since 2007, chamomile cultivation has been severely affected due to an unknown complex disease, which could be a combination of fungal infection (*Septoria* coupled with larval infestation of an unidentified insect [Gärber et al. 2013]). These parasites and the disease symptoms in chamomile plants are tabulated in Table 2.5.

2.8.15 Effect of Biopesticides

Considering the huge array of pests affecting chamomile, it becomes important to investigate safe pest control measures for chamomile. A study on the effect of biopesticides on chamomile plants was carried out by Sánchez-Govín et al. (2010). Biopesticides made of *Melia azedarach* L., *Nicotiana tabacum* L., *Beauveria brassiana*, *Verticillium lecanii*, and *Trichoderma viridis* with *Bacillus thuringiensis* were sprayed on the plants. The treatment with biopesticides showed a favorable response to the essential oil content and water-soluble substances.

TABLE 2.5
Parasites on Chamomile Plants and Symptoms of Disease

S. No.	Parasite Type	Name	Symptoms
1.	Viruses	Lettuce big-vein virus	Not visible on chamomile plants. Chamomile is the host plant that harbors these viruses. These viruses are transmitted by aphids
2.	Mollicutes	Phytoplasmas	Reduced elongation of shoot tip internode is observed. Flower malformations occur and numerous leaves and branches appear at the site of infection
3.	Plant parasite	*Orobanche* spp.	In southern Europe, this plant parasite is found to be growing on chamomile plants
4.	Fungi	*Fusarium* spp.	Plant growth is stunted and it turns yellow. The roots turn dark and decay. Basal stem rots are also observed
		Erysiphe cichoracearum, Erysiphe polyphaga	These fungi cause powdery mildew. White powdery patches appear on the plants, which soon cover the entire plant. Newly emerging flowers are dwarfed as a result and seed yield is reduced
		Oidium spp.	It was observed in Cuba that the chamomile plants were infected with *Oidium* species. The symptoms are similar to powdery mildew
		Ascochyta	*Ascochyta* causes leaf blight, where the tip of the leaves gets bleached and starts dying from the tips to the base. This fungus was found to infect chamomile plants in the Ruhr basin of Germany
		Peronospora radii, Plasmopara leptosperma	These fungi cause downy mildew. Infections appear as bleached patches on leaves and the leaves turn yellow to brown and die
		Puccinia matricariae	This fungus causes chamomile rust. The plants growing in temperate and cold climates get infected by this fungus. Pale brown rusty pustules form on the leaves and stem. Black powdery telia are also found but these are restricted to the stems
		Stemphylium botryosum	This fungus can be found on chamomile plants after long periods of wet weather. The symptoms include spherical brown, grey, or black spots on the leaves. The midribs eventually collapse

Chamomile: Botany

TABLE 2.5 (*Continued*)
Parasites on Chamomile Plants and Symptoms of Disease

S. No.	Parasite Type	Name	Symptoms
5.	Insects	*Melolontha* spp., *Phyllopertha* spp., *Rhizotrogus* spp., *Pales* spp., *Tipula* spp., *Agriotes* spp., *Athous niger, Melanotus brunnipes, Sciota* spp., *Delia* spp., *Gryllotalpa vulgaris, Blaniulus guttulatus, Cylindroiulus teutonicus*	These insects chew on the chamomile plants. Found in the temperate to warm climates, mostly the larvae of these insects chew on the roots, stem bases, and leaves close to the ground. This results in stunted growth of the plants and they eventually die. Most of the infected plants can be easily pulled out of the soil as the stem bases have detached from the roots
		Meloidogyne hapla, Meloidogyne incognita (Kofoid and White) Chitwood race 3	These insects form root galls. At the point of infection, the roots swell and form galls in which the females lay eggs. The infected plants are stunted and are more susceptible to drought stress
		Cucullia tanaceti, Phalonia implicate, Helix spp., *Arianta* spp., *Arion* spp., *Deroceras reticulatum*	These insects feed on the leaves and shoots causing considerable damage to chamomile plants
		Aphis spp., *Cerosipha gossypii, Myrus persicae, Brachycaudus* spp., *Cicadinae* spp., *Eupteryx atropunctata, Empoasca pteridis, Empoasca flavescens, Chlorita viridula, Lygus lucorum, Lygus pubescens, Exolygus pratensis, Plagiognathus chrysamthemi, Adelphocoris lineolatus, Calocoris norvegicus, Trialeurodes vaporariorum*	Some species of these insects, such as *Aphis*, are found on chamomile plants all over the world. These insects feed on the sap of the stem and the leaves. The tip of the leaves curls up and the leaves turn yellow. Finally, the leaves die. Deformation of the leaves and stems is also observed
		Ceutorhyncus rugulosus, Apion confluens, Phytomyza atricornis, Phytomyza matricariae, Liriomyza strigata, Typetha zoe	These insects bore and mine the leaves and central stem pith of the stem causing damage to the chamomile plants

(*Continued*)

TABLE 2.5 (*Continued*)
Parasites on Chamomile Plants and Symptoms of Disease

S. No.	Parasite Type	Name	Symptoms
		Meligethes, Olibrus aeneus, Pseudostyphlus pilumnus, Trupanea stellata	These insects feed on mature pollen released by the anthers of chamomile flowers. As a result of the infestation, although no damage is done, the flower color is affected, which in turn affects drug quality
		Thrips physapus, Thrips tabaci	These insects feed on the tubular flowers of chamomile, which wither and turn brown
		Coleoptera	These insects are ladybugs, which colonize on aphid infested chamomile plants. The harvested plant material contains the bright wings of the insects, which are difficult to remove. The quality of the drug is affected
		Aceria matricariae	These insects have been found on the chamomile plants growing in Serbia. These infest the flowers of chamomile and cause considerable damage

2.9 BIOCHEMISTRY

The mechanisms of different biochemical reactions involved in the formation of the secondary metabolites are being explored and the enzymes catalyzing the reactions in different biochemical pathways are being discovered.

2.9.1 Enzymes

A preliminary biochemical investigation has resulted in the isolation of a protein—P33 of 33 kDa molecular weight. The protein has been found to be associated with the oil bodies in the cultured chamomile shoot primordia. The oil bodies are normally found in the mature seeds. The protein shows carboxypeptidase activity and in an assay, it acted by liberating the carboxyl terminal amino acid residues of angiotensin I (Hirata et al. 1996). The protein showed maximum activity at 6.5 pH. It was inactive below 5 pH or above 8.5 pH. The activity was inhibited by specific inhibitors—iodoacetamide, *p*-chloromercuribenzoic acid, and L-3-*trans*-carboxyoxiran-2-carbonyl-L-leucyl-agmatine (E-64). It appears that the P33 protein has a cell-protective function.

An enzyme, flavone glycoside that cleaves β-glucosidase was isolated and purified from the freeze-dried ligulate florets of chamomile (Maier et al. 1991a, 1993). This enzyme is thought to be involved in the formation of the pharmacologically active apigenin from its glucosides namely apigenin-7-glucoside and its other derivatives. The enzyme was observed to have a molecular weight of 500 kDa. Only one subunit of the enzyme of 60 kDa has been isolated using sodium dodecyl sulfate polyacrylamide gel electrophoresis. The enzyme showed specificity by cleaving the sugar linkage of apigenin-7-*O*-glucosides, preferring over apigenin-3-*O*-glucosides.

Chamomile: Botany

It showed maximum activity at 5.6 pH with half maximum activity at 4.3 pH and 6.7 pH. Its isoelectric point was at 4.6 pH. The enzyme acted at an optimum temperature of 36°C and the activity decreased to 50% at 26°C and 45.5°C. The energy of activation of this enzyme was estimated as 32.9 kJ/mol (Maier et al. 1991b, 1993). In a separate study, apigenin-7-O-β-glucoside was hydrolyzed by β-glucosidase extracted from almonds (Pekic et al. 1994).

2.9.2 Polysaccharides and Pectic Acid

Several polysaccharides with possible immunostimulating effects were isolated from the capitula of chamomile (Fuller et al. 1991; Fuller 1992; Fuller and Franz 1993). Yakovlev and Gorin (1977) isolated pectic acid and studied it using enzyme hydrolysis, periodate oxidation, partial acid hydrolysis, and methylation. The main polysaccharide chain of the pectic acid was isolated and identified as 2,3,6-tri-O-methyl-D-galactose. This suggested the presence of D-galacturonic acid residues in the pyranose form linked by 1–4 glycosidic bonds. Single branchings consisting of the neutral monosaccharides galactose, arabinose, and xylose are possible. L-Rhamnose was also isolated, which could be included in the main polysaccharide chain.

2.9.3 Chamazulene Synthesis

Chamazulene is responsible for the blue color of chamomile oil (Kaul et al. 1993). It was reported that chamazulene itself is not present in the capitula, but it is formed from the precursor matricine, a sesquiterpene lactone, during the process of distillation (Mann and Staba 1986). Azulene and guaiazulene have also been identified as the colored compounds in chamomile oil. Like chamazulene, azulene and guaiazulene are also formed from the precursors during distillation. The process of conversion of matricine to chamazulene during hydrodistillation seems to involve hydrolysis of matricine to the intermediate hydroxycarboxylic acid, which loses water to form chamazulenic acid, which is then converted into chamazulene by decarboxylation of the carboxyl group.

2.9.4 Biosynthesis of Other Sesquiterpenes

The essential oil of chamomile contains α-bisabolol, bisabolol oxide A, bisabolol oxide B, and matricine. These compounds are known as sesquiterpenes. The sesquiterpenes are a class of terpenes that are made of three isoprene units. The sesquiterpene structure could be acyclic, monocyclic, bicyclic, or tricyclic. α-Bisabolol is an unsaturated monocyclic sesquiterpene alcohol. It is a viscous oil with a sweet floral aroma. α-Bisabolol and its oxides, namely bisabolol oxide A and B, and matricine are synthesized from a basic molecule known as isopentenyl diphosphate (IPP). It has been shown that IPP is formed by two pathways, namely the acetate–mevalonate pathway and the nonmevalonate pathway. The acetate–mevalonate pathway works in the cytoplasm and the starting compound is acetyl CoA, which is converted to mevalonate through a series of steps. The mevalonate is converted to mevalonate phosphate by the enzyme mevalonate kinase, and the enzyme mevalonate diphosphate decarboxylase converts the mevalonate diphosphate into IPP (Dubey et al. 2003).

The nonmevalonate pathway works in the plastids and the starting compound is pyruvate or glyceraldehyde-3-phosphate. These are converted to 1-deoxy-D-xylulose-5-phosphate (DOXP) by the enzyme DOXP synthase. After several steps, dimethylallyl diphosphate is produced, which gets converted to IPP (Dubey et al. 2003).

Labeling studies have shown that two of the isoprene building blocks of bisabolol oxide A and chamazulene are formed via the nonmevalonic pathway (triose pyruvate pathway [TPP]) and the third isoprene is derived via both mevalonic acid pathway and the nonmevalonic pathway (TPP) (Holz 1979; Holz and Meithing 1994; Adam and Zapp 1998).

2.9.5 Biosynthesis of Coumarins

The coumarin herniarin (7-methoxycoumarin) is synthesized from (Z)- and (E)-GMCA. By the action of an enzyme, or by the action of UV light, (E)-GMCA gets isomerized to (Z)-GMCA. (Z)-GMCA on the action of β-glucosidase gets hydrolyzed and spontaneously lactonizes to herniarin (Eliašová et al. 2004). Umbelliferone (7-hydroxycoumarin) is produced from L-phenylalanine via the shikimate pathway.

2.10 DESCRIPTORS OF CHAMOMILE

2.10.1 Need of Descriptors

The biodiversity of chamomile plants needs to be conserved. The variety of genotypes or accessions that are found in different climates needs to be documented as descriptors. These documented and conserved accessions will help in better understanding how different climates, soils, and other factors affect the combinations of genes that are found in plants, and plant breeders and other users of germplasm collections will be able to more quickly identify traits that are best for improving crops (Steiner and Greene 1995). The authors had opined that at that time, a description of the collection site was required when collecting accessions, but the kinds and amount of information recorded that described the natural environmental features of the collection site varied greatly depending on the collector. They emphasized the need for standardized and detailed ecological descriptors for collected accessions. The guidelines for preparing the descriptors had been started by the International Bureau of Plant Genetic Resources in the 1970s (Biodiversity International 1995).

2.10.2 Definition and Types of Descriptor

A descriptor is defined as an attribute, characteristic, or measurable trait that is observed in an accession of a gene bank. It is used to facilitate data classification, storage, retrieval, exchange, and use. Over the years, the format of the descriptors has developed considerably. The different types of descriptors are described below in brief:

1. Minimum descriptors: A minimum number of characteristics of a crop were provided. However, several useful descriptors lacked internationally accepted definitions of the description of the characteristics.
2. Comprehensive list of descriptors: Comprehensive lists that included all descriptions of a crop plant were used for characterization and evaluation. These also included details of site environment where the plant was found and management of the drop plant.
3. Highly discriminating descriptors for international harmonization: The comprehensive descriptor lists were revised to harmonize according to international requirements.

The crop descriptors begin with an introduction to the crop, which highlights the economical and nutritional value of the crop. Each descriptor consists of a descriptor name, a descriptor state, and a descriptor method. Descriptor names are frequently composed of an object or item, and a characteristic or attribute name, such as species name and leaf color. For describing colors, a standard color chart should be used. A descriptor state is a clearly definable state of expression to define a characteristic and harmonize descriptions, such as cordate, oblong, or ovate leaf state. A descriptor method describes in detail how and under what conditions a descriptor is measured or scored. For example, the method requires that an object, such as a leaf, is to be described. In this case, the measuring points have to be also described, such as from the base of the petiole to the leaf tip. The method also requires the definition of the environmental conditions under which the measurements had been made. The method further requires the number of samples based on which the observations have been made.

2.10.3 Chamomile Descriptors: Morphology, Yield of Flower and Oil, and Chemical Constituents of Oil

Although chamomile is a very popular medicinal crop, its descriptors have not been found in the literature studied. In 1995–1996, Das and Rajeshwari (Das 1999) prepared minimum descriptors to characterize the chamomile plants growing in the experimental field of Central Institute of Medicinal Plants (CIMAP), Lucknow, Uttar Pradesh, India to facilitate documentation and conservation in the gene bank of CIMAP. As these were minimum descriptors, and the sample size varied considerably, a more detailed work needs to be carried out to enhance the information in the descriptors. However, this could serve as a basis for developing highly discerning descriptors for international harmonization in the future. To prepare the descriptors (morphological and yield-related descriptors), the accessions growing in the field experimental conditions were evaluated. Their descriptions along with the methods used are presented here. The morphological descriptors are presented in Table 2.6. Forty morphological characteristics of the plant were evaluated using specific methods and each was provided a rank or unit of measurement or description.

TABLE 2.6
Morphological Descriptors of Chamomile

S. No.	Characters	No. of Samples	Methods	Rank or Measurement Unit	
1.	Habit	335 Plants	Observation	Erect Semi erect Prostrate	In the year 1995–1996, a large variation was observed in the morphological characters and it was decided to record them for 335 plants
2.	Plant pigmentation	335 Plants	Observation	Light purple Purple Dark purple Dark ashy purple Brownish purple Brown ashy purple Peach Peachish green No anthocyanin	A color chart was used
3.	Plant height	335 Plants	Measurement	≤50 cm 51–70 cm 71–90 cm 91–110 cm ≥111 cm	Base to shoot tip (tip of the first capitulum) was measured in centi meter
4.	Canopy	335 Plants	Measurement	≤315 cm^2 316–615 cm^2 –915 cm^2 ≥916 cm^2	Two measurements of the radius was taken just above the base and calculated using the formula πr^2
5.	Plant Hairiness	25 Plants	Observation	Yes – No	The stem and branches were observed using dissection microscope. The plants were chosen at random from the experimental blocks in the field
6.	Number of branches	335 Plants	Measurement	≤500 501–1000 1001–1500 1501–2000 ≥2001	The primary branches were counted

TABLE 2.6 (*Continued*)
Morphological Descriptors of Chamomile

S. No.	Characters	No. of Samples	Methods	Rank or Measurement Unit	Remarks
7.	Number of leaves	335 Plants	Observation	Yes – No	The stem and branches were observed using dissection microscope. The plants were chosen at random from the experimental blocks in the field
8.	Shape of leaf	335 Plants	Measurement	Linear Linear–Lanceolate Ovate–Oblong	
9.	Number of leaves	335 Plants	Measurement	≤1000 1001–2500 2501–4000 4001–5500 ≥5501	The primary branches were counted
10.	Length of leaf	335 Plants	Measurement	≤5 cm 6–10 cm 11–15 cm ≥16 cm	The leaf at the base of the fifth branch was measured
11.	Width of leaf	335 Plants		≤1.5 cm 1.6–3.5 cm 3.6–5.5 cm ≥5.6 cm	The leaf at the base of the fifth branch was taken. At its maximum width the measurement was taken
12.	Area of leaf	335 Plants		≤10 cm^2 11–15 cm^2 16–20 cm^2 21–25 cm^2 ≥26 cm^2	The leaf at the base of the fifth branch was taken. The area was calculated from the length and width of the leaf
13.	Coating of leaf	25 Plants	Observation	Waxy Non-waxy	Hand lens and microscopic examination of the stem was done
14.	Nature of leaf surface	25 Plants	Observation	Smooth Rough	A dissection microscopic examination of the stem was done

(*Continued*)

TABLE 2.6 (*Continued*)
Morphological Descriptors of Chamomile

S. No.	Characters	No. of Samples	Methods	Rank or Measurement Unit	Remarks
15.	Tip of leaf	25 Plants	Observation	Acute Obtuse	A dissection microscopic examination of the stem was done
16.	Thickness of stem (diameter)	25 Plants	Measurement	≤0.5 cm 0.6–1.0 cm 1.1–1.5 cm ≥1.5 cm	A Vernier caliper was used
17.	Hairiness of peduncle	25 Plants	Observation	+ Present − Absent	A dissection microscopic examination of the stem was done
18.	Length of peduncle	25 Plants	Measurement	≤2 cm 3–5 cm ≥6 cm	
19.	Color of peduncle	25 Plants	Observation	Green Light green Yellowish green	A color chart was used
20.	Number of capitula per plant	335 Plants	Measurement	≤1500 1501–3500 3501–5500 5501–7500 7501–9000 ≥9001	The counting was done during the first harvest of flowers
21.	Diameter of capitulum	25 Plants	Measurement	≤0.7 cm 0.8–1.7 cm 1.8–2.7 cm 2.8–3.7 cm ≥3.8 cm	A Vernier caliper was used. A visual observation led to the selection of the samples
22.	Number of bracts	25 Plants	Measurement	≤10 11–20 21–30 ≥31	A dissection microscope was used
23.	Size of outer bracts (length)	25 Plants	Measurement	≤1.5 mm 1.6–2.5 mm 2.6–3.5 mm 3.6–4.5 mm ≥4.6 mm	A light microscope with stage and ocular micrometer was used
24.	Size of inner bracts (length)	25 Plants	Measurement	≤1 mm 2–3 mm ≥4 mm	A light microscope with stage and ocular micrometer was used

TABLE 2.6 (*Continued*)
Morphological Descriptors of Chamomile

S. No.	Characters	No. of Samples	Methods	Rank or Measurement Unit	Remarks
25.	Shape of bracts	25 Plants	Observation	Lanceolate Ovate Ovate–Lanceolate Linear Ovate–Oblong Linear–Lanceolate Oblong	Dissection microscope was used
26.	Margins of bracts	25 Plants	Observation	Scarious Hyaline Herbaceous	Dissection and light microscopes were used
27.	Number of ligulate florets	335 Plants	Measurement	≤10 11–15 16–20 21–25 ≥26	
28.	Shape of ligulate florets	25 Plants	Observation	Oblong Elliptic Elliptic–oblong Broadly–ovate Ovate Linear Linear–lanceolate	A dissection microscope was used
29.	Number of lips	25 Plants	Measurement	≤1 2–3 ≥4	A dissection microscope was used
30.	Bracteole tip	25 Plants	Observation	Entire Fringed	A dissection microscope was used
31.	Length of appendage	25 Plants	Measurement	≤0.5 mm 0.6–1.0 mm 1.1–1.15 mm ≥1.6 mm	A light microscope was used
32.	Diameter of receptacle	25 Plants	Measurement	≤5 mm 6–10 mm 11–15 mm ≥16 mm	A dissection microscope was used
33.	Tip of receptacle	25 Plants	Observation	Acute Obtuse Flat	A dissection microscope was used
34.	Shape of receptacle	25 Plants	Observation	Conical Dome Round	A dissection microscope was used

(*Continued*)

TABLE 2.6 (*Continued*)
Morphological Descriptors of Chamomile

S. No.	Characters	No. of Samples	Methods	Rank or Measurement Unit	Remarks
35.	Disc floret size	25 Plants	Measurement	≤1 mm 1.1–1.5 mm 1.6–2.0 mm 2.1–2.5 mm ≥2.6 mm	A dissection microscope and light microscope were used
36.	Tip of petals	25 Plants	Observation	Acute Obtuse Obtuse–Acute	
37.	Average size of five anthers	25 Plants	Measurement	≤0.2 mm 0.3–0.5 mm 0.6–0.8 mm ≥0.9 mm	A light microscope was used
38.	Size of seeds	25 Plants		≤0.4 mm 0.5–0.7 mm ≥0.8 mm	A light microscope was used
39.	Length of style	25 Plants		≤1.0 mm 1.1–3.5 mm 3.6–6.0 mm 6.1–8.5 mm 8.6–10.0 mm ≥10.1 mm	A dissection microscope and a light microscope were used
40.	Color of style	25 Plants	Observation	White Light Yellowish	
41.	Ribs of the seed	25 Plants	Measurement	≤3 4–5 6–7 ≥8	Dissection and light microscopes were used

The characteristics of the yield of chamomile, such as total fresh weight of plant and capitula per plant, and oil yield were studied and the descriptors were prepared as presented in Table 2.7.

The oil quality was determined in two growing seasons of 1995–1996 and 1996–1997. To evaluate the compounds of the essential oil, the oils were subjected to column chromatography and gas chromatography–mass spectrometry. The compounds were identified by peak enrichment with standard compounds, comparisons of retention indices relative to C9–C20 alkanes with literature values, and comparison of mass spectral fragments with those reported in the literature. Area percentages were computed without applying correction factors. The descriptors of the quality determining compounds in the chamomile essential oil is presented in Table 2.8.

TABLE 2.7
Descriptors of Yield and Essential Oil

S. No.	Characters	No. of Samples	Methods	Rank or Measurement Unit	Remarks
1.	Biological yield (total fresh plant weight)	335 Plants	Measurement	≤500 g 501–750 g 751–1000 g 1001–1250 g ≥1251 g	A weighing balance was used to measure fresh plants. It includes weight of the capitula
2.	Yield of capitula (fresh weight)	335 Plants	Measurement	≤25 g 26–75 g 76–125 g 126–175 g ≥176 g	
3.	Harvest index	335 Plants	Measurement	≤10% 11%–25% 26%–40% 41%–55% ≥56%	The percentage weight of capitula per total plant weight was calculated
4.	Yield of capitula essential oil	84 Plants	Measurement	≤0.15% v/w 0.16%–0.25% v/w 0.26%–0.35% v/w 0.36%–0.45% v/w ≥0.46% v/w	Oil of 45 genotypes were evaluated in 1995–1996 and 39 genotypes in 1996–1997. 100 g of fresh flowers were distilled in clevenger-type apparatus for 7–8 hours and the oil was collected
5.	Color of the capitula oil	84 Plants	Observation	White Yellow Light brown Light blue Blue Ink blue	A color chart was used
6.	Clarity of oil	84 Plants	Observation	Clear Turbid	The observation was carried out just after distillation
7.	Leaf and stem oil	20 Plants	Measurement	≤0.01% v/w 0.02%–0.04% v/w 0.05%–0.07% v/w ≥0.08% v/w	100 g of fresh herb (without) flowers were distilled in clevenger-type apparatus for 7–8 hours and the oil was collected

(Continued)

TABLE 2.7 (Continued)
Descriptors of Yield and Essential Oil

S. No.	Characters	No. of Samples	Methods	Rank or Measurement Unit	Remarks
8.	Leaf oil color	20 Plants	Observation	Greenish light blue Greenish blue Brownish	A color chart was used
9.	Root oil	20 Plants	Measurement	$\geq$0.003% v/w 0.004%–0.008% v/w $\leq$0.01% v/w	100 g of fresh roots were distilled in clevenger-type apparatus for 7–8 hours and the oil was collected
10.	Root oil color	20 Plants	Observation	White Yellow Light brown	A color chart was used

TABLE 2.8
Descriptors of Flower Essential Oil Quality Determining Compounds

S. No.	Characters	No. of Samples	Methods	Rank or Measurement Unit	Remarks
1.	(E)-β-farnesene	84 Plants	Measurement	$\leq$5 6–15 16–25 26–35 $\geq$36	Oil of 45 genotypes were evaluated in 1995–1996 and 39 genotypes in 1996–1997
2.	Chamazulene	84 Plants	Measurement	$\leq$5 6–10 11–15 16–20 $\geq$21	Oil of 45 genotypes were evaluated in 1995–1996 and 39 genotypes in 1996–1997
3.	α-Bisabolol	84 Plants	Measurement	$\leq$10 22–25 26–40 41–55 $\geq$56	Oil of 45 genotypes were evaluated in 1995–1996 and 39 genotypes in 1996–1997
4.	α-Bisabolol oxide B	84 Plants	Measurement	$\leq$15 16–30 31–45 46–60 $\geq$61	Oil of 45 genotypes were evaluated in 1995–1996 and 39 genotypes in 1996–1997

TABLE 2.8 (Continued)
Descriptors of Flower Essential Oil Quality Determining Compounds

S. No.	Characters	No. of Samples	Methods	Rank or Measurement Unit	Remarks
5.	α-Bisabolol oxide A	84 Plants	Measurement	≤10 11–25 26–40 42–55 ≥56	Oil of 45 genotypes were evaluated in 1995–1996 and 39 genotypes in 1996–1997
6.	Cis-en-yn-dicycloether	84 Plants	Measurement	≤5 6–10 11–15 16–20 ≥21	Oil of 45 genotypes were evaluated in 1995–1996 and 39 genotypes in 1996–1997

REFERENCES

Abd El-Twab, M. H., Mekawy, A. M., and Saad El-Katatny, M. 2008. Karyomorphological studies of some species of *Chrysanthemum sensu lato* in Egypt. *Chromosome Botany* 3: 41–47.

Acosta, L., Fuentes, V., Durand, D., Martin, G., Rodriguez, C., and Ramos, R. 1986. Cultivation of chamomile (*Matricaria chamomilla* L.) in two locations of the country. *Revista Cubana de Farmacia* 20(2): 169–175.

Adam, K. P. and Zapp, J. 1998. Biosynthesis of isoprene units of chamomile sesquiterpenes. *Phytochemistry* 48(6): 953–959.

Ahmad, S., Koukab, S., Razzaq, N., Islam, M., Rose, A., and Aslam, M. 2011. Cultivation of *Matricaria recutita* L. in highlands of Balochistan, Pakistan. *Pakistan Journal of Agricultural Research* 24(1–4): 35–41.

Ale-Agha, N., Feige, G. B., and Dachowski, M. 2002. Microfungi on compositae in the Ruhr Basin. *Mededelingen* 67(2): 217–226.

Aly, M. S. and Hussien, M. S. 2007. Egyptian chamomile—Cultivation and industrial processing. *ISHS Acta Horticulturae* 749: 81–91.

Andreucci, A. C., Ciccarelli, C., Desideri, I., and Pagni, A. M. 2008. Glandular hairs and secretory ducts of *Matricaria chamomilla* (Asteraceae): Morphology and histochemistry. *Annales Botanici Fennici* 45: 11–18.

Anonymous. 2012. Linnaeus and Hortus Cliffortianus. Natural History Museum. http://www.nhm.ac.uk/research-curation/research/projects/clifford-herbarium/linnaeus.html (Accessed October 30, 2012).

Applequist, W. L. 2002. A reassessment of the nomenclature of *Matricaria* L. and *Tripleurospermum* Sch. Bip. (Asteraceae). *Taxon* 51: 757–761.

Azizi, M., Bos, R., Woerdenbag, H. J., and Kayser, O. 2007. A comparative study of four chamomile cultivars cultivated in Iran. *ISHS Acta Horticulturae* 749: 93–96.

Bączek-Kwinta, R., Koziel, A., and Seidler-Łożykowska, K. 2011. Are the fluorescence parameters of German chamomile leaves the first indicators of the anthodia yield in drought conditions? *Photosynthetica* 49(1): 87–97.

Bara, I. I. 1979. The karyotype of some plant species, II: The study of mitotic chromosomes in *Matricaria chamomilla* [*Chamomile recutita* (L.) Rauschert]. variety Zloty Lan. *Biologie Vegetale* 31: 73–75.

Bharat, N. K. 2003. Powdery mildew of matricaria in Himachal Pradesh. *Plant Disease Research Ludhiana* 18(2): 202.

Biodiversity International. 1995. Developing crop descriptor lists: Guidelines for developers. *Biodiversity Technical Bulletin* (13), 1–3. http://www.bioversityinternational.org/uploads/tx_news/Developing_crop_descriptor_lists_1226.pdf (Accessed October 30, 2012).

BM-000647192. 2003. *Matricaria chamomilla*. The Linnaean Plant Name Typification Project.

Carle, R., Jung-Heiliger, H., and Schroder, M. B. 1993. Determination of ploidy in *Chamomilla recutita by* cytomorphological methods and by flow cytometry. *Planta Medica* 59 (Suppl. 1): A 699.

Chand, S., Pandey, A., and Patra, D. D. 2012. Influence of vermicompost on dry matter yield and uptake of Ni and Cd by chamomile (*Matricaria chamomilla*) in Ni- and Cd-polluted soil. *Water, Air, & Soil Pollution* 223(5): 2257–2262.

Chandra, V., Misra, P. N., and Singh, A. 1979. *Matricaria for blue oil*. Extension bulletin No.4. National Botanical Research Institute, Lucknow, India.

Cinc, A. S., Cinc, M., and Tomazin, E. P. 1984. Preliminary trials of the cultivation of chamomile in Kenya. *Farmacevtski vestnik* 35(3): 1973–1978.

Claphman, A. R., Tutin, T. G., and Warburg, E. F. (eds.). 1962. *Flora of the British Isles*. Second Edition. Cambridge: Cambridge University Press.

Correa, C. Jr. 1995. Mandirituba: New Brazilian chamomile cultivar. *Horticultura Brasiteira* 13(1): 61.

Dadkhah, A. R. 2010. Effect of salt stress on growth and essential oil of *Matricaria chamomilla*. *Research Journal of Biological Sciences* 5(10): 643–646.

Dagar, J. C., Minhas, P. S., and Kumar, M. 2011. Cultivation of medicinal and aromatic plants in saline environments. *Perspectives in Agriculture, Veterinary Science, Nutrition and Natural Resources* 6(009): 1–11.

Daniel, C., Kersten, T., Kehraus, S., König, G. M., and Knöss, W. 2008. Metabolic fingerprinting for the identification and classification of medicinal plants. *Zeitschrift für Phytotherapie* 29(6): 270–274.

Darlington, C. D. and Wylie, A. P. 1955. *Chromosome Atlas of Flowering Plants*. London, United Kingdom: Allen and Unwin, p. 268.

Das, M. 1999. Analysis of the genetic variation in Chamomile (*Matricaria recutita*) for the selection of improved breeding types. PhD Thesis submitted for the award of Doctor of Philosophy in Botany to the Lucknow University, Lucknow, India, p. 296.

de Santayana, M. P. and Morales, R. 2010. Chamomiles in Spain: The dynamics of plant nomenclature. In: M. Pardo de Santayana, A. Pieroni, and R. K. Puri (eds.) *Ethnobotany in the New Europe: People, Health and Wild Plant Resources*. New York, NY: Berghahn Books, pp. 282–306.

Dellacecca, V. 1996. Five years of research into chamomile (*Chamomilla recutita* L. Rauschert). In: *Atti convegno internazionale: Cultivazione emiglioramento di piante officinali*. Trento, Italy: Instituto Sperimentale per I' Assesmento Forestale e per Alpicoltura, giugno 2–3, 1994, pp. 27–45.

Donalisio, M. G. R. 1986. Preliminary determination of essential oil content in chamomile growing in Brazil. *Bragantia* 44(1): 407–410.

Dragland, S. 1996. Inhold av Kadmium og bly i kamille (*Chamomilla recutita* L.) og matrem (*Tanacetum parthenium* L.) dryket pa ulike steder i Norge [Content of cadmium and lead in chamomile (*Chamomilla recutita* L.) and feverfew (Tanacetum parthenium L.) grown in different parts of Norway]. *Norsk Landbruksforking* 10(3–4): 181–188.

Dragland, S., Paulsen, B. S., Wold, J. K., and Aslaksen, T. H. 1996. Flower yield and the content and quality of essential oil of chamomile, *Chamomilla recutita* (L.) Rauschert, grown in Norway. *Norwegian Journal of Agricultural Sciences* 10(4): 363–370.

Dubey, V. S., Bhalla, R., and Luthra, R. 2003. An overview of the non-mevalonate pathway for terpenoid biosynthesis in plants. *Journal of Biosciences* 28(5):637–646.
Eliasová, A., Poracká, V., Pal'ove-Balang, P., Imrich, J., and Repčák, M. 2012. Accumulation of tetracoumaroyl spermine in *Matricaria chamomilla* during floral development and nitrogen deficiency. *Zeitschrift für Naturforschung C* 67(1–2): 58–64.
Eliašová, A., Repčák, M., and Pastírová, A. 2004. Quantitative changes of secondary metabolites of *Matricaria chamomilla* by abiotic stress. *Zeitschrift für Naturforschung C* 59: 543–548.
Emongor, V. E., Chweya, J. A., Keya, S. O., and Munavi, R. M. 1990. Effect of nitrogen and phosphorus on the essential oil yield and quality of chamomile (*Matricaria chamomilla* L.) flowers. *East African Agricultural and Forestry Journal* 55(5): 261–264.
Falistocco, E., Menghini, A., and Veronesi, F. 1994. Osservazioni cariologiche in *Chamomilla recutita* (L.) Rauschert [Karyological observations on *Chamomilla recutita* (L.) Rauschert]. In: *Atti del convegno internazionale*: Coltivazione miglioramento di piante officinalo. Trento, Italy, giugno 2–3.
Farkoosh, S. S., Ardakani, M. R., Rejali, F., Darzi, M. T., and Faregh, A. H. 2011. Effect of mycorrhizal symbiosis and *Bacillus coagolance* on qualitative and quantitative traits of *Matricaria chamomilla* under different levels of phosphorus. *Middle-East Journal of Scientific Research* 8(1): 1–9.
Farzadfar, S. and Zarinkamar, F. 2012. Morphological and anatomical responses of *Matricaria chamomilla* plants to cadmium and calcium. *Advances in Environmental Biology* 6(5): 1603–1609.
Farzadfar, S. and Zarinkamar, F. 2013. Exogenously applied calcium alleviates cadmium toxicity in *Matricaria chamomilla* L. plants. *Environmental Science and Pollution Research* 20(3): 1413–1422.
Fatma, R., Abd El-Waheed., M. S. A., and Gamal El Din, K. M. 2010. Effect of indole acetic acid, gibberellic acid and kinetin on vegetative growth, flowering, essential oil pattern of chamomile plant (*Chamomile recutita* L. Rausch). *World Journal of Agricultural Sciences* 6(5): 595–600.
Fatma, R., Shahira, T., Abdel-Rahim, E. A., Afify, A. S., and Ayads, H. S. 1999. Effect of low temperature on growth, some biochemical constituents and essential oil of chamomile. *Annals of Agricultural Science* 44(2): 741–760.
Franke, R. 2005. Plant sources. In: R. Franke and H. Schilcher (eds.) *Chamomile: Industrial Profiles*. Boca Raton, FL: CRC Press, pp. 40–41.
Franke, R. and Schilcher, H. 2007. Relevance and use of chamomile (*Matricaria recutita* L.). *ISHS Acta Horticulturae* 749: 29–43.
Fránová, J., Petrzik, K., Paprštein, F., Kukerová, J., Navrátil, M., Válová, P., Nebesárová, J., and Jakešová, H. 2007. Experiences with phytoplasma detection and identification by different methods. *Bulletin of Insectology* 60(2): 247–248.
Franz, C. 1980. Content and composition of the essential oil in flower heads of *Matricaria chamomilla* L. during its ontogenetical development. *ISHS Acta Horticulturae* 96: 317–321.
Franz, C., Fritz, D., and Schroeder, F. J. 1975. Influence of ecological factors on the essential oils and flavonoids of two chamomile cultivars. 2. Effect of light and temperature. *Planta Medica* 27(1): 46–52.
Franz, C. and Isaac, O. 1984. A process for the production of a new tetraploid bisabololreichen chamomile and having improved properties. German patent number DE 3446220C2.
Fuller, E. 1992. Struktur und biologische aktivitat wasserloslicher polysaccharide aus *Chamomilla recutita* (L.) Rauschert. PhD Dissertation, University of Regensburg.
Fuller, E., Blaschek, W., and Franz, G. 1991. Characterization of water soluble polysaccharides from chamomile flowers. *Planta Medica* 57(Suppl. 2): A40.

Fuller, E. and Franz, G. 1993. News von den kamillen polysacchariden vergleich der Droge Kamillenbuten mit einem Fertigarzneimittel. *Deutsche Apotheker Zeitung* 133(45): 4225–4227.

Fuss, G., Geiser, E., and Patzner, R. 2005. On the host plants of several leaf beetles of Central Europe – the problem of fame and evidence (Coleoptera: Chrysomelidae). *Koleopterologische Rundschau* 75: 359–371.

Galambosi, B., Marczal, K., Litkey, K., Svab, M. J., and Petri, G. 1988. Comparative examination of chamomile varieties grown in Finland and Hungary. *Herba Hungarica* 27(2/3): 45–55.

Galambosi, B., Szeheri-Galambosi, Z., Repcak, M., and Cernaj, P. 1991. Variation in the yield and essential oil of four chamomile varieties grown in Finland in 1985–1988. *Journal of Agricultural Science in Finland* 63: 403–410.

Gärber, U., Plescher, A., and Hagedorn, G. 2013. Occurrence of diseases and damage when cultivating chamomile (*Matricaria recutita* L.). *Zeitschrift für Arznei- & Gewürzpflanzen*, 18(3): 124–131.

Ghanavati, M., Houshmand, S., Zainali, H., and Ejlali, F. 2011. Salinity effects on germination and growth of chamomile genotypes. *Journal of Medicinal Plants Research* 5(30): 6609–6614.

Ghasemi, M., Jelodar, N. B., Modarresi, M., and Bagheri, N. 2013. Morphological response of German chamomile to heat stress accompanies salicylic acid-mediated under field conditions. *International Journal of Agriculture and Crop Sciences* 5(7): 756–760.

Ghanavati, M. and Sengul, S. 2010. Salinity effect on the germination and some chemical components of *Chamomilla recutita* L. *Asian Journal of Chemistry* 22(2): 859–866.

Ghosh, M. L. 1989. Introduction and scientific growing of some essential oil yielding plants in the Gangetic plains of Hoogly district West Bengal India. *Indian Perfumer* 33(4): 286–290.

Gosztola, B., Sárosi, S., and Németh, É. 2010. Variability of the essential oil content and composition of chamomile (*Matricaria recutita* L.) affected by weather conditions. *Natural Product Communications* 5(3): 465–470.

Grejtovský, A., Markušová, K., and Eliašová, A. 2006. The response of chamomile (*Matricaria chamomilla* L.) plants to soil zinc supply. *Plant, Soil and Environment* 52(1): 1–7.

Grejtovský, A., Markušová, K., and Nováková, L. 2008. Lead uptake by *Matricaria chamomilla* L. *Plant Soil and Environment* 54(2): 47–54.

Grejtovský A., Repčák M., Eliašová A., Markušová K. 2001. Effect of cadmium on active principle contents of *Matricaria recutita*. L. *Herba Pol.*, 47(3): 203–208.

Hansen, H. V. and Christensen, K. I. 2009. The common chamomile and the scentless mayweed revisited. *Taxon* 58: 261–264.

Herb. Clifford: 415, *Matricaria* 1 (BM). 2003. http://www.nhm.ac.uk//resources/research-curation/projects/clifford-herbarium/lgimages/BM000647192.JPG (Accessed September 15, 2012).

Heywood, V. H. and Harborne, J. B. (eds.). 1977. Anthemideae—Systematic review. In: *The Biology and Chemistry of Compositae*. Volume II. London, United Kingdom: Academic Press, pp. 851–898.

Hirata, T., Izumi, S., Akita, K., Fukuda, N., Katayama, S., Taniguchi, K., Dyas, L., and Goad, L. J. 1996. Lipid constituents of oil bodies in the cultured shoot primordia of *Matricaria chamomilla*. *Phytochemistry* 41(5): 1275–1279.

Hitchcock, A. and Green, M. 1935. Species lectotypicae generum Linnaei. In: *International Rules of Botanical Nomenclature*. Third Edition (Suppl. II), Volume 1, 139–143. Jena, Germany: Gustav Fischer.

Hojati, M., Modarres-Sanavy, S. A. M., Ghanati, F., and Panahi, M. 2011. Hexaconazole induces antioxidant protection and apigenin-7-glucoside accumulation in *Matricaria chamomilla* plants subjected to drought stress. *Journal of Plant Physiology* 168(8): 782–791.

Holubář, J. 2005. Growing varieties of chamomile in the Czech Republic. In: R. Franke and H. Schilcher (eds.) *Chamomile: Industrial Profiles*. Boca Raton, FL: CRC Press, pp. 139–141.

Holz, J. 1979. Investigations on the synthesis of sesquiterpenes and spiroethers of *Matricaria chamomilla* L. Abstract of paper presented at the 27th Annual Meeting of the Society for Medicinal Plant Research. Budapest, Hungary, July 16–22. *Planta Medica* 36(3): 226.

Holz, W. and Meithing, H. 1994. Active substances in aqueous chamomile infusions. 2. Communication: Release of essential oil components. *Pharmazie* 49(1): 53–55. http://archive.org/stream/mobot31753000036829#page/379/mode/2up/search/matricaria; http://www.nhm.ac.uk/research-curation/research/projects/linnaean-typification/database/detailimage.dsml?ID=558400 (Accessed October 20, 2012).

Huțanu-Bashtawi, L., Toma, C., and Ivănescu, L. 2009. Considerations to the Histo-Anatomical Study of the *Chamomilla Recutita* (L.) Rauschert Strain Treated with Thiophanate Methyl (Topsin M). *Analele ştiinţifice ale Universităţii 'Al. I. Cuza' Iaşi Tomul LV, fasc. 1, s.II a. Biologie vegetală*, 21–30.

Inceer, H. and Ozcan, M. 2011. Leaf anatomy as an additional taxonomy tool for 18 taxa of *Matricaria* L. and *Tripleurospermum* Sch. Bip. (Anthemideae-Asteraceae) in Turkey. *Plant Systematics and Evolution* 296: 205–215.

The International Code of Botanical Nomenclature (ICBN). 2006. Appendix IV Nomina Specifica Conservanda et Rejicienda. Online Version. http://home.kpn.nl/klaasvanmanen/icbn/0106AppendixIIINGSp00302.htm#Matricaria (Accessed October 29, 2012).

Jeffery, C. 1979. Note on the lectotypification of the names *Cacalia* L., *Matricaria* L. and *Gnaphalium* L. *Taxon* 28(4): 349–351.

Juárez-Rosete, C. R., Rodríguez-Mendoza, M. N., Trejo-Téllez, L. I., Aguilar-Castillo, J. A., Gómez-Merino, F. C., Trejo-Téllez, L. I., and Rodríguez-Mendoza, M. N. 2012. Inorganic and organic fertilization in biomass and essential oil production of *Matricaria recutita* L. *ISHS Acta Horticulturae* 947: 307–311.

Kapoor, L. D. 1982. My experiences in the cultivation of medicinal plants. *International Journal of Crude Drug Research* 20(4): 187–191.

Kaul, V. K., Singh, B., and Sood, R. P. 1993. Volatile constituents of the essential oil of *Tanacetum longifolium* Wall. *Journal of Essential Oil Research* 5: 597–601.

Kováčik, J. 2013. Hyperaccumulation of cadmium in *Matricaria chamomilla*: A never-ending story? *Acta Physiologiae Plantarum* 35(5): 1721–1728.

Kováčik, J. and Bačkor, M. 2007. Changes of phenolic metabolism and oxidative status in nitrogen-deficient *Matricaria chamomilla* plants. *Plant and Soil* 297(1–2): 255–265.

Kováčik, J., Grúz, J., Bačkor, M., Strnad, M., and Repčák, M. 2009. Salicylic acid-induced changes to growth and phenolic metabolism in *Matricaria chamomilla* plants. *Plant Cell Reports* 28(1): 135–143.

Kováčik, J., Grúz, J., Klejdus, B., Štork, F., and Hedbavny, J. 2012c. Accumulation of metals and selected nutritional parameters in the field-grown chamomile anthodia. *Food Chemistry* 131(1): 55–62.

Kováčik, J., Klejdus, B., Bačkor, M., and Repčák, M. 2007. Phenylalanine ammonia-lyase activity and phenolic compounds accumulation in nitrogen-deficient *Matricaria chamomilla* leaf rosettes. *Plant Science* 172(2): 393–399.

Kováčik, J., Klejdus, B., Hedbavny, J., Mártonfi, P., Štork, F., and Mártonfiová, L. 2011c. Copper uptake, physiology and cytogenetic characteristics in three *Matricaria chamomilla* cultivars. *Water, Air, & Soil Pollution* 218(1–4): 681–691.

Kováčik, J., Klejdus, B., Hedbavny, J., Štork, F., and Grúz, J. 2012b. Modulation of copper uptake and toxicity by abiotic stresses in *Matricaria chamomilla* plants. *Journal of Agricultural and Food Chemistry* 60(27): 6755–6763.

Kováčik, J., Klejdus, B., Hedbavny, J., and Zoń, J. 2011b. Significance of phenols in cadmium and nickel uptake. *Journal of Plant Physiology* 168(6): 576–584.

Kováčik, J., Klejdus, B., Štork, F., and Hedbavny, J. 2011a. Nitrate deficiency reduces cadmium and nickel accumulation in chamomile plants. *Journal of Agricultural and Food Chemistry* 59(9): 5139–5149.

Kováčik, J., Štork, F., Klejdus, B., Grúz, J., and Hedbavny, J. 2012a. Effect of metabolic regulators on aluminium uptake and toxicity in *Matricaria chamomilla* plants. *Plant Physiology and Biochemistry* 54: 140–148.

Kováčik, J., Tomko, J., Bačkor, M., and Repčák, M. 2006. *Matricaria chamomilla* is not a hyperaccumulator, but tolerant to cadmium stress. *Plant Growth Regulation* 50(2–3): 239–247.

Král'ová, K., Masarovičová, E., Kubová, J., and Svajlenová, O. 2007. Response of *Matricaria recutita* plants to some copper (II) chelates. *ISHS Acta Horticulturae* 749: 237–243.

Krstic-Pavlovic, N. and Dzamic, R. 1984. Contribution to the investigation of the influence of fertilization on the yield and quality of cultivated chamomile (*Matricaria chamomilla* L.) in the region of northern Banat (Yugoslavia, stimulant plants and crops). *Agrohemija* 3: 207–215.

Kuberappa, G. C., Shilpa, P., Shwetha, B. V., and Nataraju, M. S. 2012. Pollinator fauna, abundance, foraging activity and impact of modes of pollination in increasing the productivity of chamomile, *Matricaria chamomilla* L. *Mysore Journal of Agricultural Sciences* 46(4): 772–777.

Kuberappa, G. C., Shilpa, P., Vishwas, A. B., and Vasundhara, M. 2007. Floral biology and phenology of chamomile (*Matricaria chamomilla* L.). *Biomed* 2(3): 257–259.

Kummerova, M., Zezulka, S., Kral'ova, K., Masarovicova, E. 2010. Effect of zinc and cadmium on physiological and production characteristics in *Matricaria recutita*. *Biologia Plantarum* 54(2): 308–314.

Lal, R. K., Sharma, J. R., Mishra, O. P., and Singh, S. P. 1993. Induced floral mutants and their productivity in German chamomile (*Matricaria recutita*). *Indian Journal of Agricultural Science* 63: 27–33.

Lesíková, J., Král'ová, K., Masarovičová, E., Kubová, J., and Onderjkovičová, I. 2007. Effect of different cadmium compounds on chamomile plants. *ISHS Acta Horticulturae* 749: 223–229.

Letchamo, W. 1993. Nitrogen application affects yield and content of the active substances in camomile genotypes. In: J. Janick and J. E. Simon (eds.) *New Crops*. New York: Wiley, pp. 636–639.

Letchamo, W., Marquard, R., and Friedt, W. 1995. Alternative methods for determination of ploidy level in chamomile (*Chamomilla recutita* (L.) Rausch.) breeding. *Journal of Herbs Spices & Medicinal Plants* 2(4): 19–25.

Leto, C., Carruba, A., and Cibella, R. 1994a. Correlazione tra fattori producttivinella camomilla (*Chamomilla recutita* Rausch) nell' ambiente caldo-arids Siciliano [Correlation among yield components in chamomile (*Chamomilla recutita* Rausch) in a semi arid Sicilian environment. In: *Atti convegno internationale: Coltivazione miglioramento de piante officinali* (Proceedings of the international meeting cultivation and improvement of medicinal and aromatic plants). MinisterodelleRisorseAgricole, Alimentari e Forestali (Ministry of Agricultural, Food and Forestry Policies), Trento, Italy, giugno 2–3.

Leto, C., Carruba, A., and Cibella, R. 1994b. Effecti della densita di semina sulla Chamomilla [*Chamomilla recutita* (L.) Rausch.] nell' ambiente caldo arido Siciliano [Effects of sowing density on chamomile (*Chamomilla recutita* Rausch) in a semi arid Sicilian environment]. In: *Atti convegno internazionale: Coltivazione e miglioramento di piante officinali* (Proceedings of the international meeting cultivation and improvement of medicinal and aromatic plants). MinisterodelleRisorseAgricole, Alimentari e Forestali (Ministry of Agricultural, Food and Forestry Policies), Trento, Italy, giugno 2–3.

Leto, C., Carruba, A., Cibella, R. 1994c. Resultati di un quadriennio di coltivazione di camomilla (*Chamomilla recutita* Rausch.) nell' ambiente caldo-arido Siciliano [Results of a four year trial period of chamomile (*Chamomilla recutita* Rausch.) Cultivation in semi-arid Sicilian environment]. In: *Atti convegno internazionale: Coltivazione e miglioramentdi piante officinali*. Trento, Italy, giugno 2–3.

Linnaei, C. 1745. *Matricaria. Flora Suecica: Exhibens Plantas Per Regnum Sveciae Crescentes, Systematice Cum Differentiis Specierum, Synonymis Autorum, Nominibus Incolarum, Solo Locorum, Usu Pharmacopaeorum.* First Edition. Stockholmiae: Sumtu & literis Laurentii Salvii, pp. 251–252. http://www.botanicus.org/title/b12069474 (Accessed September 30, 2012).

Linnaei, C. 1753. *Species Plantarum Exhibentes Plantas Rite Cognitas Ad Genera Relatas, Cum Differentiis Specificis, Nominibus Trivialibus, Synonymis selectis, Locis natalibus, Secundum Systema Sexuale Digestas*. Volume 2, pp. 890–891. Holmiae: Impensis Laurentii Salvii. http://biodiversitylibrary.org/page/358911; http://biodiversitylibrary.org/page/358912 (Accessed September 30, 2012).

Linnaei, C. 1755. Matricaria. In: *Flora Suecica*. Second Edition, pp. 296–297. http://books.google.co.in/books?id=wUo-AAAAcAAJ&printsec=frontcover&source=gbs_ge_summary_r&cad=0#v=onepage&q=matricaria&f=false (Accessed October 4, 2012).

Linnaei, C. 1754. Matricaria-Genera plantarum: Eorumque characteres naturales secundum numerum, figuram, situm, et proportionem omnium fructificationis partium, p. 380. http://www.botanicus.org/title/b11659750 (Accessed October 4, 2012).

Linnaei, C. 1763. Species Plantarum Exhibentes Plantas Rite Cognitas Ad Genera Relatas, Cum Differentiis Specificis, Nominibus Trivialibus, Synonymis Selectis, Locis Natalibus, Secundum Systema Sexuale Digestas. Volume 2, pp. 1255–1256. http://biodiversitylibrary.org/page/11834642; http://biodiversitylibrary.org/page/11834643 (Accessed October 10, 2012).

Linnaeus, C. 1775. A generic and specific description of British plants Translated from Genera et Species Plantarum of the Celebrated Linnaeus with notes and Observations by James Jenkinson. Kendal, pp. 206–207.

Linnaeus, C. and Ehret, G. D. 1737. Matricaria. In: *Hortus Cliffortianus*, pp. 415–416. http://www.biodiversitylibrary.org/item/13838#page/4/mode/1up (Accessed October 5, 2012).

Mackova, A. 1992. Testing of sensitivity of Tripleurospermum inodorum and Matricaria recutita to dichlorprop. *Zahradnictvi-UVTIZ* 19(2): 101–108.

Madhusoodanan, K. J. and Arora, O. P. 1979. Induced autotetraploidy in *Matricaria chamomilla* L. *Cytologia* 44: 227–232.

Maier, R., Carle, R., Kreis, W., and Reinhard, E. 1993. Purification and characterization of a flavone 7-*O*-glucoside-specific glucosidase from ligulate florets of *Chamomilla recutita*. *Planta Medica* 59: 436–441.

Maier, R., Kreis, W., Carle, R., and Reinhard, E. 1991a. Partial purification and substrate specificity of a β-glucosidase from *Chamomilla recutita*. *Planta Medica* 57(Suppl. 2): A84–A85.

Maier, R., Kreis, W., Carle, R., and Reinhard, E. 1991b. A flavone-glucoside-cleaving beta-glucosidase from *Chamomilla recutita*. *Planta Medica* 57(3): 297–298.

Mann, C. and Staba, E. J. 1986. The chemistry, pharmacognosy and chemical formulations of chamomile. *Journal of Herbs, Spices, and Medicinal Plants* 1: 236–280.

Mostafa, M., El-Fouly, M., Meawad, A, and Afifi, A. 1983. The combined effects of nitrogen fertilization and growth regulators on chamomile plants [Egypt]. *Zagazig Journal of Agricultural Research* 10(2): 61–75.

Muravenko, O. V., Samatadze, T. E., and Zelenin, A. V. 1999. Image and visual analyses of G-banding patterns of camomile chromosomes. *Membrane & Cell Biology* 12(6): 845–855.

Nasrin, F., Nasibi, F., and Rezazadeh, R. 2012. Comparison of the effects of nitric oxide and spermidin pretreatment on alleviation of salt stress in chamomile plant (*Matricaria recutita* L.). *Journal of Stress Physiology & Biochemistry* 8(3): 214–223.
NCBI. 2013. Species *Matricaria chamomilla*. http://www.uniprot.org/taxonomy/98504 (Accessed October 16, 2012).
Nikolova, A., Kozhuharova, K., Zheljazkov, V. D., and Craker, L. E. 1999. Mineral Nutrition of Chamomile (*Chamomilla recutita* (L.) K. *ISHS Acta Horticulturae* 502: 203–208:II WOCMAP Congress Medicinal and Aromatic Plants, Part 3: Agricultural Production, Post Harvest Techniques, Biotechnology.
Ohe, C., Sugino, M., Minami, M., Hasegawa, C., Ashida, K., Ogaki, K., and Kanamori, H. 1995. Studies on the cultivation and evaluation of chamomilae flos. Seasonal variation in production of the head (capitula) and accumulation of glycosides in the capitula of *Matricaria chamomilla* L. *Yakugaku Zasshi* 115(2): 130–135.
Okoń, S. and Surmacz-Magdziak, A. 2011. The use of RAPD markers for detecting genetic similarity and molecular identification of chamomile (Chamomilla recutita (L.) Rausch.) genotypes. *Herba Polonica* 57(1): 38–47.
Omer, E. A., Said–Al-Ahl, H. A. H., El-Gendy, A. G., Shaban, K. A., and Hussein, M. S. 2013. Effect of amino acids application on production, volatile oil and chemical composition of chamomile cultivated in saline soil at Sinai. *Journal of Applied Sciences Research* 9(4): 3006–3021.
Pavlovič, A., Masarovicová, E., Kráľova, K., and Kubová, J. 2006. Response of chamomile plants (*Matricaria recutita* L.) to cadmium treatment. *Bulletin of Environmental Contamination and Toxicology* 77(5): 763–771.
Pekic, B., Zekovic, Z., and Lapojevic, Z. 1994. Investigation of apigenin-7-O-β-glucoside hydrolysis by β-glucoside from almonds. *Biotechnology Letters* 16(3): 229–234.
Pérez Bocourt, Y., López Manes, D., and López Mesa, M. O. 2005. New host plants for the family Erysiphaceae in Cuba. *Fitosanidad* 9(1): 73.
Petanović, R. U., Boczek, J., and Shi, A. 2000. Four new Aceria species (Acari: Eriophyoidea) from Serbia. *Acta Entomologica Serbica* 5(1–2): 119–129.
Pirkhezri, M., Hassani, M. E., and Hadian, J. 2010. Genetic diversity in different populations of Matricaria chamomilla L. growing in southwest of Iran, based on morphological and RAPD markers. *Research Journal of Medicinal Plant* 4(1): 1–13.
Pirzad, A., Shakiba, M. R., Zehtab-Salmasi, S., Mohammadi, S. A., Sharifi, R. S., and Hassani, A. 2011. Effects of irrigation regime and plant density on essential oil composition of German chamomile (*Matricaria chamomilla*). *Journal of Herbs, Spices & Medicinal Plants* 17(2): 107–118.
The Plant List. 2012. http://www.theplantlist.org/ (Accessed October 29, 2012).
Plescher, A. 2005. Abiotic and biotic stress affecting the common chamomile (*Matricaria recutita* L.) and the Roman chamomile (*Chamaemelum nobile* L. syn. *Anthemis nobilis* L.). In: R. Franke and H. Schilcher (eds.) *Chamomile: Industrial Profiles*. Boca Raton, FL: CRC Press, pp. 167–172.
Pourakbar, L. 2013. Effect of static magnetic field on germination, growth characteristics and activities of some enzymes in chamomile seeds (*Matricaria chamomilla* L.). *International Journal of Agronomy and Plant Production* 4(9): 2335–2340.
Purbrick, P. and Blessing, P. 2007. Chamomile demand, cultivation and use in Australia. *ISHS Acta Horticulturae* 749: 65–70.
Rashid, M. A. and Ahmad, F. 1994. Pharmacognostical studies on the flowers of *Matricaria chamomilla* Linn. *Hamdard Medicus* 37(2): 73–81.
Rauschert, S. 1974. Nomenklatorische Probleme in der Gattung *Matricaria* L. *Folia Geobotanica et Phytotaxonomica* 9(3): 249–260.
Repčák, M., Imrich, J., and Franeková, M. 2001. Umbelliferone, a stress metabolite of *Chamomilla recutita* (L.) Rauschert. *Journal of Plant Physiology* 158(8): 1085–1087.

Repčák, M., Smajda, P., Cernaj, P., Honcariv, R., and Podhradhs, D. 1980. Diurnal rhythms of certain sesquiterpenes in wild camomile (*Matricaria chamomilla* L.). *Biologia Plantarum* 22(6): 420–427.

Repčák, M. and Suvák, M. 2013. Methyl jasmonate and *Echinothrips americanus* regulate coumarin accumulation in leaves of *Matricaria chamomilla*. *Biochemical systematics and Ecology* 47: 38–41.

Salamon, I. 1992. Production of chamomile, *Chamomilla recutita* (L.) Rauschert, in Slovakia. *Journal of Herbs, Spices and Medicinal Plants* 1(1–2): 37–45.

Salamon, I. 1993. Chamomile. The Modern Phytotherapist. *Mediherb*. Australia. pp. 13–16.

Salamon, I. 1996. Large scale cultivation of chamomile in Slovakia and its perspectives. In: *Atti convegno internazionale: Cultivazione e miglioramento di piante officinali*. Trento, Italy: Institute Sperimentale per I' Assessmento Forestale e per I' Alpicoltura, giugno 2–3, 1994, pp. 413–416.

Salamon, I. and Honcariv, R. 1994. Growing conditions and breeding of chamomile (*Chamomilla recutita* (L.) Rauschert regarding the essential oil qualitative-quantitative characteristics in Slovakia. *Herba Polonica* 40(1–2): 68–74.

Salamon, I., Král'ová, K., and Masarovičová, E. 2007a. Accumulation of cadmium in chamomile plants cultivated in Eastern Slovakia regions. *ISHS Acta Horticulturae* 749: 217–222.

Salamon, I., Labun, P., Skoula, M., and Fabian, M. 2007b. Cadmium, lead and nickel accumulation in chamomile plants grown on heavy metal-enriched soil. *ISHS Acta Horticulturae* 749: 231–237.

Saleh, M. 1971. The effect of air temperature and thermoperiod on the quality and quantity of *Matricaria chamomilla* L. oil. *Meded LandboirwhoResch Wagenigen* 70: 1–17.

Saleh, M. 1973. Effects of light upon quantity and quality of *Matricaria chamomilla* oil. III. Preliminary study of light intensity effects under controlled conditions. *Planta Medica* 24(4): 337–340.

Salimi, F., Shekari, F., Azimi, M. R., and Zangani, E. 2012. Role of methyl jasmonate on improving salt resistance through some physiological characters in German chamomile (*Matricaria chamomilla* L.). *Iranian Journal of Medicinal and Aromatic Plants* 27(4): 700–711.

Samatadze, T. E., Muravenko, O. V., Konstantin, M., Popov, V., and Zelenin, A. V. 2001. Genome comparison of the *Matricaria chamomilla* L. varieties by the chromosome C- and OR-banding patterns. *Caryologia* 54(4): 299–306.

Sánchez Govín, E., Rivera Amita, M. M., and Carballo Guerra, C. 2010. Biopesticides influence on the quality parameters of *Calendula officinalis* L. and *Matricaria recutita* L. *Revista Cubana de Plantas Medicinales* 15(2): 60–65.

Schilcher, H. 1978. Influence of herbicides and some heavy metals on growth of *Matricaria chamomilla* L. and the biosynthesis of the essential oils. *ISHS Acta Horticulturae* 73: 339–341.

Seidler-Lozykowska, K. 2007. Chamomile cultivars and their cultivation in Poland. *ISHS Acta Horticulturae* 749: 111–114.

Sell, P. and Murrell, G. 2006. *Flora of Great Britain and Ireland: Campanulaceae— Asteraceae*. Volume 4. Cambridge: Cambridge University Press, pp. 483–484. http://books.google.co.in/books? id = Ulm5mCqAVZUC&pg = PA484&lpg = PA484 &dq = Akylopsis+suaveolens+%28Pursh%29+Lehm&source = bl&ots = bA5E zp9vSG&sig = kTa7HsYnEcqfjVhQqbwEkOCcfNI&hl = en&sa = X&ei = rbFy UO-dG9CGrAeW34CQDA&ved = 0CDcQ6AEwAw#v = onepage&q = chamomilla&f = false (Accessed October 28, 2012).

Sharafzadeh, S. and Bayatpoor, N. 2012. Effect of naphthaleneacetic acid and spermidine on essential oil constituents of German Chamomile. *International Journal of Agricultural and Crop Sciences* 4(23): 1803–1806.

Sharafzadeh, S., Bazrafshan, F., and Dastgheibif, N. 2012. Essential oil components of German chamomile as affected by vermicompost rates. *Technical Journal of Engineering and Applied Sciences* 2(9): 275–278.
Sharifi-Tehrani, M. and Ghasemi, N. 2011. Matricaria L. (Anthemideae, Asteraceae) in Iran: A chemotaxonomic study based on flavonoids. *Taxonomy and Biosystematics* 3(8): 25–34.
Siahsar, B. A., Sarani, S., and Allahdoo, M. 2011. Polypeptide electrophoretic pattern of *Matricaria chamomilla* and *Anthemis nobilis* under salt and Fe-deficiency stress. *African Journal of Biotechnology* 10(54): 11182–11185.
Šiljkovic, Ž. and Rimanić, A. 2005. Geographic aspects of medicinal plants organic growing in Croatia. *Geoadria* 10(1): 53–68.
Skaria, B. P., Joy, P. P., Mathew, S., Joseph, A., and Joseph, R. 2007. Chamomile. In: K. V. Peter (ed.) *Aromatic Plants, Horticulture Science Series*. Volume 1. New Delhi, India: New India Publishing Agency, pp. 68–70.
Solouki, M., Mehdikhani, H., Zeinali, H., and Emamjomeh, A. A. 2008. Study of genetic diversity in Chamomile (*Matricaria chamomilla*) based on morphological traits and molecular markers. *Scientia Horticulturae* 117(3): 281–287.
Steiner, J. J. and Greene, S. L. 1995. Proposed ecological descriptors and their utility for plant germplasm collections. *Crop Science* 36(2): 439–451.
Stevanovič, Z. D., Vrbničanin, S., and Jevdjovič, R. 2007. Weeding of cultivated chamomile in Serbia. *ISHS Acta Horticulturae* 749: 149–155.
Sulborska, A. 2011. Micromorphology of flowers, anatomy and ultrastructure of *Chamomilla recutita* (L.) Rausch. (Asteraceae) nectary. *Acta Agrobotanica* 64(4): 23–34.
Tai, Y., Yang, X., Yuan, Y., Jiang, L., Yu, D., and Hu, F. 2013. Effects of NaCl stress on seed germination and seedling growth, physiological index and anatomical structure of Matricaria chamomilla. *Journal of Plant Resources and Environment* 22(2): 78–85.
Tutin T. G., Heywood V. H., Burges N. A., and Valentine D. H. (eds.) 2010. *Flora Europaea: Plantaginaceae to Compositae (and Rubiaceae)*. Volume 4. Cambridge: Cambridge University Press, pp. 166–167. http://books.google.de/books?id=QXRooltqAVMC&printsec=frontcover&hl=de&source=gbs_ge_summary_r&cad=0#v=onepage&q=chamomilla&f=false (Accessed October 28, 2012).
Tutin, T. G., Heywood, V. H., Burges, N. A., Valentine, D. H., Walters, S. M., and Webb, D. A. 1964. *Flora Europea*. Volume 4. Cambridge: Cambridge University Press, p. 167.
Upadhyay, R. K. and Patra, D. D. 2011. Influence of secondary plant nutrients (Ca and Mg) on growth and yield of chamomile (*Matricaria recutita* L.). *Asian Journal of Crop Science* 3(3): 151–157.
Uroviche, R. C., Volpini, A. F. N., Dias, D. C., Lopes, A. R., Zaghi Jr., L. L., Souza, S. G. H. de., and Alberton, O. 2012. Mycorrhization and microbial activity from a soil cultivated with coriander and chamomile. *Arquivos de Ciências Veterinárias e Zoologia da UNIPAR* 15(2): 121–125.
WHO. 1999. *WHO Monographs on Selected Medicinal Plants*. Volume 1. World Health Organisation, p. 86. http://apps.who.int/medicinedocs/pdf/s2200e/s2200e.pdf (Accessed October 30, 2012).
Woo, S. 1989. Biosystematics and life history strategies of scentless chamomile (*Matricaria perforata* Merat) in Canada. MSc Dissertation, University of Saskatchewan, Saskatoon, Saskatchewan, Canada. http://library.usask.ca/theses/available/etd-12152010-085844/unrestricted/Woo_Sheridan_Lois_sec_1989.pdf (Accessed October 30, 2012).
Yakovlev, A. I. and Gorin, A. G. 1977. Structure of the pectic acid of *Matricaria chamomilla*. *Chemistry of Natural Compounds* 13(2): 160–162.
Young, T. K., Young, P. J., Chankim, O., Hoikim, Y., Chang, H., and Ra, D. Y. 1992. Volatile components of chamomile (*Matricaria chamomilla* L.) cultivated in Korea. *Journal of the Korean Agricultural Chemical Society* 35(2): 122–125.
Zehtab-Salmasi, S. 2008. The influence of salinity and seed pre-treatment on the germination of German chamomile (*Matricaria chamomilla* L.). *Research Journal of Agronomy* 2(2): 28–30.

3 Chamomile
Medicinal Properties

3.1 INTRODUCTION

Chamomile is used as a folk remedy for various diseases by many people across the world. This is evident from the ethnobotanical studies carried out in the different regions of the world. Raal et al. (2013) found that in Estonia about 56% patients of common cold and flu treated themselves with medicinal plants, such as chamomile. Tudor and Georgescu (2011) reported the ethnobotanical use of chamomile in Budapest. Snežana et al. (2007) reported that wild chamomile was used most commonly for medicinal purposes in Serbia. Silvia et al. (2007) reported the ethnobotanical use of chamomile in the western Pyrenees. Smitherman et al. (2005) reported that the black community in the United States extensively used chamomile-based medication to treat colic and teething in children. Similarly, Michael et al. (2000) reported the extensive use of chamomile among the Latino healers in New York.

The chemical constituents of the chamomile drug (dried flower) possess anti-inflammatory, spasmolytic, carminative, antiseptic, sedative, and ulcer-protecting properties. The researches have revealed that chamazulene possesses antioxidant and anti-inflammatory properties and provides protection against liver damage. Bisabolol protects against ulcers. Apigenin has anti-inflammatory and spasmolytic properties. The polysaccharides are immunostimulating. A pilot study indicates encouraging results on the usefulness of luteolin and quercetin in reducing autism spectrum disorders (Taliou et al. 2013).

Chamomile is used in Unani medicine to treat congestive dysmenorrhea (Arif Zaidi et al. 2012). Chamomile extracts are also used to treat periodontal diseases (Nimbekar et al. 2012). The efficacy of chamomile extracts in ameliorating diabetes in streptozotocin-induced diabetic rats has been investigated with promising results (Emam 2012). The hepatoprotective effect of chamomile is well known (Kumar 2012; Kumar et al. 2012). Research data on the usefulness of chamomile in cancer therapy are increasing everyday. Matić et al. (2013) evaluated the antitumor activity of chamomile tea on various malignant cell lines and found that it indeed had antitumor potential. Guimarães et al. (2013) also reported that chamomile methanolic extract could inhibit the growth of several cancer cell lines.

It is well evidenced by numerous studies that chamomile essential oil and extracts have antimicrobial properties. Roby et al. (2013) found that there is a dose-dependent antimicrobial activity of chamomile extracts. The other studies are enumerated in Section 3.24.

In Sections 3.2 through 3.26, a brief overview is provided on the diverse pharmacological properties of this plant. Most of the investigations on the therapeutic activity of the drug or specific chemical compounds have been carried out on animal models or in vitro systems. A few clinical trials have also been carried out on humans. In Sections 3.2 through 3.26, specific pathophysiological conditions are described, followed by the description of the therapeutic activity of the drug or specific chemical compound.

3.2 ANTI-INFLAMMATORY EFFECTS OF CHAMOMILE

Inflammation is the body's protective response to harmful stimuli, such as pathogens or injury and wound. The symptoms are pain, heat, redness, swelling or edema, and loss of function. The inflammatory response is mediated through the cellular response of the white blood cells (leukocytes and granulocytes) and humoral response (antibodies, cytokines, growth factors, reactive oxygen species [ROS], hydrolytic enzymes, lipids, and so on). The inflammation process can be divided into two phases—acute and chronic. Acute inflammation is a rapid response to a stimulus and mostly the cellular system is involved. It lasts for a few days. In the case of chronic inflammation, a delayed response may start after some time has elapsed and may last for many months or even years. Chronic inflammation results in disorders such as atherosclerosis, allergy and asthma, inflammatory bowel disease, and even cancer and rheumatoid arthritis. Inflammation is mediated through the phospholipases; prostaglandins; leukotrienes; platelet-activating factors; interleukins; interferons; histamines; complements; substance P; polymorphonuclear leukocytes; kallikrein–kininogen–kinin systems; cell adhesion molecules such as vascular cellular adhesion molecule-1 (VCAM-1), intracellular adhesion molecule-1 (ICAM-1), and selectins; and transcription factors such as nuclear factor-κB (NF-κB) (Moses et al. 1964; Sharma and Mohsin 1990; Strober and Fuss 2011; Pearlman 1999; O'Connor et al. 2004).

The prescribed drugs for inflammation are the corticosteroids, such as dexamethasone; the nonsteroidal anti-inflammatory drugs (NSAIDs), such as aspirin and ibuprofen; and the immune-selective anti-inflammatory derivatives (ImSAIDs) (Imulan BioTherapeutics 2008). The corticosteroids bind to the glucocorticoid receptors, which activates the nucleus to express anti-inflammatory proteins or to reduce cytosolic expression of the proinflammatory proteins (Saklatvala 2002). The NSAIDs inhibit the enzyme cyclooxygenase (COX) from producing prostaglandins from arachidonic acid. Prostaglandins are mediators and bind to the mast cells to release histamine that causes inflammation.

The side effects of the steroids and the NSAIDs are well known, which include Cushing's syndrome, hypertension, and central serous retinopathy. The ImSAIDs are a relatively new class of drugs and their effects need to be studied thoroughly.

Candidate drugs with minimum side effects are being investigated and several studies have been carried out to determine the anti-inflammatory activity of the constituent compounds of chamomile essential oil and extracts. Many studies have reported that chamazulene and bisabolol in the essential oil have anti-inflammatory properties. Similarly, several studies report that the flavonoids of chamomile extract, such as apigenin, quercetin, and luteolin, have anti-inflammatory properties.

3.2.1 ANTI-INFLAMMATORY EFFECT OF CHAMOMILE EXTRACT

Chamomile extract and essential oil have been found to have anti-inflammatory effect. The various compounds in the extract and oil probably are the active agents that produce the anti-inflammatory effect. The anti-inflammatory activity of chamomile extract after topical application was evaluated by Tubaro et al. (1984). The hydroalcoholic extract contained (−)-α-bisabolol (0.05 mg/mL), bisabolol oxides (0.45 mg/mL), apigenin and its glucosides (0.4 mg/mL), en-in-dicycloethers (0.8%), and azulenes (0.02%). The researchers used Swiss albino mice and induced edema in their ears using croton oil. They then applied the hydroalcoholic extract to the ears and measured edema. Benzydamine and hydrocortisone were used as standard drugs. They found that the chamomile extract showed a reduction in the edema similar to benzydamine. They concluded that chamomile preparation acts at a vascular level to reduce inflammation.

Al-Hindawi et al. (1989) evaluated the anti-inflammatory property of chamomile ethanolic extract on carrageenan-induced paw edema. Chamomile extract was found to reduce inflammation considerably.

Della Loggia et al. (1990) evaluated the anti-inflammatory activity of chamomile preparations on mouse ear edema. They obtained hydroalcoholic extracts from fresh and dried flowers and separated essential oil, and isolated compounds (bisabolol, matricin, chamazulene, and apigenin). They induced dermatitis in mouse ear by croton oil. Benzydamine (4%, 600 μg) was used as standard drug. The reduction of edema was taken as a measure of the anti-inflammatory activity of the substances. They found that fresh chamomile flower extract (15 μL) reduced edema by 31%, which was similar to the action of benzydamine. They concluded that fresh flower extracts had more effective anti-inflammatory activity than the dried flower extracts, which was because of higher concentrations of matricine in the extracts. They also found that apigenin was 10 times more effective than matricine, whereas isolated chamazulene was 10 times less effective.

Mazokopakis and Ganotakis (2007) reviewed and provided evidence-based information on the anti-inflammatory and mucosal protective properties of chamomile.

The anti-inflammatory effects of chamomile essential oil in mice were studied by Fabian et al. (2011). At the beginning of the experiment, the mice were fed with chamomile essential oil in three concentrations (1250, 2500, and 5000 ppm). Inflammation was induced in rat paw by injection of carrageenan and colitis was induced by intrarectal administration of trinitrobenzene sulfonic acid. The results showed that chamomile oil at a concentration of 5000 ppm significantly reduced edema and weight of mice paws. Chamomile essential oil of the same concentration (i.e., 5000 ppm) showed protective effect on the colonic mucosa and a reduction in bacterial translocation to mesenteric lymph nodes. The researchers concluded that chamomile essential oil in the diet was able to reduce edema and colonic inflammation.

Chandrashekhar et al. (2011) studied the antiallergic effect of chamomile methanolic extract on mast cells. The extract showed inhibitory effects on anaphylaxis induced in mast cells by compound 48/80. It also exhibited antipruritic activity by inhibiting mast cell degranulation in a dose-dependent manner. Mast cell membrane stabilization was also observed. The extract also reduced histamine release and lowered nitric oxide (NO) levels in serum and bronchoalveolar lavage fluid (BALF). These results showed the antiallergic properties of chamomile extracts. Drummond et al. (2013) also reported the anti-inflammatory properties of chamomile aqueous extract.

Bulgari et al. (2012) investigated the anti-inflammatory effect of chamomile infusion and individual flavonoids on AGS cells and human neutrophil elastase. The AGS cell line is a culture collection that was obtained from stomach epithelial cells of a human with gastric adenocarcinoma. The researchers treated the AGS cells and neutrophil elastase with phorbol 12-myristate 13-acetate (PMA) to stimulate the cells and then added the chamomile infusion and flavonoids separately. They found that there was an inhibition in activities of neutrophil elastase and metalloproteinase-9. Further, the secretion of metalloproteinase-9 was also decreased, which was attributed to the inhibition of NF-κB-driven transcription. The researchers also speculated that the flavonoid-7-glycosides were perhaps responsible for the observed anti-inflammatory action.

In all the studies mentioned earlier, although the anti-inflammatory activities of chamomile essential oil or extracts were established, the mechanism of action could not be elucidated. Because the essential oil and extracts are a mixture of several chemical compounds, the therapeutic activity could not be attributed to any specific compound. However, several other studies on the anti-inflammatory activities of chamomile have been carried out using specific chamomile compounds. This makes it easy to pinpoint the anti-inflammatory effect of chamomile to a single compound. Some of these studies carried out using chamazulene, (−)-α-bisabolol, *trans*-en-yn-dicycloether, apigenin, quercetin, and luteolin are described in Sections 3.2.2 through 3.2.5.

3.2.2 Anti-Inflammatory Effect of Chamazulene

Jakovlev et al. (1983) investigated the anti-inflammatory effect of chamazulene. Rats weighing 98–138 g were injected with 0.1 mL of 1% carrageenan in the paws to induce edema. Chamazulene and standard salicylamide (which controls edema) were administered orally. After 2.3 and 4 hours, the effect was noted by comparing the inhibition of edema between control and chamazulene-treated rats. The researchers found that chamazulene reduced edema effectively but was comparatively less effective than salicylamide and matricin. They concluded that chamazulene and matricin might be having different anti-inflammatory activities.

The anti-inflammatory effect of chamazulene was investigated by Safahyi et al. (1994), who studied the effect of matricine and its product chamazulene on the leukotriene production in neutrophilic granulocytes. The granulocytes were treated with glycogen and calcium ionophore A23187 to increase the level of endogenous arachidonic acid. These were then incubated with different concentrations of chamazulene in the presence of dimethyl sulfoxide (DMSO, 0.5%) for 5 minutes. The cells were extracted and the extract was analyzed using reversed-phase high-performance liquid chromatography and ultraviolet detector. A series of chamazulene concentrations were used. Chamazulene at 10 and 15 μM showed 50% inhibition (half-maximal inhibitory concentration [IC_{50}]) of leukotriene B4 formation in intact cells. Chamazulene also blocked the peroxidation of arachidonic acid. Matricin was found to have no effect. Safayhi et al. (1994) concluded that chamazulene, and not matricin, contributed to the anti-inflammatory activity of chamomile extracts by inhibiting leukotriene synthesis and other antioxidative effects, such as blocking the chemical peroxidation of arachidonic acid.

Miller et al. (1996) studied the effect of chamazulene on the release of histamine from rat mast cells. In fact, they concluded that chamazulene did not show any significant reduction in histamine release. In fact, chamazulene at higher concentrations (10^{-4} M and 10^{-5} M) stimulated histamine release.

The decomposition of matricin yields chamazulene carboxylic acid (CCA), which on further decomposition yields chamazulene. CCA is a very unstable compound. However, researchers (Ramadan et al. 2006) have found that CCA is anti-inflammatory just like a profen, such as ibuprofen. CCA was shown to inhibit COX-2, but not COX-1, and to have anti-inflammatory activity in several animal models with local and systemic application.

Ogata et al. (2005) studied the effect of azunol (sodium azulene sulfonate—a derived form of azulene) on postoperative sore throat (POST) that occurs after endotracheal intubation. Azulene is used to treat inflammations, chronic gastritis, and ulcers. They found that gargling with azulene effectively attenuated POST in the patients and no adverse side effects were observed.

3.2.3 Anti-Inflammatory Effect of (−)-α-Bisabolol

Early investigators found that (−)-α-bisabolol reduced inflammation in rat paw edema and adjuvant arthritis (AA) in rats. They also said that (−)-α-bisabolol reduced the production of mucopolysaccharides in cell cultures. Experiments also showed that (−)-α-bisabolol was more effective than (+)-α-bisabolol and the synthetic racemic bisabolol (Jakovlev et al. 1979). Therefore, these researchers investigated the antiphlogistic (anti-inflammatory) effect of (−)-α-bisabolol and bisabolol oxides in rats. The paws of rats weighing 95–130 g were injected with 0.1 mL of 1% carrageenan to induce edema. The test substance containing (−)-α-bisabolol was administered to one group and another group was kept as control and treated with carboxymethyl cellulose. After 2 hours of (−)-α-bisabolol administration, the rats were killed to measure edema. The inhibition of edema by the test substance and control was given in percentage. (−)-α-Bisabolol was more effective (ED_{50} = 1465 mg/kg) as compared to dragostanol (ED_{50} = 300 mg/kg) in controlling the inflammation.

Miller et al. (1996) studied the effect of (−)-α-bisabolol on histamine release from rat mast cells. (−)-α-Bisabolol did not show any significant reduction in histamine release. The researchers reported that in fact (−)-α-bisabolol, at higher concentrations (10^{-4} M and 10^{-5} M), stimulated histamine release.

Kim et al. (2011) investigated the anti-inflammatory effects of (−)-α-bisabolol and its mechanisms of action. They found that (−)-α-bisabolol inhibited lipopolysaccharide (LPS)-induced production of NO and prostaglandin E2 (PGE2) in RAW264.7 macrophage cell line. They also found through the Western blot and luciferase reporter assays for COX-2 and inducible nitric oxide synthase (iNOS) genes that the expression of COX-2 and iNOS genes was reduced. They also found that (−)-α-bisabolol reduced the LPS-induced activation of activating protein-1 (AP-1) and NF-κB promoters, reduced the LPS-induced phosphorylation of IκBα, and attenuated LPS-induced phosphorylation of the extracellular-regulated kinase (ERK) and p38 proteins. They concluded that (−)-α-bisabolol exerts anti-inflammatory effects by downregulating the expression of iNOS and COX-2 genes through inhibition of NF-κB and AP-1 (ERK and p38) signaling.

Rocha et al. (2011) tested the effect of (−)-α-bisabolol at doses of 100 and 200 mg/kg in rodent models of paw edema. Edema was induced in various groups of mice by carrageenan, dextran, 5-hydroxytryptamine (HT), and histamine. The mice were then treated with (−)-α-bisabolol. The researchers found that the mice with carrageenan- and dextran-induced edema, when treated with (−)-α-bisabolol, showed smaller edemas compared to those treated only with the vehicle. It was also found that (−)-α-bisabolol reduced paw edemas induced by 5-HT but not those induced by histamine. In addition, the researchers also concluded that (−)-α-bisabolol has antinociceptive (painkilling) activity.

Leite et al. (2011) examined the anti-inflammatory and antinociceptive effects of (−)-α-bisabolol in mice. The groups of mice were pretreated with (−)-α-bisabolol and then dermatitis was induced in the ear by croton oil, arachidonic acid, phenol, and capsaicin. (−)-α-Bisabolol was found to inhibit inflammation in all the cases of dermatitis except that induced by capsaicin. The researchers concluded that (−)-α-bisabolol is an anti-inflammatory agent.

3.2.4 ANTI-INFLAMMATORY EFFECT OF *TRANS*-EN-YN-DICYCLOETHER

Miller et al. (1996) studied the effect of *trans*-en-yn-dicycloether on histamine release from rat mast cells. They found that it strongly inhibited protamine sulfate-mediated histamine release from rat mast cells. They concluded that *trans*-en-yn-dicycloether may exert the effects through membrane-directed and exocytosis processes.

3.2.5 ANTI-INFLAMMATORY EFFECT OF FLAVONOIDS

To investigate the role of flavonoids in the anti-inflammatory activity of chamomile, Della Loggia et al. (1986) tested the hydroalcoholic extract, a pentane extract containing essential oil components, and an ethyl acetate extract containing the flavonoids (apigenin, quercetin, luteolin, rutin, and apigenin-7-glucoside) in croton oil-induced dermatitis in mouse ear. Indomethacin and hydrocortisone were used as standard drugs. The researchers found that the flavonoids were more active than the essential oil components. Apigenin was shown to be effective, followed by luteolin. The effect of apigenin was found to be the same as that of indomethacin. The effect of apigenin and luteolin was tested for leukocyte infiltrate inhibition measured as the myeloperoxidase activity. Apigenin and luteolin were found to strongly inhibit leukocyte infiltration. The researchers suggested that while luteolin might be inhibiting inflammation by inhibiting the arachidonic acid metabolism pathway, apigenin might be working through a different mechanism such as inhibition of histamine release or free radical scavenging activity.

Ammon et al. (1996) investigated the anti-inflammatory effects of flavonoids of chamomile extract. Neutrophil granulocytes of rats were stimulated with calcium ionophore A23187 to produce arachidonic acid. The so-prepared cells were then treated with varying concentrations of flavonoids. The researchers found that apigenin at $IC_{50} = 8$ μM prevented the formation of leukotriene B_4 and at $IC_{50} = 8$ μM prevented the formation of 5-hydroxyeicosatetraenoic acid, thereby showing anti-inflammatory effects.

Sawatzky et al. (2006) reported that apigenin reduced inflammation in a rat carrageenan-induced pleurisy model. Apigenin was found to inhibit molecular signaling via the ERK1/2 pathway.

Lee et al. (2007) studied the anti-inflammatory activity of apigenin on the RAW 264.7 macrophage cells. The cells were activated by LPS to produce NO and COX. Apigenin inhibited the NO and COX production significantly. The researchers also studied the effect of apigenin on the adhesion molecules of the cells. In addition, apigenin significantly reduced the tumor necrosis factor-α (TNF-α)-induced adhesion of monocytes to human umbilical vein endothelial cells (HUVECs) monolayer. Apigenin significantly suppressed the TNF-α-stimulated upregulation of VCAM-1, ICAM-1, and E-selectin-mRNA to the basal levels.

Courtney (2009), while working on the molecular and physiological mechanisms of apigenin, found that apigenin inhibits transcriptional activation of NF-κB (a protein complex that controls the transcription of DNA, which is involved in expression of cytokines and free radicals) and subsequent release of proinflammatory cytokines TNF-α, interleukin (IL)-1β, and IL-8 in response to LPS stimulation. He also found that in a mouse model of sepsis-related acute lung injury, intraperitoneal apigenin reduced LPS-induced mortality, improved cardiac function, and reduced TNF-α in serum. Apigenin failed to modulate cell death in spleens, which is traditionally shown to correlate with sepsis-related death in humans and mice. However, apigenin decreased neutrophil infiltration, cell death, expression of chemotactic proteins in mouse lungs, and chemotactic response in human neutrophils in vitro.

Funakoshi-Tago et al. (2011) found that apigenin and luteolin significantly inhibited TNF-α-induced NF-κB transcriptional activation through inhibition of the transcriptional activity of GAL4/NF-κB p65 fusion protein. TNFα-induced expression of CCL2/MCP-1 (a cytokine) and CXCL1/KC (a cytokine) was significantly inhibited by apigenin and luteolin. Furthermore, the administration of apigenin and luteolin markedly inhibited acute carrageenan-induced paw edema in mice demonstrating the anti-inflammatory effect of the flavonoids.

Drummond et al. (2013) studied the anti-inflammatory effect of the chamomile flavonoids apigenin and quercetin on the THP1 macrophages. After treating the macrophages with the flavonoids, the researchers measured the levels of inflammatory cytokines IL-1β, IL-6, and TNF-α. They found that the flavonoids had effective anti-inflammatory properties.

González et al. (2011) reviewed the effects of flavonoids on inflammation and concluded that their potential therapeutic application in inflammation extends to sites and conditions, including arthritis, asthma, encephalomyelitis, and atherosclerosis.

3.2.5.1 Anti-Inflammatory Effect of Apigenin

The anti-inflammatory activity of apigenin has been investigated by many researchers. Doseff et al. (2011) investigated the anti-inflammatory effect of apigenin on a mouse model of LPS-induced airway inflammation. They also analyzed human monocytes and macrophage cell lines to determine the molecular basis of anti-inflammatory activity of apigenin. When they analyzed the BALF of the apigenin-treated mice, they found reduced levels of chemokines. Further they found reduced levels of LPS-induced TNF and neutrophil lung infiltration. They reported to find a novel link between the apigenin and the master regulators of mRNA metabolism, which are involved in the regulation of NF-κB.

Man et al. (2012) investigated the effect of topical application of apigenin on inflammation. They used mouse models of acute allergic contact dermatitis and acute irritant contact dermatitis established by topical application of oxazolone and PMA, respectively. The effect of apigenin on stratum corneum function in a murine subacute allergic contact dermatitis model was assessed with an MPA5 physiology monitor. The researchers found that topical application of apigenin was effective in reducing both acute irritant contact dermatitis and allergic contact dermatitis models. In addition, it was found that apigenin significantly reduced transepidermal water loss, lowered skin surface pH, and increased stratum corneum hydration. The results indicated that a lower dose (1.23 mg/kg body weight) of apigenin could improve acute dermatoses, and topical applications of apigenin at dosage of 4 mg/kg body weight per day for 7 days could relieve subacute dermatitis. The researchers concluded that apigenin is a safe and effective anti-inflammatory agent, especially for topical use.

Topical formulations of apigenin as substitutes for corticosteroid therapy have been developed by Arsić et al. (2011). These formulations are both liposomal and non-liposomal.

3.2.5.2 Anti-Inflammatory Effect of Quercetin

Several investigators have worked on quercetin to elucidate its anti-inflammatory action. Lindahl and Tagesson (1993) studied the effect of quercetin on the phospholipase A_2 (PLA_2). PLA_2 plays a very important role in inflammation by catalyzing the formation of arachidonic acid that further produces leukotrienes and prostaglandins. They found that quercetin was a potent inhibitor of PLA_2.

Chan et al. (2000) studied the effect of quercetin on the iNOS gene expression. They used macrophage cell line RAW 264.7 stimulated with LPS and interferon-γ (IFN-γ) to induce iNOS expression. They found that quercetin suppressed iNOS gene expression and NO production. Quercetin was also found to be a scavenger of NO in an acellular system using sodium nitroprusside under physiological conditions. They concluded that quercetin acts on the iNOS pathway.

Guardia et al. (2001) studied the anti-inflammatory activity of quercetin in rats using a model of acute and chronic inflammation. Quercetin was given intraperitoneally. They found that a dose equivalent to 80 mg/kg quercetin inhibited both acute and chronic phases of inflammation.

Cho et al. (2003) investigated the effect of quercetin on the LPS-stimulated RAW 264.7 cells. They found that quercetin strongly reduced the activation of phosphorylated ERK and p38 mitogen-activated protein kinase (MAPK) but not c-Jun N-terminal kinase (JNK) and MAPK induced by the LPS treatment. In addition, quercetin treatment inhibited NF-κB activation through stabilization of the NF-κB/IκB complex and also inhibited IκB degradation, proinflammatory cytokines, and NO/iNOS expression.

Morikawa et al. (2003) examined the effect of quercetin on the inflammatory response induced by carrageenan in the rat. The rats were treated with either vehicle or quercetin at a dose of 10 mg/kg/h. Forty-eight hours after carrageenan challenge, the air pouches were removed and analyzed. The volume, protein amounts, and cell counts in the exudation obtained from the quercetin-treated animals were significantly reduced compared to those from vehicle-treated animals. The levels of PGE2, TNF-α, RANTES (regulated on activation, normal T-cell expressed, and secreted), MIP-2, and the mRNA for COX-2 were also suppressed in these rats.

Rotelli et al. (2003) investigated the anti-inflammatory activity of quercetin in animal models of acute and chronic inflammation. They found that quercetin significantly decreased carrageenan-induced rat paw edema.

O'Leary et al. (2004) reported that quercetin reduced COX-2 mRNA expression in both unstimulated and IL-1β-stimulated colon cancer (Caco2) cells, suggesting that quercetin possessed anti-inflammatory activity.

Comalada et al. (2005) reported that quercetin was able to downregulate the inflammatory response of bone marrow-derived macrophages in vitro. Moreover, they found that quercetin inhibits cytokine and iNOS expression through inhibition of the NF-κB pathway without modification of JNK activity (both in vitro and in vivo).

Boots et al. (2008) studied the in vitro and ex vivo anti-inflammatory activity of quercetin in healthy volunteers. Quercetin inhibited in vitro LPS-induced tumor TNF-α production in the blood of healthy volunteers. Their study indicated that quercetin increased antioxidant capacity in vivo and displayed anti-inflammatory effects in vitro, but not in vivo or ex vivo, in the blood of healthy volunteers.

Mamani-Matsuda et al. (2006) investigated in vivo effects of oral or intracutaneous quercetin in chronic rat AA. The researchers reported that injection of quercetin (5 × 160 mg/kg), intracutaneous injections of quercetin in lower doses (5 × 60 mg/kg), and injection of relatively low quercetin doses (5 × 30 mg/kg) prior to AA induction to arthritic rats resulted in a clear decrease of clinical signs of AA. The analysis of ex vivo macrophage response to quercetin was evaluated and a significant decrease of inflammatory mediators produced by peritoneal macrophages was found, ex vivo and in vitro, which correlated with anti-arthritic effects of quercetin.

Rogerio et al. (2007) investigated the anti-inflammatory effect of quercetin in a murine model of asthma. They found that in the quercetin-treated mice, the eosinophil counts were lower in the BALF, blood, and lung parenchyma. The researchers suggested that quercetin was able to suppress eosinophil-mediated inflammation.

Min et al. (2007) reported that when human mast cells (HMC-1) were stimulated with PMA and calcium ionophore A23187 (PMACI) to produce inflammatory cytokines, and treated with quercetin, it was found that quercetin decreased the gene expression and production of TNF-α, IL-1β, IL-6, and IL-8. Quercetin also attenuated PMACI-induced activation of NF-κB and p38 MAPK.

González-Segovia et al. (2008) analyzed the effect of oral administration of pure quercetin on inflammation and lipid peroxidation induced by *Helicobacter pylori* in the gastric mucosa of the guinea pig. The guinea pigs were orally infected with *Helicobacter pylori* and after 60 days received 200 mg/kg of quercetin daily orally for 15 days. They found that quercetin significantly reduced the mononuclear cell and neutrophil infiltration, and bacterial colonization in the pyloric antrum and corpus. Further, the lipid hydroperoxide concentration was significantly decreased in infected animals treated with quercetin. The researchers concluded that the in vivo oral quercetin administration decreased *Helicobacter pylori* infection in the gastric mucosa and reduced both the inflammatory response and lipid peroxidation.

Ying et al. (2009) studied the effect of quercetin on the ICAM-1 expression in pulmonary epithelial cell line A549. They pretreated the A549 cells with quercetin and then stimulated them with IL-1β to express ICAM-1. It was found that quercetin effectively inhibited the ICAM-1. This activity was carried out by quercetin through the inhibition of the degradation of I-κB (inhibitor of NF-κB), inhibition of the activity of NF-κB, and sequential attenuation of the c-fos and c-jun mRNA expressions. The researchers suggested that quercetin negatively modulating ICAM-1 was partly dependent on MAPK pathways.

Chuang et al. (2010) examined the extent to which quercetin prevented inflammation in primary cultures of human adipocytes treated with TNF-α. They found that quercetin reduced the TNF-α-induced expression of inflammatory genes such as IL-6, IL-1β, IL-8, and monocyte chemoattractant protein-1 (MCP-1) and the secretion of IL-6, IL-8, and MCP-1. Quercetin was also found to attenuate TNF-α-mediated phosphorylation of extracellular signal-related kinase and JNK and prevent the degradation of inhibitory κB protein. Quercetin was also reported to decrease the TNF-α-induced NF-κB transcriptional activity.

Boots et al. (2011) investigated the effect of quercetin on sarcoidosis, an inflammatory disease. A group of untreated sarcoidosis patients were given 4 × 500 mg quercetin ($n = 12$) orally within 24 hours. Oxidative damage was measured as the levels of plasma malondialdehyde and inflammation was measured as the plasma ratios of TNFα/IL-10 and IL-8/IL-10. Quercetin was found to increase the total plasma antioxidant capacity, indicating its antioxidant property. It was also found to reduce the markers of oxidative stress and inflammation in the blood of sarcoidosis patients.

MicroRNAs (miRNAs) have been identified as powerful posttranscriptional gene regulators of the genes involved in inflammation. The effect of quercetin on miRNA regulation in vivo was investigated by Boesch-Saadatmandi et al. (2012). Mice were fed for 6 weeks with control or quercetin-enriched diets. The biomarkers of inflammation as well as hepatic levels of miRNAs previously involved in inflammation (miR-125b) and lipid metabolism (miR-122) were determined. They found that quercetin increased the hepatic miR-122 and miR-125b concentrations. Quercetin was also found to lower the mRNA steady-state levels of the inflammatory genes IL-6, C-reactive protein, MCP-1, and acyloxyacyl hydrolase. In addition, they found evidence for an involvement of redox factor 1, a modulator of NF-κB signaling, on the attenuation of inflammatory gene expression mediated by quercetin.

Weng et al. (2012) studied the effect of quercetin on cultured HMCs and on two clinical trials on dermatitis patients. Quercetin (100 µM) effectively inhibited secretion of histamine PGD_2 and leukotrienes. It also effectively inhibited IL-8, and reduced IL-6 TNF release from the mast cells. Moreover, it was found to inhibit cytosolic calcium level increase and NF-κB activation. Interestingly, it was found to be effective prophylactically and significantly decreased contact dermatitis and photosensitivity, skin conditions that do not respond to conventional treatment.

3.2.5.3 Anti-Inflammatory Effect of Luteolin

Luteolin has demonstrated significant anti-inflammatory activity in well-established models of acute and chronic inflammation, such as xylene-induced ear edema in mice (ED_{50} = 107 mg/kg), carrageenan-induced swelling of the ankle, acetic acid-induced pleurisy, and croton oil-induced gaseous pouch granuloma in rats. Luteolin has been found to antagonize the inflammatory effects of the slow-reacting substance of anaphylaxis or *SRS-A* (a mixture of the leukotrienes LTC4, LTD4, and LTE4) and also histamine. Luteolin was also reported to have immunostimulating activities, such as the production of hemolysin in mice, experimental allergic encephalomyelitis in guinea pigs, and the proliferative response of mouse splenic lymphocytes in vitro (Min-Zhu et al. 1986).

Luteolin has been found to inhibit protein kinase C (PKC, an enzyme actively involved in the inflammation process) by blocking the ATP binding site of PKC (Middleton et al. 2000).

Shimoi et al. (2000) evaluated the chemopreventive role of luteolin in inflammatory responses involved in the pathogenesis of atherosclerosis and cancer. The effect of luteolin on the cell surface expression of adhesion molecules in HUVECs was also examined. The researchers found that luteolin suppressed the TNF-α-induced ICAM-1 expression significantly.

Xagorari et al. (2001) examined the ability of luteolin to modulate the production of proinflammatory molecules from LPS-stimulated mouse macrophages and investigated their mechanism(s) of action. They reported that luteolin inhibited protein tyrosine phosphorylation, NF-κB-mediated gene expression, and proinflammatory cytokine production in the macrophages.

Kotanidou et al. (2002) found that luteolin pretreatment reduced LPS-stimulated ICAM-1 expression in the liver and abolished leukocyte infiltration in the liver and lung. They concluded that luteolin protects against LPS-induced lethal toxicity, possibly by inhibiting proinflammatory molecule (TNF-α, ICAM-1) expression in vivo and reducing leukocyte infiltration in tissues.

Kim et al. (2003) investigated the effects of luteolin on NF-κB activation. They found that luteolin showed the most potent inhibition on LPS-stimulated NF-κB transcriptional activity in Rat-1 fibroblasts. Luteolin prevented LPS-stimulated interaction between the p65 subunit of NF-κB and the transcriptional coactivator CREB Binding Protein (CBP). They concluded that inhibition of NF-κB transcriptional activity by luteolin may occur through competition with transcription factors for coactivator that is available in limited amounts.

Das et al. (2003) investigated the antiasthmatic effect of luteolin on a mouse model of asthma. They found that luteolin, at a dose of 0.1 mg/kg body weight, both during and after sensitization of the mice, significantly modulated ovalbumin (OVA)-induced airway bronchoconstriction and bronchial hyperreactivity ($p < 0.05$). Luteolin also reduced OVA-specific IgE levels in the sera, increased IFN-γ levels, and decreased IL-4 and IL-5 levels in the BALF.

Hirano et al. (2004) investigated the inhibition of IL-4 and IL-13 production by anti-IgE antibody alone or anti-IgE antibody along with IL-3-stimulated basophils in the presence of luteolin. They also investigated whether luteolin suppressed leukotriene C4 synthesis by basophils and IL-4 synthesis by T cells in response to anti-CD3 antibody. Luteolin was found to inhibit IL-4 and IL-13 production by basophils but did not affect leukotriene C4 synthesis. At higher concentrations, luteolin was found to suppress IL-4 production by T cells.

Kim et al. (2005) investigated the role of luteolin, a major flavonoid of *Lonicera japonica*, on TNF-α-induced IL-8 production in human colonic epithelial cells. Luteolin suppressed TNF-α-induced IL-8 production in dose-dependent manner. In addition, it inhibited TNF-α-induced phosphorylation of p38 MAPK and ERKs, degradation of IκB, and activation of NF-κB. The researchers concluded that luteolin inhibited TNF-α-induced IL-8 production through blocking the phosphorylation of MAPKs, following IκB degradation and NF-κB activation.

Harris et al. (2006) reported that pretreatment of RAW 264.7 macrophage-like cells with 25, 50, or 100 μmol/L concentrations of luteolin inhibited LPS-induced COX-2 protein expression. Luteolin also suppressed LPS-induced PGE2 formation. Luteolin inhibited xanthine/xanthine oxidase-generated superoxide formation and also reduced LPS-induced hydroxyl radical formation significantly.

Wen-hui and Mao (2006) studied the effects of luteolin on airway remodeling in asthmatic mice. They found that luteolin was able to remodel the airways through a decrease in the airway wall thickness and smooth muscle thickness possibly by the way of decreasing the levels of IL-5 and increasing the levels of IFN-γ in the lung tissue.

Chiu-Yuan et al. (2007) examined the effects of luteolin on the production of NO and PGE2, as well as the expression of iNOS, COX-2, TNF-α, and IL-6 in mouse alveolar macrophage MH-S and peripheral macrophage RAW 264.7 cells. Luteolin dose dependently inhibited the expression and production of these inflammatory mediators. Luteolin also reduced the DNA binding activity of NF-κB and blocked the degradation of IκB-α and nuclear translocation of NF-KB p65 subunit. In addition, luteolin significantly inhibited the DNA binding activity of AP-1 and also attenuated the PKB and If-kappa B kinase (IKK) phosphorylation, as well as ROS production. It was thus shown that luteolin was able to block the production of many mediators and genes involved in inflammation.

Kempuraj et al. (2008) reported that luteolin pretreatment inhibited mast cell activation, Jurkat cell activation, and mast cell–dependent Jurkat cell stimulation. This finding led the researchers to suggest that luteolin may be useful in the treatment of autoimmune diseases.

Lopez-Lazaro (2009) summarized the biological activities of luteolin and suggested that modulation of ROS levels, inhibition of topoisomerases I and II, and reduction of NF-κB and AP-1 activity are some of the possible mechanisms involved in the biological activities of luteolin.

Kang et al. (2010) studied the effect of luteolin on PMA along with A23187-induced mast cell activation. They reported that luteolin suppressed the expression of PMA and A23187-induced TNF-α, IL-8, IL-6, GM-CSF, and COX-2 through a decrease in the intracellular Ca^{2+} levels, and also showed a suppression of the ERK1/2, JNK1/2, and NF-κB activation.

Hänler et al. (2012) examined the influence of luteolin on the inflammatory response in ARPE-19 (a cell line from retinal epithelium) cells. The cells were exposed to the lipid peroxidation end product 4-hydroxynonenal to induce the inflammatory response and luteolin was added simultaneously. Luteolin significantly decreased the production of proinflammatory cytokines IL-6 and IL-8. The researchers concluded that these results indicated the luteolin could reduce inflammation ARPE-19 cells and could be useful in the therapy of age-related macular degeneration.

3.3 CHAMOMILE AS AN ANTIOXIDANT

A free radical is any atom or group of atoms that has an unpaired electron in its outer orbit. The ROS are free radicals and the types of ROS are superoxide ($^{\bullet}O_2^-$), hydroxyl ($^{\bullet}OH$), peroxyl ($ROO^{\bullet}$), alkoxyl ($RO^{\bullet}$) radicals, radicals of NO ($^{\bullet}NO$), nitrogen dioxide ($^{\bullet}NO_2$). The ROS are mostly generated by cells during inflammation or injury and may cause damage to the DNA, lipid, protein, and carbohydrate molecules. The ROS, thus, may be involved in processes such as mutagenesis, carcinogenesis, membrane damage, lipid peroxidation, as well as carbohydrate damage (Datta et al. 2000).

3.3.1 ANTIOXIDANT PROPERTY OF CHAMOMILE

The antioxidant property of chamomile has been investigated by many researchers.

3.3.1.1 Chamazulene

Rekka et al. (1996) investigated the effect of chamazulene on free radical processes. They induced lipid peroxidation in the cell membrane by Fe^{2+}/ascorbate and assessed as the 2-thiobarbituric acid reactive material. They studied the hydroxyl radical scavenging activity by allowing the competition of chamazulene and DMSO to scavenge for the HO radical generated by Fe^{3+}/ascorbate. Finally, the interaction of chamazulene with 2,2-diphenyl-1-picrylhydrazyl (DPPH) (an inorganic compound composed of stable free radicals) was estimated photometrically (517 nm). They found that chamazulene inhibited lipid peroxidation in a concentration- and time-dependent manner presenting IC_{50} of 18 μM after 45 minutes of incubation. Chamazulene was also reported to inhibit the autoxidation of DMSO (33 mM) by 76% at 25 mM, and had a weak capacity to interact with DPPH. It was concluded that chamazulene could play an important role in controlling free radicals.

In another experiment, heat-inactivated rat liver microsomes were exposed to different concentrations of chamazulene (5–50 μM) dissolved in DMSO. In addition, the reaction mixture contained ascorbic acid 0.2 μM in tris buffer. The mixture was incubated at 37°C for 30 minutes. Lipid peroxidation by ascorbic acid was determined photometrically by detecting barbituric acid at 535 nm in the mixture. Chamazulene inhibited lipid peroxidation completely at 50 μM (IC_{50} = 18 μM). This activity was comparable to quercetin (IC_{50} = 10 µM) and propyl gallate (IC_{50} = 10 µM). Chamazulene had more potent hydroxyl scavenging activity than mannitol (Rekka et al. 1996). Chamazulene was thus found to have antioxidant properties.

3.3.1.2 (−)-α-Bisabolol

Braga et al. (2009) investigated the antioxidant property of (−)-α-bisabolol in human polymorphonuclear cells. They stimulated the cells with *Candida albicans* and *N*-formyl-methionyl-leucyl-phenylalanine (fMLP) to produce ROS through respiratory bursts. The ROS produced was measured using luminol-amplified chemiluminescence (LACL). The effect of (−)-α-bisabolol on the ROS production was measured as inhibition of LACL. In addition to the polymorphonucelar cells, cell-free systems (SIN-1 and H_2O_2/HOCl(−) systems) were also used to study the ability of bisabolol for its ROS scavenging activity. After *Candida albicans* stimulation, significant concentration-dependent LACL inhibition was observed at bisabolol concentrations ranging from 7.7 to 31 µg/mL. Similarly, after the fMLP stimulus, significant LACL inhibition was observed at bisabolol concentrations ranging from 3.8 to 31 µg/mL. A similar effect was observed in the SIN-1 and H_2O_2/HOCl(−) systems. The researchers concluded that (−)-α-bisabolol findings may reduce oxidative stress and restore the redox balance.

de Siqueira et al. (2012) investigated the pharmacological effects of different concentrations of (−)-α-bisabolol on various smooth muscle preparations from rats, such as duodenal strips, aortic rings, urinary bladder strips, and tracheal and colonic tissues. They found that (−)-α-bisabolol relaxed intact aortic rings and urinary bladder strips at higher concentrations (600–1000 µmol/L) and also on tracheal and colonic tissues. (−)-α-Bisabolol completely decreased spontaneous contractions in duodenum. They concluded that (−)-α-bisabolol is biologically active in smooth muscle and it may be an inhibitor of voltage-dependent Ca^{2+} channels.

3.3.1.3 Chamomile Extract and Essential Oil

Luqman et al. (2009) studied the effect of chamomile extracts for their protective activity against oxidative stress in erythrocytes induced by hydrogen peroxide (2 mM) and *tert*-butyl hydroperoxide (0.01 mM). They found that chamomile extract protected the erythrocytes from oxidative stress.

Abdoul-Latif et al. (2011) tested the antioxidant activities of chamomile oil and chamomile methanolic extract using DPPH free radical scavenging assay and β-carotene–linoleic acid assay. In the β-carotene–linoleic acid assay, oxidation was effectively inhibited by chamomile.

Muñoz-Velázquez et al. (2012) studied the phenolic content, antioxidant capacity, and anti-inflammatory activity of a commercial infusion of chamomile. They tested the infusion on the inhibition of COX-2 enzyme. They found that chamomile infusion inhibited the COX-2 enzyme, suggesting its anti-inflammatory potential.

On the basis of the observation that the various components of chamomile essential oil and extracts have anti-inflammatory activities, Ghavimi et al. (2012) put forth the hypothesis that chamomile could provide relief from gouty arthritis.

3.3.1.4 Polysaccharides

Füller et al. (1991) reported that the polysaccharides of chamomile, namely fructosan, pectinic polymer, and 4-0-methylglucuronoxylan, have inflammatory activities, when applied topically.

3.4 ANTINOCICEPTIVE EFFECT OF CHAMOMILE

Nociception is the sense of pain. Nociceptive pain is a kind of early warning system that tells us about the dangers of the environment we are in. The nociceptor neurons are nerve cells that are involved in this kind of pain. These nociceptors sense any environmental stimuli, such as heat (high or low), pressure (high or low), taste, olfaction, and chemicals (in addition to external chemicals, such as acids and capsaicin in chilies, chemicals those are generated during tissue injury, such as histamines and prostaglandins). The nociceptors convey these simuli to the spinal cord or brain from any part of the body through electrical impulses. Nociceptors work through receptors (specialized proteins) that get activated under extreme conditions of heat, pressure, or chemicals. When activated, the receptors change their conformation and form ion channels in the cell membrane through which positive ions, such as sodium and calcium move into the cells thereby lowering the voltage across the membrane. In other cases, the positive potassium ions may move out of the cell. When the stimulus is extreme, this change in the voltage further gets lowered through other proteins in the cell and an impulse is generated, which travels across the neuron to the spinal cord. Thus, these cationic channels are responsible for the excitability of the neurons. In addition to nociceptive pain, there is another kind of pain classified as clinical pain. Clinical pain occurs in response to tissue injury and inflammation (inflammatory pain), damage to the nervous system (neuropathic pain), and alterations in the normal function of the nervous system (functional pain). During an inflammatory response, several intracellular pathways are activated that signal pain. It is possible to control the excitability of the cationic channels by regulating the expression or modulation of the cationic channels (Julius and Basbaum 2001; Le bars et al. 2001; Wood 2004; Lee et al. 2005; Patel 2010).

3.4.1 PAIN-RELIEVING PROPERTIES OF CHAMOMILE

Abad et al. (2011a) studied the antinociceptive effect of chamomile extract on mice that were treated with formalin in two phases to induce pain. The mice were then treated with cisplatin, an anticancer drug that causes pain. It was found that the chamomile decreased the second phase of the cisplatin-induced pain significantly.

Alves Ade et al. (2010) studied the effect of bisabolol on mice sciatic nerves by observing the changes on the compound action potential characteristics. They found that (−)-α-bisabolol decreased the neuronal excitability in a concentration-dependent manner, although such effects were not reversed when the nerve was submitted to wash out. They suggested that decreased nervous excitability elicited by (−)-α-bisabolol might be caused by an irreversible blockade of voltage-dependent sodium channels.

Leite et al. (2011) examined the antinociceptive effects of (−)-α-bisabolol in mice. The mice were treated with cyclophosphamide intraperitoneally or mustard oil intracolonically to induce pain. (−)-α-Bisabolol was found to significantly reduce pain in a dose-unrelated manner. The researchers concluded that (−)-α-bisabolol is an anti-inflammatory and visceral antinociceptive agent.

Nouri and Abad (2012) evaluated the effect of chamomile hydroalcoholic extract on vincristine-induced peripheral neuropathy in Naval Medical Research Institute (NMRI) mice. The mice received intraperitoneal and intravenous injections of different doses of chamomile extract for 12 days. The pain responses were treated by formalin. Chamomile extracts were found to significantly decrease the pain induced by vincristine. The researchers concluded that chamomile extracts may be used for treating vincristine-induced peripheral neuropathic pain.

3.5 CHAMOMILE AND WOUND HEALING

Wound healing occurs through four phases: hemostasis, inflammation, proliferation, and remodeling. For a wound to heal successfully, all four phases must occur in the proper sequence and time (Guo and DiPietro 2010). The complete healing response was summarized by Flynn and Rovee (1982) as follows:

- Platelet aggregation at the site of injury followed by blood clotting to arrest bleeding.
- An inflammatory response with the infiltration of the leukocytes and associated inflammatory mechanisms at the site of injury. The neutrophils move in and begin the process of phagocytosis to remove damaged cells, foreign matter, and bacteria. Then the macrophages come in and release growth factors and cytokines.
- Fibroblast accumulation with enhanced collagen synthesis at the injury site.
- Epithelial cell migration, division, and differentiation.
- Contraction of the wound, scarring, and remodeling of the scar.

This is a highly ordered process that occurs naturally. However, in pathological conditions, such as severe infection or ulcers, the process becomes disorderly. The inflammatory process becomes chronic with the presence of excessive neutrophils and ROS at the injury site. Sometimes there is an excessive deposition of collagen matrix leading to fibrosis (Diegelmann and Evans 2004).

3.5.1 Wound Healing and Ulcer-Protective Property of Chamomile

Chamomile extracts and oil have been found to have ulcer-protective capabilities. The active compound appears to be (−)-α-bisabolol.

3.5.1.1 Chamomile Extract

Presibella et al. (2006, 2007) studied the effect of chamomile extracts on human leukocyte chemotaxis. Leukocytes were extracted and a 1% casein gradient was used for leukocyte migration in a Boyden system. The effect of different concentrations of chamomile extract (0.1–10^3 µg/mL) was tested on leukocyte migration. Dexamethasone was used as the standard drug. It was found that chamomile extracts (0.1–10 µg/mL) significantly reduced the leukocyte migration. Presibella et al. concluded that by inhibiting leukocyte migration during inflammation of injury sites, chamomile could alleviate the symptoms of inflammation and improve wound healing.

Karbalay-Doust and Noorafshan (2009) carried out a study to explore the antiulcerogenic effect of chamomile extract in Balb/c mice. The mice were pretreated with 400 mg/kg extract and then gastric ulceration was induced. After 14 days the gastric lesions were measured. It was found that chamomile extract prevented gastric ulceration.

The effect of chamomile extract on wound healing was also studied by Jarrahi et al. (2010). Wistar rats were subjected to linear wounds on the skin of the back. Chamomile extract dissolved in olive oil was applied and significant improvement in wound healing was observed indicating that chamomile extract applied topically has wound healing potential.

Cemek et al. (2010) studied the effect of chamomile hydroalcoholic extracts on ethanol-induced gastric mucosal injury in rats. During the experiment, rats were pretreated with different doses of chamomile extract and then were induced with gastric mucosal injury using ethanol. It was found that chamomile extract significantly reduced gastric lesions and malondialdehyde levels. It was also found that the glutathione, serum β-carotene, and retinol levels were significantly higher in rats that were administered 200 mg/kg chamomile extract. The researchers concluded that the gastroprotective effect was due to the reduction in lipid peroxidation and also due to the augmentation in antioxidant activity.

Aqueous extracts of chamomile studied by Pizarro Espín et al. (2012) were found to have gastroprotective properties in Sprague Dawley rats. Indomethacin (50 mg/kg) was used to induce ulcer in the rats, ranitidine (100 mg/kg) was used as a positive control, and distilled water was used as negative control. The aqueous extracts of chamomile at doses 125, 250, and 500 mg/kg were administered to the rats. Chamomile at all the doses showed gastroprotective activity but the most effective doses were 250 and 500 mg/kg.

Zaidi et al. (2012) studied the anti-inflammatory and cryoprotective effects of chamomile extracts on *Helicobacter pylori*-infected gastric epithelial cells. They also investigated the secretion of IL-8 and the presence of ROS in these cells. They found that chamomile did not significantly affect IL-8 secretion, but its effect was significantly high in aiding ROS generation. They concluded that this result demonstrated the potential of chamomile as a candidate drug for chemoprevention against peptic ulcer or gastric ulcer.

3.5.1.2 (−)-α-Bisabolol

The studies on the wound healing effect of (−)-α-bisabolol were reviewed by Isaac (1979). In this review, he reported that in guinea pigs exposed to UV light, treatment with (−)-α-bisabolol led to a decrease in the skin temperature. In cutaneous burns of guinea pigs, there was a significant shortening of the healing time. (−)-α-Bisabolol was found to promote epithelization and granulation.

Szelenyi et al. (1979) studied the ulcer-protective effect of chamomile. Albino rats (190–250 g) were starved for 48 hours. Ulceration was caused by 20 mg/kg indomethacin in 1% methyl cellulose or by ethanol. Metiamid was used as control. Different concentrations of bisabolol were administered orally followed 1 hour later by indomethacin. This was repeated for 3 days. The extent of ulcer inhibition was estimated according to ulcer index given by Munchow. The healing properties of the tested substance were measured comparatively with control and expressed in percentage. (−)-α-Bisabolol was found to inhibit indomethacin- or ethanol-induced ulcerations at ED_{50} = 145 mg/kg as compared to ED_{50} = 595 mg/kg of metiamid.

Bezerra et al. (2009) studied the effect of chamomile and α-bisabolol against acute gastric lesions induced by absolute ethanol in rats. They also studied the role of prostaglandins, NO, and K_{ATP}^+ channels in the mechanism of gastroprotection induced by α-bisabolol. They found that chamomile reduced the gastric lesions. α-Bisabolol (50 and 100 mg/kg) significantly reduced the lesions. Further, they found α-bisabolol was effective even after pretreatment with N-nitro-L-arginine methyl ester (10 mg/kg), a NO antagonist, and indomethacin, an inhibitor of COX. When the rats were pretreated with glibenclamide, an inhibitor of K_{ATP}^+ channel activation, the effect of α-bisabolol was significantly reduced. This led the researchers to conclude that α-bisabolol reduces gastric damage by the activation of K_{ATP}^+ channels.

The effect of α-bisabolol on acute gastric mucosal lesions in mice was studied by Rocha et al. (2010). They induced gastric ulcers in mice using ethanol and indomethacin. Different concentrations of α-bisabolol of 100 and 200 mg/kg were administered to the mice. The results revealed that α-bisabolol was able to heal and protect the gastric mucosa from ulcer. Administration of Ofl-NAME, glibenclamide, or indomethacin on rats treated with α-bisabolol (200 mg/kg) was not able to revert the gastroprotection. In addition, it was observed that ethanol and indomethacin treatment had decreased the levels of nonprotein sulfhydryl groups whereas α-bisabolol had decreased the reduction of these sulfhydryl groups. This observation suggested an increase in the bioavailability of gastric sulfhydryl groups through α-bisabolol activity leading to a reduction of gastric oxidative injury induced by ethanol and indomethacin.

3.5.1.3 Chamomile Oil

Rezaie et al. (2012) carried out a comparative study to assess chamomile and zinc oxide on healing of skin wounds in rats. Incisional wound was made on Wistar rats and these were treated with two different concentrations of chamomile oil (10% and 20%). It was found that the 10% concentration of chamomile oil has significant wound healing properties.

3.5.1.4 Others

The wound healing potential of chamomile was also examined in farm animals. Ahmad et al. (1995) took farm animals and two wounds were made on each animal. One wound served as a control. To the other wound, chamomile lotion and ointment were applied topically and clinical, histopathological, histochemical, and microbial studies were carried out. They reported that quick healing was observed with chamomile lotion.

Langhorst et al. (2013) reported that in a clinical trial, they found that a herbal mixture containing chamomile was extremely safe and effective in the treatment of ulcerative colitis.

3.6 CHAMOMILE AS AN IMMUNOMODULATOR

Immunomodulation is the therapeutic process by which the immune system is normalized. This normalization is brought about through biological or chemical substances called immunomodulators that are intrinsic (cytokines, growth factors, etc.) or extrinsic (plant polysaccharides, vaccines, etc.). An immunomodulator has the ability to stimulate, suppress, or modulate any of the components of the immune system, influencing both the innate and the adaptive immune responses. The immunomodulators could be immunostimulants (stimulating the immune response in a weak immune system) or immunosuppressants (reducing the abnormal immune response). The immunostimulants enhance a body's resistance against infections, and in a healthy person, these are expected to enhance the basal levels of immune response. The immunosuppressants can be used to control the inflammatory response to allergy, hypersensitivity, graft rejections, and so on (Lebish and Moraski 1987; Agarwal and Singh 1999). Botanical polysaccharides have been found to have the ability to modulate innate and adaptive immunity (Schepetkin and Quinn 2006) through interactions with T cells, monocytes, macrophages, and polymorphonuclear lymphocytes (Tzianabos 2000).

3.6.1 IMMUNOMODULATING EFFECT OF CHAMOMILE

Laskova and Uteshev (1992) studied the immunomodulating effect of chamomile heteropolysaccharides. Two separate groups of rats (active) or swimming and nonactive or nonswimming) were injected with sheep erythrocytes, followed by heteropolysaccharides extracted from chamomile. Both the groups showed induction of helper factors by the spleen cells. The helper factors or T cells are critical in modulating the immune response.

The effect of chamomile extract on phytohemogglutinin-induced proliferation of human lymphocytes and on mixed lymphocyte reaction was studied by Amirghofran et al. (2000). Their results indicated that the chamomile extract showed no stimulatory effect on the human lymphocytes. However, when mixed lymphocytes were treated with 10 μg/mL of chamomile extract, cell proliferation significantly increased. This observation suggested that chamomile enhances the proliferation of lymphocytes, thus suggesting its immunomodulatory properties.

Ghonime et al. (2011) evaluated the immunomodulatory effect of chamomile methanolic extract in Balb/c mice. The mice were given intraperitoneal injections of five doses of chamomile extract. The researchers found that chamomile significantly increased the white blood cells. Bone marrow cellularity and spleen weight also increased significantly when infected with *Candida albicans*, the chamomile-treated mice resistance against *Candida* confirming the immunomodulatory activity of chamomile.

3.7 ANTIMUTAGENIC EFFECT OF CHAMOMILE

Mutagenesis is the process by which mutations or changes in the DNA occur in an organism. Mutations could be spontaneous or induced through chemicals or UV light (Friedberg et al. 1995). These lead to many diseases and conditions, such as cancer. The mutation in the DNA sequence leads to the production of an altered protein, which may alter the path of growth and development with undesirable effects, or a defective protein, which may be ineffective and stop the growth and development of the organism. With the change in the environment, we are becoming increasingly susceptible to harmful chemicals that are reported to be mostly mutagenic or genotoxic and carcinogenic.

3.7.1 Antigenotoxic and Tumoricidal Effect of Chamomile

The research into using natural plant products to reduce genotoxic and/or carcinogenic effects is continuously growing (Demir et al. 2013). Several researchers have found that chamomile essential oil inhibits the genotoxic damage produced by mutagens in mice.

Hernández-Ceruelos et al. (2002) investigated the effect of chamomile oil on the mutagenic activity of daunorubicin and on the sister chromatid exchanges in mouse bone marrow cells. The groups of mice were treated with different doses of chamomile oil (5, 50, and 500 mg/kg) followed by 10 mg/kg daunorubicin (a drug used to treat some types of cancer). Another group of mice were treated with chamomile oil (250, 500, and 1000 mg/kg) and then with 25 mg/kg methyl methanesulfonate. The results showed a dose-dependent inhibition of sister chromatid exchanges by chamomile oil. In case of daunorubicin, there was a 75% inhibition by chamomile oil at 500 mg/kg concentration. There was no change in cellular proliferation kinetics, but a reduction in the mitotic index was detected. In case of methyl methanesulfonate (a carcinogen), a 60% inhibition by chamomile oil (1000 mg/kg) was observed. The cellular proliferation kinetics and mitotic index were unaffected. The researchers suggested that chamomile could be an effective mitogen.

Hernández-Ceruelos et al. (2010) further investigated the antigenotoxic effect of chamomile oil in mouse spermatogonial cells. Daunorubicin was used to induce genotoxic damage. The effect of chamomile oil on the sister chromatid exchange frequency after treatment with daunorubicin only and after treatment with daunorubicin and chamomile oil was noted. They reported a dose-dependent inhibition with even a 95% inhibition of genotoxicity. To better understand the antigenotoxic mechanism of action of the chamomile oil, they further determined the antioxidant capacity of the oil using DPPH free radical scavenging method and ferric thiocyanate assays. Although the radical scavenging capacity of chamomile oil was found to be moderate, the antioxidant capacity was found to be similar to vitamin E. They concluded that the chemoprotective effect of chamomile oil might be attributed to its antioxidant effects.

Anter et al. (2011) tested the genotoxic, antigenotoxic, tumoricidal, and apoptotic effect of apigenin and bisabolol. The genotoxic and antigenotoxic studies were carried out by the drosophila wing-spot test. To assess the tumoricidal activity, human

Chamomile: Medicinal Properties

leukemia cells were used and to assess apoptosis, a DNA fragmentation test was used. The researchers found that apigenin and bisabolol did not show any genotoxic effect. These compounds however showed antigenotoxic and tumoricidal activity, and induced apoptosis in the leukemia cells.

Delarmelina et al. (2012) studied the cytotoxic, genotoxic, and mutagenic effects of chamomile tincture in mice. They subjected the mice to 0.02 and 0.1 µL/g/d doses that were equivalent to 20 and 100 drops daily, respectively, for an adult. The treatment was given to the mice for 5 days and they were killed on the sixth day. Micronucleus assay of the bone marrow was carried out and it was found that the chamomile tincture at 0.02 and 0.1 µL/g/d doses did not have any cytotoxic, genotoxic, and mutagenic effects.

Arango et al. (2012) studied the effect of apigenin on DNA damage and apoptosis in leukemic cells. They found that apigenin induced the phosphorylation of the ataxia-telangiectasia mutated kinase and histone H2AX, two key regulators of the DNA damage response. This was mediated by p38 and PKCδ. Apigenin was found to delay cell-cycle progression at G1/S and increased the number of apoptotic cells. A genome-wide mRNA analysis showed that apigenin-induced DNA damage led to downregulation of genes involved in cell-cycle control and DNA repair. As per the researchers, this activity of apigenin contributes to its anticarcinogenic activities.

3.8 ANTICANCER EFFECT OF CHAMOMILE

Cancer is caused due to abnormal cell division. The cells continue to grow and undergo metastasis. Cancer is not one disease but a combination of many diseases. The disease symptoms are manifested as pain, swelling, bleeding, fever, weight loss, and tiredness (American Cancer Society 2012). In many cases, the abnormally grown cancerous cells release substances in the body that cause these symptoms. According to an estimate, there are more than 100 types of cancer (National Cancer Institute 2013). Cancer is treated by several methods, such as surgery, chemotherapy, radiation therapy, and immunotherapy. Chemotherapy and radiation therapy are known to cause undesirable side effects. To reduce these side effects, efforts are being made to identify candidate drugs, and encouraging results are being obtained from medicinal plants. The plant-based drugs are being advocated for adjuvant therapy (treatment given after primary therapy to increase the chance of long-term survival) and neoadjuvant therapy (treatment given before primary therapy).

3.8.1 ANTICANCER PROPERTY OF CHAMOMILE

Chamomile has also been evaluated for its anticancer property. Evans et al. (2009) evaluated the effect of a botanical supplement (TBS-101) on invasive prostate cancer in animal models. TBS-101 contains seven standardized botanical drugs including chamomile. Evans et al. implanted mice with PC-3 cells and after sufficient growth of the tumor they treated the mice with different doses of TBS-101 to study apoptosis. They found an induction of apoptosis and significant inhibition of tumor growth.

Srivastava and Gupta (2009) reported that the apigenin glucosides inhibited cancer cell growth through deconjugation of glycosides to produce aglycone, apigenin.

Zu et al. (2010) studied the effect of chamomile essential oil on three human cancer cell lines: A-549 (human lung cancer), PC-3 (human prostate cancer), and MCF-7 (human breast cancer). Cell viability was determined by the (3-[4,5-dimethylthiazol-2-yl]-2,5 diphenyl tetrazolium bromide) (MTT) assay. Chamomile at a concentration of 0.002% (v/v) showed cytotoxic effects toward PC-3. The viable MCF-7 cells were reduced to 6.9% by chamomile oil treatment. These results suggested that chamomile oil could possess potential anticancer activity.

Ogata et al. (2010) studied the effect of bisabolol oxide A on rat thymocytes by using fluorescent dyes. They found that bisabolol oxide A was able to induce apoptosis in rat thymocytes. When the cells were incubated with bisabolol oxide A for 24 hours, the bisabolol oxide A at concentrations of 30 µM or more significantly increased populations of dead cells, shrunken cells, and cells with phosphatidylserine exposed on membrane surface. Phosphatidylserine is a phospholipid component of the cell membrane, which normally is exposed to the cyotosolic portion of a living cell. When apoptosis occurs, it flips over to the surface of the cell. In addition, bisabolol oxide A significantly increased the population of cells containing hypodiploid DNA, and the increase was completely attenuated by Z-VAD-FMK, a pan inhibitor for caspases, indicating an involvement of caspase activation. Caspases (cysteine-dependent aspartate-directed proteases) are important enzymes that play a role in apoptosis. The researchers concluded that bisabolol oxide A is likely to induce cell death by apoptosis.

Ogata-Ikeda et al. (2011) studied the cytotoxic effect of bisabolol oxide A on human leukemia K562 cells. They also studied the combined effect of bisabolol oxide A and 5-fluorouracil (5-FU), an anticancer agent, on the K562 cells. They found that bisabolol oxide A at the concentrations 5–10 µM exerted a cytotoxic action on the K562 cells. Further they found that bisabolol oxide A in combination with 5-FU inhibited the growth of K562 cells. The researchers suggested that a simultaneous application of chamomile could reduce the dosage of 5-FU.

Cavalieri et al. (2011) had found that α-bisabolol may induce apoptosis in cells. Therefore, they studied the activity of α-bisabolol in acute leukemia cells to find out if α-bisabolol would induce apoptosis in cancer cells. α-Bisabolol was added in different doses to the leukemia cells and kept for 24 hours. They found that α-bisabolol induced apoptosis in these cells by damaging the mitochondria, resulting in a decrease in NADH-supported respiration and a disruption in the mitochondrial membrane potential. They concluded that α-bisabolol is a proapoptotic agent for primary human acute leukemia cells.

Kogiannou et al. (2013) analyzed the ability of chamomile phenols to (1) scavenge free radicals, (2) inhibit proliferation, (3) decrease IL-8 levels, and (4) regulate NF-κB in epithelial colon cancer (HT29) and prostate (PC3) cancer cells. They found that chamomile phenols had the ability to scavenge the free radicals and also to reduce cell proliferation. The chamomile phenols were found to reduce the IL-8 levels in the PC3 cells. In HT29 cells, the molecular target of chamomile phenols was found to be NF-κB.

3.9 CHAMOMILE FOR TREATING INSOMNIA

Insomnia or sleep disorder is known to be the most common disorder affecting millions of people. It can be understood as a disorder of the sleep–wake schedule and has been defined variously. It is thought to result from a state of hyperarousal or an elevated state of alertness. As a result it is difficult to sleep. The symptoms of insomnia are attributed to many underlying conditions such as depression, stress, or abuse of substance. In a review (Mai and Buysse 2008), it was reported that chronic insomniacs reported more heart disease, hypertension, chronic pain, as well as increased gastrointestinal, neurological, urinary, and breathing difficulties. The converse was also shown to be true, in which subjects with hypertension, chronic pain, breathing, gastrointestinal, and urinary problems complained of insomnia more often than non-insomniacs. Thus, insomnia is both a disorder and a symptom.

The medications used in insomnia are generally sedative hypnotics, such as benzodiazepines and non-benzodiazepines. Both benzodiazepines and non-benzodiazepines target γ-aminobutyric acid type A ($GABA_A$) receptor sites in the brain that modulate the effects of the neurotransmitter GABA. GABA regulates the neuronal excitability. When GABA binds with $GABA_A$ receptors on the neurons, chloride ions (Cl^-) move out of the cell causing excitability in the neurons. This causes alertness leading to sleep disorder. The sedative hypnotics bind to the $GABA_A$ receptors and block the site so that GABA cannot bind, thus blocking the development of the symptoms.

Benzodiazepines are potentially dangerous when combined with alcohol. Side effects include morning sedation, allergies, depression, incontinence, and so on. Complementary and alternative therapies are increasingly used by the people to treat insomnia. Many people, including doctors, recommend using chamomile for the treatment of insomnia due to its hypnotic and sedative effects (Gould et al. 1973; UMMS 2012).

3.9.1 Chamomile as a Sedative

The sedative effect of chamomile has been reported by several investigators. Della Loggia et al. (1982) studied the effect of chamomile on the central nervous system (CNS) of mice. They prepared lyophilized infusion of chamomile flowers according to the *European Pharmacopoeia* and it was injected into the female mice that had been starved for 16 hours. After intraperitoneal injection of the infusion, the mice were subjected to several tests of long-term motility, motor coordination, exploratory activity, and barbiturate-induced sleeping time. A significant sleeping time potentiating effect was noted at the doses of 160 and 320 mg/kg of chamomile. It was concluded that chamomile infusion administered intraperitoneally had a depressive effect on the CNS and therefore chamomile could be a potent sedative.

Viola et al. (1995) reported that apigenin competitively inhibited the binding of flunitrazepam to $GABA_A$ receptors. They also found that a 10-fold increase in dosage produced a mild sedative effect. Apigenin also showed a clear anxiolytic activity in mice in the elevated plus-maze test.

Avallone et al. (1996, 2000) studied the pharmacological effect of apigenin, the flavonoid constituent of chamomile extract, on the cultured cerebellar granule cells. They reported that apigenin reduced GABA-activated Cl^- currents in a dose-dependent manner and the effect was blocked by coapplication of Ro 15-1788, a specific benzodiazepine receptor antagonist.

Shinomiya et al. (2005) studied the effect of chamomile extract on a model of sleep-deprived rats. Wistar rats were treated with chamomile water extract for 7 days. Thereafter the animals were anesthetized with pentobarbital sodium (35 mg/kg, i.p.) and subjected to electroencephalogram recording and electromyogram. The sleep–wake states were also recorded. Sleep latency (i.e., the length of time that it takes to accomplish the transition from full wakefulness to sleep) was also measured. A significant shortening of sleep latency was observed with chamomile extract at a dose of 300 mg/kg. The researchers concluded that the shortening of sleep latency induced by chamomile extract may be caused by apigenin through its binding with benzodiazepine receptors. Viola et al. (1995) reported that apigenin induced anxiolytic action in an elevated plus-maze test of mice. From these results, the researchers concluded that the effect of chamomile extract may be due to not only hypnotic activities but also anxiolytic activities of the herb.

Kakuta et al. (2007) investigated the effect of chamomile on men and women. They gave them cold and warm chamomile jelly to eat and tested their skin temperature and a test with Mood Checklist followed by a survey next day to determine changes in sleep consciousness. They found that eating warm chamomile increased the blood flow and blood temperature. The mood was also enhanced as indicated by an increased relaxed feeling score of the examinees. The survey next day showed that the male group who ate warm chamomile jelly had a significant increase in the scores of the improved sleep consciousness.

Zick et al. (2011) carried out a preliminary examination of the efficacy and safety of a standardized chamomile extract for chronic primary insomnia. They treated insomnic patients with chamomile (270 mg) twice daily for 28 days. They concluded that chamomile improved the daytime functioning and its symptoms in the insomnic patients.

3.10 CHAMOMILE FOR RELIEVING STRESS

Stress has become very common today. It is estimated that in the United States about 75%–90% of the patients are reported to visit primary care physicians (Kathleen and Kelly 2009). During stress, the stress hormone norepinephrine is released from the nerve endings. The adrenal glands also release the stress hormones epinephrine and norepinephrine into the bloodstream. This happens within a few seconds of the stressful event. Subsequently, the adrenal glands secrete extra cortisol, another stress hormone, into the bloodstream. During the stress response, the endocrine glands, such as the hypothalamus, secrete corticotropin-releasing factor, which in turn stimulates the pituitary gland to release adrenocorticotropic hormone (ACTH). The ACTH stimulates the adrenalin glands to release more cortisol. Stress may cause serious incidents including myocardial infarction and cardiac arrest (Kathleen and Kelly 2009).

3.10.1 Anxiolytic and Stress-Reducing Properties of Chamomile

Chamomile, because of its calming effects, has been used traditionally as an anxiolytic and stress reducer. Roberts and Williams (1992) subjected a group of 22 human volunteers to chamomile oil vapors and a placebo as control. They found that chamomile oil vapors shifted the mood rating and frequency of judgment of the volunteers toward a more positive direction. It has also been proved through animal experiments that chamomile oil reduces the ACTH levels in animals. When Schmidt and Vogel (1992) subjected the ovariectomized rats under restriction stress to vapors of chamomile oil, the plasma ACTH levels decreased in the animals. This finding was corroborated when Yamada et al. (1996) also found that inhalation of chamomile essential oil vapors decreased the stress-induced increased levels of plasma ACTH in ovariectomized rats.

Reis et al. (2006) reported that the cortisol levels in the blood of Nelore calves fed with chamomile were significantly less. According to the researchers, this indicated that chamomile was instrumental in reducing the cortisol levels thereby decreasing stress in the animals.

3.11 NEUROPROTECTIVE EFFECT OF CHAMOMILE

Neuroprotection is the mechanism of protection of neurons of the CNS and the brain. The concept of neuroprotection can be traced to the studies of the pathology of ischemic brain injury but it is now used for disease conditions such as Alzheimer's disease, Parkinson's disease, traumatic injury, and epilepsy (seizure) as well. During injury there is a deprivation of oxygen and glucose in the neurons that leads to a cascade of events or pathways that eventually lead to neuronal death. Some of the events are the excessive activation of glutamate receptors, accumulation of calcium ions in the cells, heavy recruitment of inflammatory cells, generation of free radicals, and apoptosis (Cheng et al. 2004; Halász and Rásonyi 2004). The neuroprotective agents are expected to stop one or more of these events and pathways to successfully provide neuroprotection. Several plant phenolics have been identified as potential neuroprotective agents. Most of these phenolic compounds have anti-inflammatory and antiapoptotic activity and work by reducing oxidative stress in the cells (Kim 2010).

3.11.1 Neuroprotective Properties of Chamomile

To explore the neuroprotective properties of chamomile, Heidari et al. (2009) studied the antiseizure effect of chamomile hydromethanolic extract in mice. Animals were pretreated with 100, 200, and 300 mg/kg extract. Phenobarbital (40 mg/kg) was used as the standard drug. After 20 minutes of treatment with chamomile extract, the animals were injected with picrotoxin (12 mg/kg) to induce seizure. The results showed that the latency of the beginning of the seizure increased in chamomile-treated mice, the most effective dose being 200 mg/kg. The researchers concluded that chamomile could have antiseizure properties.

Chandrashekhar et al. (2010) studied the neuroprotective effect of chamomile methanolic extract against global cerebral ischemia/reperfusion injury-induced oxidative stress in Sprague Dawley rats. They found that chamomile showed neuroprotective activity in a dose-dependent manner. There was a significant decrease in lipid peroxidation and increase in the levels of superoxide dismutase, catalase, glutathione, and total thiol. In addition, cerebral infarction area was significantly reduced. They concluded that chamomile showed potent neuroprotective activity.

Abad et al. (2011b) studied the effect of chamomile hydrochloric extract on the pentylenetetrazole-induced seizure in mice. They found that chamomile extract significantly increased the threshold of T2-induced seizure. Further, it significantly increased seizure threshold in vincristine-treated groups of mice.

Ranpariya et al. (2011) studied the neuroprotective activity of chamomile methanolic extracts in Sprague Dawley rats. The rats were subjected to aluminum fluoride, causing oxidative stress in them. Chamomile extracts were administered to the rats at concentrations of 100, 200, and 300 mg/kg. The results showed a significant decrease in lipid peroxidation and increase in the levels of superoxide dismutase, catalase, glutathione, and total thiol in the rats. Histopathological observations showed that there was a dose-dependent protection of brain damage. The researchers concluded that chamomile extracts showed a dose-dependent neuroprotective activity.

Abad et al. (2011a) studied the effect of chamomile extract on cisplatin-induced neuropathy in NMRI mice. Mice were treated with formalin in two phases to induce pain. The effect of chamomile was studied on cisplatin-treated mice. It was found that the chamomile decreased the second phase of the cisplatin-induced pain significantly. They found that chamomile was able to decrease cisplatin-induced pain and inflammation better than morphine.

Sofiabadi et al. (2012) studies the effect of chamomile ethanolic extract on seizure. Then they injected different concentrations of chamomile ethanolic extract in these male rats. After 30 minutes, they induced seizure in the rats by pentylenetetrazole. They found that injecting the rats with chamomile extract at the doses of 100, 200, and 500 mg/kg significantly reduced the onset of tonic seizures in a dose-dependent manner.

3.12 EFFECT OF CHAMOMILE ON MORPHINE WITHDRAWAL SYNDROME

Morphine is given as an antinociceptive in many disease conditions such as cancer. The long-term prescription of morphine leads to its dependence. When morphine is stopped, it leads to severe withdrawal symptoms, which include the somatic/physical and cognitive/affective components. Morphine withdrawal has been found to change the cerebellar neurotransmission and make the dopaminergic, noradrenergic, and serotonergic neurotransmitter systems dysfunctional (Sekiya et al. 2004; Lelevich et al. 2009). Treatment includes the use of opioid antagonists, such as naloxone, and N-methyl-D-aspartic acid (NDMA) antagonists. Naloxone binds to opioid receptors in the brain and thus competes with morphine. The central glutamatergic system is involved in morphine dependence and the excitatory amino acids, such as glutamate, are known to bind to the NDMA receptors in this process. The NDMA antagonist mimics glutamate and binds to the glutamate receptor. These treatments bring about morphine withdrawal symptoms that proceed through a number of stages. There are some side effects reported for these drugs, which may cause cardiac, gastric, and nervous system disorders, and to attenuate this, plant-based adjuvants are being investigated.

3.12.1 EFFECT OF CHAMOMILE EXTRACT ON MORPHINE DEPENDENCE AND ABSTINENCE

Gomaa et al. (2003) investigated the effect of chamomile extract on morphine dependence and abstinence in rats. They treated the rats with morphine to induce morphine dependence and then injected naloxone to induce morphine withdrawal signs (paw tremor, rearing, teeth chattering, body shakes, ptosis, diarrhea, and urination). Administration of chamomile extract along with morphine significantly reduced withdrawal signs and weight loss. When a single dose of chamomile extract was given before naloxone, it abolished the manifestation of the withdrawal signs.

Esmaeili et al. (2007) studied the effect of chamomile extract on morphine withdrawal symptoms in mice. They treated groups of mice with morphine for 4 days and then treated one group with different doses of chamomile extract (10, 20, or 30 mg/kg) intraperitoneally. Naloxone was injected in all groups of mice after 3 hours to induce withdrawal symptoms. Treatments with the different doses of chamomile extract were found to significantly attenuate the morphine withdrawal syndrome signs (jumping, climbing, writhing, and weight loss) compared with morphine group.

Kesmati et al. (2008) studied the effect of chamomile on morphine withdrawal syndrome in Wistar rats. The rats were subjected to morphine dependence and then naloxone was injected for the induction of morphine withdrawal syndrome. The withdrawal signs of climbing, jumping, and face washing were measured. In chamomile-treated rats, the number of climbing decreased significantly. The researchers concluded that chamomile extract has a sedative effect on morphine withdrawal syndrome and this is probably related to its benzodiazepine-like components that act on benzodiazepine receptors.

Kesmati et al. (2009) studied the effect of chamomile hydroalcoholic extract on the morphine withdrawal symptoms in Wistar rats in the presence and absence of tamoxifen, an estrogen antagonist. The parameters measured in the rats were rearing, jumping, and grooming. Chamomile hydroalcoholic extract containing apigenin and chrisin (phytoestrogens) induced sedative effects on some of the morphine withdrawal symptoms such as rearing and grooming. The researchers suggested that these phytoestrogens probably act with some neurochemical systems of dependence in the CNSs and reduce the symptoms of dependence.

Similar results were reported by Safari (2010) who suggested that chamomile might be helpful in the treatment of morphine withdrawal syndrome.

3.13 ANXIOLYTIC EFFECT OF CHAMOMILE

Anxiety is one of the most prevalent disorders today. The most common are generalized anxiety disorder, panic disorder, social anxiety disorder, and post-traumatic stress disorder. These disorders are physical and psychological disorders having a huge impact on the functioning and work productivity of the patients (Kroenke et al. 2007). The medicines used to treat anxiety are mostly benzodiazepines, selective serotonin reuptake inhibitors (SSRIs), and non-benzodiazepines, such as buspirone. The benzodiazepines act by binding to GABA receptors thereby enhancing the levels of GABA, which is a neurotransmitter in the CNS and has a calming effect (Lydiard 2003). The SSRIs block the reuptake of the neurotransmitter serotonin in the brain cells, thereby reducing its levels in the brain and reducing anxiety. Buspirone binds to the serotonin receptors in the brain cells. It reduces the levels of excessive serotonin by partially blocking the serotonin receptor 5-HT1A (Brennan 2008).

3.13.1 ANXIOLYTIC PROPERTIES OF CHAMOMILE

Viola et al. (1995) isolated apigenin-containing fraction from chamomile flowers and found it had an affinity to central benzodiazepine receptors (receptors that are expressed by the neurons of the CNS, e.g., $GABA_A$ receptor). They injected apigenin into the mice. These mice were then subjected to an elevated plus-maze test. Apigenin administered intraperitoneally at a dose of 3 mg/kg showed anxiolytic activity in the elevated plus-maze test. It did not have any sedation or muscle relaxant effects. They concluded that apigenin showed anxiolytic activity and slight sedative effect by binding to the central benzodiazepine receptors in the neurons of the CNS.

Amsterdam et al. (2009) carried out a randomized, double-blind, placebo-controlled trial of oral *Matricaria recutita* (chamomile) extract therapy for generalized anxiety disorder using Hamilton anxiety rating (HAM-A). A low score indicates low anxiety levels. They found that chamomile significantly lowered the HAM-A scores in the patients. The researchers suggested that chamomile might have mild anxiolytic effects on the patients with generalized anxiety disorder.

3.14 CHAMOMILE AS PSYCHOSTIMULANT AND ITS EFFECT ON ATTENTION DEFICIT HYPERACTIVITY DISORDER

Psychostimulants are those substances that temporarily improve the mental and physical function. These are used to treat clinical depression or attention deficit hyperactivity disorder (ADHD). The common psychostimulants used are amphetamines. The psychostimulants block the reuptake of catecholamines (norepinephrine and dopamine) by the neurons, thereby preventing their degradation by monoamine oxidase. In addition, amphetamine compounds release catecholamines into the synapse from the neuronal cell, thus increasing the concentration of the synaptic dopamine and norepinephrine (Wilens 2006).

3.14.1 PSYCHOPHARMACOLOGICAL EFFECT OF CHAMOMILE

Niederhofer (2009) studied the effect of chamomile on borderline cases of ADHD in three male patients. It was found that after the patients were administered chamomile, there was improvement in the Conners' hyperactivity, inattention, and immaturity factors leading to the conclusion that chamomile may improve some symptoms of ADHD. They concluded that this activity of chamomile could be attributed to its action of inhibiting serotonin and noradrenaline reuptake.

Can et al. (2012) investigated the psychopharmacological profile of chamomile essential oil in mice. They treated the mice with 25, 50, and 1000 mg/kg chamomile essential oil. They studied the spontaneous locomotor activities by activity cage measurements and motor coordination of mice by rotarod test. They also studied the open field social interactions and conducted plus-maze tests to assess the emotional state. In addition, they performed tail suspension test to determine depression levels in mice. The researchers found that at 50 and 100 mg/kg, chamomile essential oil significantly increased locomotor activities, exhibited anxiogenic effect in open field, elevated plus maze and social interaction, and decreased immobility in tail suspension tests. They concluded that these results were similar to the effects of the psychostimulant coffee.

The effect of chamomile essential oil on the CNS was investigated by Umezu (2013) in mice. They found that chamomile essential oil increased the ambulatory activity of mice thereby acting as a CNS stimulant. They opine that chamomile essential oil could be useful in treating certain kinds of mental disorders.

3.15 EFFECT OF CHAMOMILE ON SOME DERMATOLOGICAL PROBLEMS

Dermatological problems may broadly be classified as the problems of the skin, the hair, and the nails. There are many common and uncommon problems of the skin. The common problems are infection related, such as pimples, boils, and warts; injury related, such as cuts, burns, and wounds; and allergy related, such as eczema or dermatitis. The uncommon problem is cancer. The hair follicle and the nails may also have problems such as infection.

3.15.1 Treatment of Dermatological Problems with Chamomile

Chamomile has been traditionally used to treat skin problems because of its antimicrobial, anti-inflammatory, and wound healing properties. In 2003, Schmidt presented a clinical evidence that chamomile essential oil can be used to treat atopic dermatitis (Schmidt 2003).

Lee et al. (2010) examined the effect of chamomile oil on atopic dermatitis in mice. They induced atopic dermatitis in mice using 2,4-dinitrochlorobenzene. Chamomile oil was applied topically 6 days a week for 4 weeks. Thereafter, they measured the serum IgG1 and histamine levels. They found that after 2 weeks of chamomile oil application, the IgG1 and histamine levels were reduced significantly. They concluded that chamomile oil had immunoregulatory potential for alleviating atopic dermatitis through influencing of Th2 cell activation.

Shimelis et al. (2012) researched on the accessible and affordable treatments for dermatological problems in the developing countries. They tested a 10% chamomile extract cream for eczema-like lesions. They found that a large number of patients healed or improved with the use of chamomile.

Kobayashi et al. (2007) found that chamomile essential oil had the property of reducing itching or pruritus. In a study they induced sensory irritation in mice by injecting capsaicin. The coadministration of chamomile essential oil suppressed the irritation in a dose-dependent manner. In isolated guinea pig ileum, contractions were induced by injecting capsaicin, which were found to be inhibited by chamomile essential oil. They identified bisabolol oxide A as the main anti-irritation constituent in the chamomile oil.

3.16 PHYTOESTROGENIC EFFECT OF CHAMOMILE

Phytoestrogens are naturally occurring plant compounds that show estrogenic and/or antiestrogenic properties depending on the receptors they bind to. These natural compounds acting as phytoestrogens belong to flavonoids (isoflavones and coumestans) and lignans. The flavonoids found in the chamomile plant extracts are flavones (apigenin and luteolin) and flavonols (quercetin and kaempferol), which act as phytoestrogens. Flavone and isoflavone phytoestrogens have been reported to be competitive to cytochrome P450 aromatase. Aromatase converts androgen to estrogen. The phytoestrogens are therefore thought to be modifying the estrogen level in women (Gültekin and Yildiz 2006; Patisaul and Jefferson 2010).

3.16.1 CHAMOMILE AS PHYTOESTROGEN

Javaheri et al. (2009) studied the phytoestrogenic effect of chamomile extract and its relationship with endogenous estrogens on locomotor activities in NMRI mice. They treated the mice with only 50 mg/kg chamomile and another group with chamomile extract and tamoxifen (0.5 mg/kg). Tamoxifen is an estrogen receptor antagonist. They then tested the locomotor parameters such as fast and slow activity and fast and slow rearing. They found that chamomile decreased most of the locomotor activities by binding to estrogen receptors. They concluded that chamomile extract acted as phytoestrogen and bound to the estrogen receptors, and that the presence of these receptors and endogenous estrogens is essential for the phytoestrogenic components of chamomile extract.

Johari et al. (2011) evaluated the effect of chamomile hydroalcoholic extracts on the changes in ovarian tissue and on the production of follicle-stimulating hormone, luteinizing hormone, estrogen, and progesterone. Wistar rats were treated with 10, 20, and 40 mg/kg of chamomile extracts in a dose-dependent manner. After 14 days, the levels of gonadotropin, estrogen, and progesterone were determined. The researchers found that mice treated with 10 mg/kg extract showed a significant decrease in the levels of estrogen and a significant increase in the levels of progesterone. In mice treated with 20 and 40 mg/kg of chamomile extract, a significant decline in the number of primary and Graafian follicles was observed. Johari et al. concluded that the phytoestrogen present in chamomile extract causes a decrease in the estrogen levels and a subsequent increase in the progesterone levels. These hormonal changes resulted in a decrease in the numbers of ovarian follicles.

3.17 CHAMOMILE AND CARDIOVASCULAR DISEASE

The cardiovascular diseases pertain to the diseases of the heart muscles, the blood vessels, and the heart valves. The cardiovascular problems are arrhythmia; atherosclerosis; diseases of the cardiac valves, myocardium and pericardium, pulmonary circulation, and aorta; hypertension; and congenital heart disease (Selzer 1992). Needless to say, the treatment requires specialized interventions, and in most of the cases surgery is required. Medicines, such as aspirin (antiplatelet), angiotensin-converting enzyme inhibitor, β-blocker (to block the binding of epinephrine to β-receptors on heart muscles,

Chamomile: Medicinal Properties

thereby reducing stress), and statin (to reduce lipid levels) are used in the treatment of cardiovascular diseases (National Collaborating Centre for Primary Care, UK 2007).

3.17.1 Effect of Chamomile on Some Cardiovascular Disease Conditions

Gould et al. (1973) made the first attempt to understand the cardiac effects of chamomile on the hemodynamic parameters in patients with various kinds of cardiac disease. Twelve patients, who had undergone right and left ventricular catheterizations, were orally administered chamomile tea and after 30 minutes the 15 parameters were measured. The researchers did not find any change in the parameters, except a significant increase in the brachial artery pressure. However, 10 of 12 patients fell into deep sleep after ingesting chamomile tea, and the researchers suggested the role of chamomile tea as a hypnotic.

Ko et al. (1991) studied the effect of apigenin on the vasodilation of rat thoracic aorta and found that apigenin inhibited the contraction of aortic rings and induced relaxation. They concluded that apigenin relaxed the rat aorta by suppressing calcium ion influx through voltage- and receptor-operated calcium channels.

Lorenzo et al. (1996) investigated the involvement of monoamine oxidase and noradrenaline uptake in the positive chronotropic effects (those effects that change the heart rate) of apigenin in isolated spontaneously beating rat atria. They found that apigenin increased the atrial rate and decreased the uptake of noradrenalin by 60%. The exposure to 30 µM apigenin resulted in the increase of the proportion of unmetabolized noradrenaline from 11% to 45%, indicating the inactivity of monoamine oxidase. The researchers suggested that apigenin possessed the property of increasing the atrial rate and the underlying mechanism could be through reducing the noradrenaline uptake and reducing the activity of monoamine oxidase.

3.18 CHAMOMILE AND ORAL HEALTH

The oral health depends on the health of the teeth and gums. The teeth and gums are subjected to microbes and are prone to disease if proper care is not taken. Dental plaque, caries, periodontitis, and so on are some common oral diseases. In addition, chemotherapy might induce stomatitis (inflammation of the mucus lining on the inside of the mouth), which needs to be treated.

3.18.1 Effect of Chamomile on Oral Disease Conditions

Chamomile, because of its antimicrobial and wound healing properties, has been used in mouthwash formulations.

Fidler et al. (1996) gave patients undergoing 5-FU-based chemotherapy with stomatitis a chamomile mouthwash thrice daily for 14 days. They did not find any significant difference in the stomatitis levels after chamomile treatment.

Pourabbas et al. (2005) evaluated the effect of chamomile mouthwash on dental plaque and gingival inflammation in 25 patients. The patients used a chamomile mouthwash twice a day for 4 weeks. The chamomile mouthwash was found to lower both plaque and gingival inflammation significantly. No adverse reaction such as tooth staining was reported.

However, Mazokopakis et al. (2005) reported that chamomile mouthwash was effective in treating methotrexate-induced oral mucositis in a patient with rheumatoid arthritis.

Varoni et al. (2012) reviewed the existing literature on the effect of plant polyphenols on oral health. They found data on chamomile that indicated its possible use against common oral diseases such as caries, periodontitis, and candidiasis, as well as oral cancer.

On the basis of their clinical study, Lins et al. (2013) reported that chamomile mouthwash satisfactorily reduced plaque buildup and gingival inflammation.

3.19 TREATMENT OF COLICKY INFANTS WITH CHAMOMILE

Colic in infants is very common because the digestive system of the infants is not well developed. The gastrointestinal motility is also not strong enough. As a result, there could be discomfort due to gastric conditions, such as indigestion, allergy, and flatus.

3.19.1 Effect of Chamomile on Colicky Infants

Colic in infants can be treated with chamomile. Savino et al. (2005) studied the effect of chamomile preparation (ColiMil®). They found that within a week of the treatment, colic in the breast-fed infants improved.

3.20 SPASMOLYTIC EFFECT OF CHAMOMILE

A spasm is a sudden involuntary contraction of the smooth or skeletal muscles. Spasms may occur in the skeletal muscles, cardiac muscles, arteries, trachea, and the gastrointestinal tract. In cases of spasm, muscle relaxants, such as valium, hyoscyamine, and papaverine, are administered to the patients. The mechanism of action of these antispasmodic compounds is different from each other with some, such as papaverine, acting on the inhibition of phosphodiesterase enzyme (PDE10A) to increase the levels of cyclic AMP.

3.20.1 Antispasmodic Effects of Chamomile

Foster et al. (1980) investigated the antispasmodic effects of chamomile ethanolic extract in guinea pig ileum. Acetylcholine and histamine were used as spasmogens. Chamomile extract was found to significantly decrease the maximal possible contractility induced by actylcholine as well as histamine. A comparison with atropine revealed that chamomile extract was more effective at lower concentrations (0.42 µg/L) than the doses of atropine (7 µg/L) normally administered to the patients.

Achterrath-Tuckermann et al. (1980) carried out a detailed study of the antispasmodic activities of the lipophilic fraction and hydrophilic fraction of chamomile. From the lipophilic group, they studied (−)-α-bisabolol, bisabolol oxide A, bisabolol oxide B, *cis*-en-yn-dicycloether, and chamomile oil. From the hydrophilic group, they studied the flavonoids apigenin, luteolin, patuletin, quercetin, apigenin-7 (6″-0-acetylglucoside), apigenin-7-glucoside, apiin, rutin, umbelliferone, and herniarin. The antispasmodic effects of these compounds were tested in isolated guinea pig ileum. To induce the spasms, the spasmogens used were barium chloride, histamine dihydrochloride, acetylcholine, serotonin, and bradykinin. The antispasmodic activity of the chamomile compounds was compared with that of papaverine. The researchers found that the most effective lipophilic compound was (−)-α-bisabolol, which was as effective as papaverine. Bisabolol oxide A and bisabolol oxide B were half as potent as papaverine. *Cis*-en-yn-dicycloether also showed some antispasmodic activity. The flavonoids apigenin, luteolin, patuletin, and quercetin showed significant antispasmodic activity. Apigenin was found to be more potent than papaverine. The other hydrophilic flavonoids showed very small antispasmodic effect. The researchers concluded that both lipophilic and hydrophilic groups contributed to the antispasmolytic activity of chamomile.

Carle in 1990 investigated the antispasmodic activity of chamomile (Carle 1990). Intestine samples (350–450 g) from guinea pigs that had been starved for 24 hours were isolated and subjected to the spasmogen barium chloride (1×10^{-4} g/mL.) Papaverine was used as comparative spasmolytic standard against chamomile hydroalcoholic extracts. After the application of standard, the end of contractions was measured electronically. After 2 minutes of application of spasmogen, the chamomile flavonoid extract containing apigenin was injected in the intestine sample and the time required for the end of contractions was compared to the standard and expressed in percentage. Apigenin was found to be four times more effective than papaverine (Carle 1990).

Savino et al. (2008) studied the effect of ColiMil (herbal preparation containing chamomile, fennel, and lemon) and also chamomile extract separately on the upper gastrointestinal transit in mice. They found that both ColiMil and chamomile extract reduced intestinal motility significantly. To further understand how ColiMil works, Capasso et al. (2007) studied its effects on the gastrointestinal motility/transit in mice. They found that when ColiMil (0.4–0.8 mL/mouse) was administered, it delayed the gastrointestinal transit in a dose-dependent manner. They also found that chamomile extract (0.89 and 1.78 mg/mouse) significantly reduced the motility. They concluded that chamomile in ColiMil was the significant contributor to the decrease in the upper gastrointestinal tract motility and hence improved colic in breast-fed infants.

3.21 HEPATOPROTECTIVE EFFECT OF CHAMOMILE

Hepatoprotection is the ability to protect the liver from damage. The liver could get damaged due to disease conditions or addictive conditions, such as alcoholism.

3.21.1 Hepatoprotective Properties of Chamomile

Chamomile essential oil and extract have been found to possess hepatoprotective effects. Gershbein (1977) partially hepatectomized rats and gave subcutaneous injections of high levels of essential oils for 7 days or fed ad libitum for 10 days on diets supplemented with oils. It was opined that the azulenes (chamazulene and guaiazulene) stimulated regeneration when given by injection. It was also found that chamomile oil and ground chamomile flowers stimulated liver regeneration when given in the diet.

Gupta and Misra (2006) studied the hepatoprotective effect of chamomile hydroalcoholic extracts on albino mice. They fed chamomile extract (400 mg/kg body weight) and after 5 days induced hepatic damage by paracetamol (200 mg/kg body weight). After 48 hours, they estimated the different biochemical parameters such as blood glutathione, reduced liver glutathione, liver Na$^+$K$^+$-ATPase activity, serum marker enzymes (serum alanine aminotransferase [ALT], aspartate aminotransferase [AST], alkaline phosphatase [ALP]), serum bilirubin, and liver thiobarbituric acid reactive substances. Administration of chamomile extract significantly increased the concentration of glutathione in liver and in blood, and liver Na$^+$K$^+$-ATPase activity. Paracetamol increased the level of liver thiobarbituric acid reactive substances but when the mice were treated with both chamomile extract and paracetamol, the levels of these acids dropped significantly. Treatment with chamomile extract reduced the paracetamol-enhanced level of serum ALT, AST, ALP, and bilirubin. The researchers deduced that the possible mechanism for the protection of paracetamol-induced liver damage by chamomile extract could be (1) scavenging free radicals by intercepting those radicals involved in paracetamol metabolism by microsomal enzymes and (2) increasing the content of GSH in blood and liver and thus providing the liver tissue a better protection against antioxidative stress. The hepatoprotective action combined with antioxidant activity provides a synergistic effect to prevent the process of initiation and progress of hepatocellular damage. The researchers concluded that chamomile extract is a promising hepatoprotective agent.

Nwoye (2013) studied the effect of chamomile extract on alcohol-induced hepatonephrotoxicity and pancytopenia in albino rats. Chamomile extracts were fed before inducing the symptoms by alcohol as pretreatment. As posttreatment, after alcohol treatment, the rats were fed chamomile extracts for 28 days. It was found that the alcohol-induced increased levels of the enzymes ALT, AST, ALP, and γ-glutamyl transferase (GGT) were reduced significantly by chamomile extract treatment. The alcohol-induced pancytopenia with decreased levels of total red blood cells, white blood cells, and platelets, and macrocytic hypochromic anemia were reported to be ameliorated by the treatment with chamomile extract. It was also found that pretreatment with chamomile extracts produced better results than posttreatment. Nwoye deduced that one mechanism of action of chamomile extract was reducing the oxidative stress to exert the hepatoprotective function.

3.22 NEPHROPROTECTIVE EFFECT OF CHAMOMILE

Nephroprotection generally means the protection of the kidneys. The nephrons and tissues of the kidneys could get damaged due to many factors, such as inflammation, urinary tract infections, polycystic disease, and drugs and toxins.

3.22.1 Effect of Chamomile on Nephrotoxicity

Salama (2012) studied the effect of chamomile extract on the cisplatin-induced nephrotoxicity in a rat model. The rats were injected with chamomile extract intraperitoneally 5 days before experiment. The rat kidney tissue was used to determine (1) kidney function tests (serum urea, creatinine, GGT, NAG, β-gal),

(2) oxidative stress indices (NO, LPO), (3) antioxidant activities (SOD, GSH, total thiols), (4) apoptotic indices (cathepsin D, DNA fragmentation), and (5) mineral (calcium). Chamomile extract significantly increased the body weight, normalized the kidney functions, improved the apoptotic markers, reduced the oxidative stress markers, and corrected the hypocalcemia that resulted from cisplatin nephrotoxicity. It was deduced that chamomile extract could have nephroprotective properties.

3.23 ANTILEISHMANIA EFFECT OF CHAMOMILE

Leishmaniasis is caused by a protozoa parasite of the *Leishmania* sp. It is transmitted through infected sandflies. The symptoms of leishmaniasis include skin sores, breathing difficulty, and fever. The immune system is also compromised that might lead to death if untreated. According to a WHO report, leishmaniasis threatens about 350 million men, women, and children in 88 countries around the world. As many as 12 million people are believed to be currently infected, with about 1–2 million estimated new cases occurring every year (WHO 2013).

3.23.1 INHIBITORY EFFECT OF CHAMOMILE ON *LEISHMANIA*

Schnitzler et al. (1996) reported a strong inhibitory effect of chamomile extract on *Leishmania mexicana*. An ethanol extract (80%) was tested, which was found to strongly inhibit *Leishmania mexicana*. Morales-Yuste et al. (2010) also reported the antileishmania effect of chamomile with (−)-α-bisabolol playing the most important role. (−)-α-Bisabolol was found to totally inhibit *Leishmania infantum* promastigotes at the concentrations of 1000 and 500 μg/mL. The researchers suggested that (−)-α-bisabolol preparations could be a promising therapeutic solution to treat leishmaniasis caused by *Leishmania infantum*.

3.24 ANTIBIOTIC EFFECT OF CHAMOMILE

Chamomile essential oil and extracts are known to have antimicrobial effects. The oil and extracts inhibit a wide range of harmful microorganisms that include several species of bacteria and fungi.

3.24.1 ANTIBIOTIC EFFECT OF CHAMOMILE EXTRACTS

Cinco et al. (1983) conducted a study on the activity of chamomile hydroalcoholic extract. *Staphylococcus aureus*, *Streptococcus mutans*, group B Streptococcus, *Streptococcus salivarius*, *Bacillus megaterium*, *Leptospira icterohaemorrhagiae*, *Staphylococcus epidermidis*, and *Streptococcus faecalis*, cultured from vaginal and oral isolates, and a *Trichomonas* strain were used to test the activity of chamomile hydroalcoholic extract. The bacteria were incubated with different dilutions of the extract at 37°C and 0.1 mL of each sample was plated. The number of colonies recovered was taken as the measure of viability and expressed as colony-forming units (cfu/mL). Chamomile extract completely inhibited the growth of *Staphylococcus aureus*, *Streptococcus mutans*, group B Streptococcus, and *Streptococcus salivarius* and exerted a bactericidal effect on *B. megaterium* and *Leptospira icterohaemorrhagiae*, whereas the growth of *Staphylococcus epidermidis* and *Streptococcus faecalis* was significantly reduced. The cells of *Trichomonas* were cultured in the diamond medium and incubated at 37°C for 24 hours. After centrifugation, cells were resuspended in the same medium and counted in Burker counting chamber. The growth was expressed in terms of the motile cells/mL. A dose of 10 mg/mL of extract represented the minimal inhibitory concentration for *Trichomonas*.

Suganda et al. (1983) tested the effect of chamomile ethanolic extract on cell cultures infected with human herpesvirus 1 and human poliovirus 2. Chamomile extract showed an inhibitory activity on the development of the virus.

Türi et al. (1999) studied the effects of chamomile extracts on *Escherichia coli* surface structure by the salt aggregation test. They found that the constituent compounds of chamomile extract blocked the aggregation of *Escherichia coli*. They reported that these constituent compounds were thermostable.

Abdel-Hameed et al. (2008) investigated the activity of chamomile extracts on three bacteria (*B. subtilis*, *Staphylococcus aureus*, and *Escherichia coli*) and two fungi species (*Aspergillus niger* and *Candida albicans*). It was found that chamomile showed a very high microbial activity.

Fabri et al. (2011) studied the effect of leaf extracts of chamomile on *Shigella sonnei* and *Pseudomonas aeruginosa*. They found that at 0.078 mg/mL, chamomile extracts inhibited the growth of *Shigella sonnei* and *Pseudomonas aeruginosa*. The researchers concluded that the flavonoids and terpenes might be responsible for the antimicrobial activity.

Gulzar et al. (2013) reported that chamomile ethanolic extract was effective in controlling the mycelial growth and conidial germination of the fungal pathogen *Fusarium pallidoroseum* that causes twig blight in mulberry, a vital tree for raising silkworms.

3.24.2 Antibiotic Effect of Chamomile Essential Oil

Aggag and Yousef (1972) reported that chamomile oil had more inhibitory effect on gram-positive bacteria than on gram-negative bacteria. They recommended that chamomile oil could be used for treating staphylococcal infections. They also reported a marked inhibitory effect of chamomile oil on *Candida albicans*.

Kedzia (1991) tested the essential oil activity on bacterial, yeast, and fungal cultures. Inhibitory activity was stronger on gram-positive bacteria, yeasts, and fungi than on gram-negative bacteria.

Annuk et al. (1999) investigated the cell surface hydrophobicity of *Helicobacter pylori* to chamomile extract by the salt aggregation test. They found that chamomile extract could block the aggregation of *Helicobacter pylori*.

Pandey et al. (2002) investigated the mycotoxic potential of *Helminthosporium sativum* and found that chamomile oil showed more than 90% mycotoxicity at 1000 ppm. They found that chamomile oil was better than the commercial fungicides in terms of inhibiting the fungus, restricting spore germination, and restricting the formation of the inhibition zone.

Pereira et al. (2008) studied the antibacterial effect of chamomile seed oil. They found that the seed oil was able to inhibit *Pseudomonas aeruginosa*, *Escherichia coli*, and *Enterobacter aerogenes*. However, it was not found to be effective against *Salmonella choleraesuis*.

Tolouee et al. (2010) studied the mechanism of effect of chamomile essential oil on the inhibition of *Aspergillus niger*. Using electron microscope they found that chamomile oil causes swelling and disruption of hyphae, disruption of plasma membrane and intracellular organelles, detachment of plasma membrane from cell wall, and depletion of cytoplasm. The researchers suggested that these changes might be due to the effect of chamomile essential oil on the plasma membrane causing cell permeability.

Abdoul-Latif et al. (2011) tested the essential oil and extract against bacterial and fungal strains using a broth microdilution methods. They reported that chamomile oil and extract showed significant antimicrobial activity.

Alireza (2012) tested chamomile essential oil against seven strains of bacteria (both gram positive and gram negative) and the oil was found to show mild to significant antimicrobial activity.

Hoyos et al. (2012) studied the effect of chamomile oil on the fungus *Pseudocercospora griseola*. They found that chamomile oil at 0.1% and 0.5% inhibited conidia growth effectively. The main antifungal constituent in the oil was identified as *trans*-β-farnesene. They concluded that this compound can be used as an alternative to fungicide for the control of angular leaf spot in bean.

Aliheidari et al. (2013) evaluated the antimicrobial effectiveness of chamomile essential oil against *Listeria monocytogenes*, *Staphylococcus aureus*, and *Escherichia coli* and found the oil to be effective against these bacteria.

3.25 EFFECT OF CHAMOMILE AGAINST ANISAKIASIS

Anisakiasis is a parasitic infection of the gastrointestinal tract, which is caused by the nematode *Anisakis simplex*. This parasite is usually hosted in marine animals, such as fish and squids, which are transferred to humans when these animals are eaten, especially raw fish.

3.25.1 CHAMOMILE AS A POTENTIAL CURE FOR ANISAKIASIS

Chamomile essential oil has been found to be a potential cure for anisakiasis. Romero et al. (2012) investigated the effect of chamomile oil against Anisakis larvae in infected rats. The essential oil (125 μg/mL) caused death of all nematodes. Cuticle changes and rupture of intestinal wall were observed in the nematodes. The chamomile-treated infected rats showed significant decrease in gastric lesions. Chamazulene was ineffective whereas α-bisabolol was highly effective in vitro that in vivo suggesting a synergistic action of the compounds. The researchers concluded that chamomile oil could be a good biocidal agent against Anisakis type I.

3.26 ANTICULEX EFFECT OF CHAMOMILE

Diseases transmitted by the *Culex* mosquito, such as filariasis, Japanese encephalitis, West Nile virus disease, and avian malaria, cause the loss of thousands of lives every year.

3.26.1 INHIBITORY EFFECT OF CHAMOMILE AGAINST *CULEX*

Chamomile has been found to have inhibitory effects against *Culex*. Al-Khalaf (2011) investigated the effect of chamomile on the morphological features of the third larval instar and pupae of *Culex quinquefasciatus*. The results showed that exposure to chamomile extracts for 24 hours had malformed the morphological features of the adults, larvae, and pupa. The malformations included blocking of metamorphosis and ecdysis and inhibition of adults' emergence. This result indicates that chamomile is a possible candidate to fight diseases such as filariasis, Japanese encephalitis, West Nile virus, and avian malaria.

3.27 CHAMOMILE AND PREGNANCY

Chamomile has been contraindicated in pregnant women because of its emmenagogue and relaxing effect on smooth muscles. Arruda et al. (2013) investigated the effect of chamomile on pregnancy and the offspring using rats. They found that chamomile may negatively influence the weight of the mother during pregnancy and also the weight gain of the offspring after birth. Chamomile might also cause changes in the neurological reflexes of the offspring.

3.28 CHAMOMILE AND PEST MANAGEMENT

Chamomile extract could be used as a pesticide. Radwan et al. (2011) evaluated the effect of chamomile extract on the nematode *Meloidogyne incognita* that had been infecting tomato plants. They found that chamomile did not effectively suppress the nematode. They further tested the effect of chamomile seed extracts on the nematode (Radwan et al. 2012) and found that chamomile seed extracts were ineffective against the nematode.

Cui et al. (2012) studied the effect of E-β-farnesene in attracting natural enemies of aphids in cabbage fields. They found that E-β-farnesene extracted from chamomile attracted the natural enemies of aphids thereby reducing their population density in cabbage fields.

Padín et al. (2013) studied the toxicity and repellency of chamomile methanolic extracts against the red flour beetle *Tribolium castaneum*. They found that chamomile extract caused 57% mortality. The extract also showed a high insect repellency. The researchers concluded that chamomile extract could be promising as insect repellents.

3.29 CHAMOMILE USE IN POULTRY AND ANIMAL HUSBANDRY

Chamomile flowers are fed to poultry and farm animals for obtaining better yields of eggs, milk, and meat. The enhancement in the yield after feeding chamomile flowers has been demonstrated by several studies. Hamodi (2007) fed hens with chamomile flowers (0.3% and 0.6%) in the diet. They found that there was an increase in the egg production when the hens were fed with 0.6% chamomile flowers and an increase in egg mass when the hens were fed with 0.3% chamomile flowers in the diet. The researchers also found a significant decrease in the total count of microorganisms in the intestine of the chamomile-treated hens.

Poráčová et al. (2007a,b) studied the effect of chamomile essential oil on the weight of the eggs of the hens. A group of hens were fed diet enriched with chamomile essential oil (0.1%) daily for 8 weeks. They observed a significant increase in the weight of the eggs of the hens fed with chamomile oil-enriched diet.

Al-Kaisse and Khalel (2011) fed 0.25%, 0.50%, 0.75%, and 1% of chamomile flowers powder mixed with diet to the hens and measured the weight gain, feed intake, feed conversion ratio, blood parameters [RBC, Hb, Packed Cell Volume (PCV)], and carcass traits. Chamomile at 0.75% and 1% significantly increased the weight gain, feed intake, and feed conversion ratio. Chamomile also lowered the RBC, Hb, PCV, heterophil/lymphocyte, cholesterol, and glucose levels. An increase in the weights of heart, liver, and gizzard was observed in chamomile-fed hens.

El-Galil et al. (2011) studied the productive and reproductive effect of feeding chamomile flower meal to Japanese quails. They found that there was a dose-dependent increase in body weight of the quails fed with chamomile flowers. The egg shell, yolk, and egg shape showed significant improvement. The sperm concentration, sperm motility, total motile sperm, live spermatozoa, and semen quality increased when chamomile flowers were fed. The researchers concluded that using dietary chamomile flower meal at 0.50 g/kg of the diet could improve productivity without any side effects.

Simonová et al. (2007) studied the antimicrobial effect of chamomile essential oil on controlling microbes in rabbits. They found that chamomile essential oil reduced enterococci, staphylococci, *Escherichia coli*, *Clostridium*, and *Pseudomonas* in the rabbits. There was also a reduction in oocysts of *Eimeris* sp. in the feces.

Saleh et al. (2009) fed suckling lambs a feed additive comprising chamomile flowers. They found that this supplementary diet improved the physiological functions of body of the lambs.

Kholif (2010) studied the effect of chamomile flowers on the milk yield of lactating goats. They found that chamomile significantly increased the yield of milk and 4% fat corrected milk. Chamomile also increased the milk constituents. Khorshed et al. (2011) also found that feeding chamomile flowers to lactating goats increased the milk yield by 23.08% and milk fat by 28.04%.

El-Garhy (2012) studied the effect of feeding chamomile flower residue after extracting the essential oil. The results showed that buffaloes fed with chamomile flower residue shied high values of digestibility coefficients and feeding values. When fed with only chamomile flower residue, the buffaloes yielded high quantities of milk and the quality of milk was also high.

3.30 CHAMOMILE AS VETERINARY MEDICINE

Chamomile could have possible uses in veterinary medicine. However, studies are few in this area. Cassu et al. (2011) studied the effect of chamomile extract on the stress levels of female dogs after social isolation and undergoing ovariohysterectomy. The dogs were treated for 15 days with chamomile extracts and the serum cortisol levels were measured before and after the surgery. There was a significant increase in the serum levels in the dogs that were not treated with chamomile extracts, but the group treated with the extract maintained a stable level of serum cortisol. They concluded that chamomile extract could alleviate stress in dogs.

3.31 TOXICITY AND ALLERGENICITY OF CHAMOMILE COMPOUNDS

Habersang et al. (1979) studied the toxicity of (−)-α-bisabolol and reported that the acute toxicity of (−)-α-bisabolol was very low in mice and rats if administered orally (1.0–2.0 mL bisabolol/kg body weight). The toxicity was low in case of oral administration in rhesus monkeys as well as dogs. Toxicity occurred when the doses were high. Roder (1982) also reported that the chamomile drug has tolerable effects. However, Hausen et al. (1984) reported that anthecotulide present in trace amounts in the bisabolol oxide B type of chamomile of Argentine origin had been noted to cause contact dermatitis, allergy of the skin. Subiza et al. (1989) also reported that a patient of hay fever, who was previously sensitive to *Artemisia* pollen, had an anaphylactic reaction after ingesting chamomile tea. Subiza et al. (1990) reported allergic conjunctivitis in some individuals after their eyes had been washed with chamomile infusion. This effect was related to some unknown compound in the pollen. Reider et al. (2000) have opined that the incidence and risk of type I allergy to chamomile may be underestimated.

Paulsen et al. (2008) observed the reaction of chamomile pollen, tea, ointments, creams, and oil on chamomile-sensitive patients to test the elicitation of dermatitis in pre-sensitized patients. They found that when the patients were patch tested with the herbal products, the compounds elicited a positive response. In addition, the fragrances, emulsifiers, and preservatives tested positive as well. They concluded that Compositae-allergic patients should be warned against topical use of Compositae compound-containing products. Paulsen et al. (2010) tested the allergic effect of herniarin, a coumarin, on chamomile-sensitive patients and found that herniarin was one of the non-sesquiterpene lactone sensitizers. The sensitization may occur through oral ingestion of tea or topical application of chamomile-containing medicines.

Jacob and Hsu (2010) reported three cases of allergic contact dermatitis due to the use of bisabolol. They indicated that bisabolol may be involved in atopic dermatitis in children and cautioned the physicians to be aware of the effects.

Russell and Jacob (2010) recommended patch testing with bisabolol-containing products or bisabolol in patients with presumptive allergic contact dermatitis or potentially worsening atopic dermatitis. Patients sensitized to bisabolol should be counseled to avoid any bisabolol-containing products.

However, one report mentions the use of chamomile dyes for the manufacture of allergic contact dermatitis–proof clothing (Sertoli et al. 1994). Babych and Novak (1995) reported that chamomile is used for treatment of allergic dermatosis.

Pharmacodynamic interaction between warfarin and coumarin and pharmacokinetic interaction with cyclosporine has been reported in the literature (Segal and Pilote 2006). It was recommended that patient education on the potential risk of taking chamomile products while being treated with warfarin is necessary. Schulz (2009) studied the effect of chamomile on the risk of drug interaction on in vitro experiment on rats. He concluded that according to the available data, the occurrence of interactions between chamomile and cyclosporine or warfarin cannot be deduced.

Stingeni et al. (1999) evaluated the cytokine profile of the T-cell lines from a patient of allergic contact dermatitis. The cloning of the *Chamomilla recutita*–specific T-cell lines of the patient showed high amounts of IFN-γ and IL-4, which might play a role in the immunopathogenesis of allergenic contact dermatitis due to chamomile.

3.32 PHARMACOKINETIC STUDIES AND IN VIVO SKIN PENETRATIONS OF CHAMOMILE COMPOUNDS

The efficacy of the chamomile drugs based on the chamomile essential oil or extracts depends on the penetration of the compounds into the skin and maintaining stability for some time. A few studies have been conducted to elucidate the penetration of these compounds in the skin, both in vitro and in humans.

The exposed skin patches in mice were applied with 14°C-labeled bisabolol topically. After 5 hours, activity was observed by cutting the tissue and preparing autoradiogram. Densitometric measurements were taken to observe penetration of bisabolol in the skin. Half of the radioactivity was observed in skin and half in the underlying tissues. About 90% activity was in the intact bisabolol. The results showed that bisabolol penetrated into the tissue fast and remained stable for considerable period (Hahn and Hoelzl 1987).

Nine human female volunteers were applied with an aqueous alcohol solution containing apigenin (1.6 mg/100 mL), luteolin (3.2 mg/100 mL), and apigenin-7-0-β-glucoside (4.0 mg/100 mL) on their upper arms with the help of application chambers. After 60 minutes, all the chambers containing residual solutions were emptied and refilled with initial solutions and the process was repeated six times. All the residual samples were analyzed spectrometrically at 350 nm and the decline in flavonoid concentration in the residual solutions was noted. A steady state of flux was attained after 3 hours. Apigenin showed highest steady-state flux followed by luteolin. Within the first 2 hours, flavonoid was concentrated in stratum corneum but after 3 hours flavonoids diffused into deeper layers of tissue. The chamomile compounds got absorbed through the human skin and remained stable there for quite some time (Merfort et al. 1994).

REFERENCES

Abad, A. N. A., Nouri, M. H. K., Gharjanie, A., and Tavakoli, F. 2011a. Effect of *Matricaria chamomilla* hydroalcoholic extract on cisplatin-induced neuropathy in mice. *Chinese Journal of Natural Medicines* 9(2): 126–131.

Abad, A. N. A., Nouri, M. H. K., and Tavakkoli, F. 2011b. Study of *Matricaria recutita* and vincristine effects on PTZ-induced seizure threshold in mice. *Research Journal of Medical Sciences* 5(5): 247–251.

Abdel-Hameed, E. S., El-Nahas, H. A., and Abo-Sedera, S. A. 2008. Antischistosomal and antimicrobial activities of some Egyptian plant species. *Pharmaceutical Biology* 46(9): 626–633.

Abdoul-Latif, F. M., Mohamed, N., Edou, P., Ali, A. A., Djama, S. O., Obame, L. C., Bassolé, I. H. N., and Dicko, M. H. 2011. Antimicrobial and antioxidant activities of essential oil and methanol extract of *Matricaria chamomilla* L. from Djibouti. *Journal of Medicinal Plants Research* 5(9): 1512–1517.

Achterrath-Tuckermann, U., Kunde, R., Flaskamp, E., Issac, O., and Theimer, K. 1980. Pharmacological investigations with compounds of chamomile. *Planta Medica* 39(1): 38–50.

Aggag, M. E. and Yousef, R. T. 1972. Study of antimicrobial activity of chamomile oil. *Planta Medica* 22(6): 140–144.

Agarwal, S. S. and Singh, V. K. 1999. Immunomodulators: A review of studies on Indian medicinal plants and synthetic peptides. *Proceedings of the Indian National Science Association* B65(3&5): 179–204.

Ahmad, I. H., Awad, M. A., El-Mahdy, M., Gohar, H. M., and Ghanem, A. M. 1995. The effect of some medicinal plants extracts on wound healing in farm animals. *Assiut Veterinary Medical Journal* 32(64): 236–244.

Al-Hindawi, M. K., A1-Deen, I. H. S., Nab, M. H. A., and Ismail, M. A. 1989. Anti-inflammatory activity of some Iraqi plants using intact rats. *Journal of Ethnopharmacology* 26(2): 163–168.

Aliheidari, N., Fazaeli, M., Ahmadi, R., Ghasemlou, M., and Emam-Djomeh, Z. 2013. Comparative evaluation on fatty acid and *Matricaria recutita* essential oil incorporated into casein-based film. *International Journal of Biological Macromolecules* 56: 69–75.

Alireza, M. 2012. Antimicrobial activity and chemical composition of essential oils of chamomile from Neyshabur, Iran. *Journal of Medicinal Plants Research* 6(5): 820–824.

Al-Kaisse, G. A. M. and Khalel, E. K. 2011. The potency of chamomile flowers (*Matericaria chamomilla* L.) as feed supplements (growth promoters) on productive performance and hematological parameters constituents of broiler. *International Journal of Poultry Science* 10(9): 726–729.

Al-Khalaf, A. A. 2011. Effect of the LC_{50} of *Artemisia herba alba* and *Matricaria chamomilla* on morphological features of 3rd larval instar and pupa of *Culex quinquefasciatus*. *Egyptian Journal of Biological Pest Control* 21(2): 385–392.

Alves Ade, M., Gonçalves, J. C., Cruz, J. S., and Araújo, D. A. 2010. Evaluation of the sesquiterpene (−)-alpha-bisabolol as a novel peripheral nervous blocker. *Neuroscience Letters* 472(1): 11–15.

American Cancer Society. 2012. Signs and symptoms of cancer. http://www.cancer.org/cancer/cancerbasics/signs-and-symptoms-of-cancer (Accessed January 04, 2013).

Amirghofran, Z., Azadbakht, M., and Karimi, M. H. 2000. Evaluation of the immunomodulatory effects of five herbal plants. *Journal of Ethnopharmacology* 72(1–2): 167–172.

Ammon, H. P. T., Sabieraj, J., and Kaul, R. 1996. Chamomile. Mechanism of antiphlogistic activity of chamomile extracts and their contents. *Deutsche Apotheker Zeitung* 136: 17–30.

Amsterdam, J. D., Li, Y., Soeller, I., Rockwell, K., Mao, J. J., and Shults, J. 2009. A randomized, double-blind, placebo-controlled trial of oral *Matricaria recutita* (chamomile) extract therapy for generalized anxiety disorder. *Journal of Clinical Psychopharmacology* 29(4): 378–382.

Annuk, H., Hirmo, S., Türi, E., Mikelsaar, M., Arak, E. and Wadström, T. 1999. Effect on cell surface hydrophobicity and susceptibility of *Helicobacter pylori* to medicinal plant extracts. *FEMS Microbiology Letters* 172(1): 41–45.

Anter, J., Romero-Jiménez, M., Fernández-Bedmar, Z., Villatoro-Pulido, M., Analla, M., Alonso-Moraga, A., and Muñoz-Serrano, A. 2011. Antigenotoxicity, cytotoxicity, and apoptosis induction by apigenin, bisabolol, and protocatechuic acid. *Journal of Medicinal Food* 14(3): 276–283.

Arango, D., Parihar, A., Villamena, F. A., Wang, L., Freitas, M. A., Grotewold, E., and Doseff, A. I. 2012. Apigenin induces DNA damage through the PKCδ-dependent activation of ATM and H2AX causing down-regulation of genes involved in cell cycle control and DNA repair. *Biochemical Pharmacology* 84(12): 1571–1580.

Arif Zaidi, S. M., Khatoon, K., and Aslam, K. M. 2012. Role of herbal medicine in Ussuruttams (dysmenorrhoea). *Journal of Academia and Industrial Research* 1(3): 113–117.

Arruda, J. T., Approbato, F. C., Maia, M. C. S. Silva, T. M., and Approbato, M. S. 2013. Effect of aqueous extract of chamomile (*Chamomilla recutita* L.) on rat pregnancy and offspring development. *Revista Brasileira de Plantas Medicinais* 15(1): 66–71.

Arsić, I., Tadić, V., Vlaović, D., Homšek, I., Vesić, S., Isailović, G., and Vuleta, G. 2011. Preparation of novel apigenin-enriched, liposomal and non-liposomal, anti-inflammatory topical formulations as substitutes for corticosteroid therapy. *Phytotherapy Research* 25(2): 228–233.

Avallone, R., Paola, Z., Giulia, P., Matthias, K., Peter, S., and Mario, B. 2000. Pharmacological profile of apigenin, a flavonoid isolated from *Matricaria chamomilla*. *Neuroscience* 59(11): 1387–1394.

Avallone, R., Zanoli, P., Corsi, L., Cannazza, G., and Baraldi, M. 1996. Benzodiazepine-like compounds and GABA in flower heads of *Matricaria chamomilla*. *Phytotherapy Research* 10: S177–S179.

Babych, V. and Novak, B. 1995. Use of some plant species for treatment of allergic diseases. *Naturalium* 3. http://agris.fao.org/agris-search/search/display.do?f=2012%2FOV%2 FOV201208404008404.xml%3BUA19960076973 (Accessed November 10, 2012).

Bezerra, S. B., Leal, L. K., Nogueira, N. A., and Campos, A. R. 2009. Bisabolol-induced gastroprotection against acute gastric lesions: Role of prostaglandins, nitric oxide, and K_{ATP}^+ channels. *Journal of Medicinal Food* 12(6): 1403–1406.

Boesch-Saadatmandi, C., Wagner, A. E., Wolffram, S., and Rimbach G. 2012. Effect of quercetin on inflammatory gene expression in mice liver in vivo—role of redox factor 1, miRNA-122 and miRNA-125b. *Pharmacological Research* 65(5): 523–530.

Boots, A. W., Drent, M., de Boer, V. C., Bast, A., and Haenen, G. R. 2011. Quercetin reduces markers of oxidative stress and inflammation in sarcoidosis. *Clinical Nutrition* 30(4): 506–512.

Boots, A. W., Wilms, L. C., Swennen, E. L., Kleinjans, J. C., Bast, A., and Haenen, G. R. 2008. In vitro and ex vivo anti-inflammatory activity of quercetin in healthy volunteers. *Nutrition* 24(7–8): 703–710.

Braga, P. C., Dal Sasso, M., Fonti, E., and Culici, M. 2009. Antioxidant activity of bisabolol: Inhibitory effects on chemiluminescence of human neutrophil bursts and cell-free systems. *Pharmacology* 83(2): 110–115.

Brennan, B. 2008. What Is Buspirone (Buspar), How Does It Work, and How Is It Used to Treat Anxiety Disorders Œ *abc News*. http://abcnews.go.com/Health/AnxietyTreating/story?id=4660154#.UOf9eazDu-0 (Accessed March 13, 2014).

Bulgari, M., Sangiovanni, E., Colombo, E., Maschi, O., Caruso, D., Bosisio, E., and Dell'Agli, M. 2012. Inhibition of neutrophil elastase and metalloproteinase-9 of human adenocarcinoma gastric cells by chamomile (*Matricaria recutita* L.) infusion. *Phytotherapy Research* 26(12): 1817–1822.

Can, O.D., Demir Ö. U., Kıyan, H.T., and Demirci, B. 2012. Psychopharmacological profile of Chamomile (*Matricaria recutita* L.) essential oil in mice. *Phytomedicine*, 19(3-4):306–10.

Capasso, R., Savino, F., and Capasso, F. 2007. Effects of the herbal formulation ColiMil® on upper gastrointestinal transit in mice *in vivo*. *Phytotherapy Research* 21(10): 999–1001.

Carle, R. 1990. Anti inflammatory and spasmolytic botanical drugs. *British Journal of Phytotherapy* 1(1): 33–39.

Cassu, R. N., Andreazi, C. D., and Pereira, L. 2011. Effect of the *Matricaria chamomilla* CH12 in stress response in dogs. *Agrariae* 7(2): 1–7.

Cavalieri, E., Rigo, A., Bonifacio, M., Carcereri de Prati, A., Guardalben, E., Bergamini, C., Fato, R., Pizzolo, G., Suzuki, H., and Vinante, F. 2011. Pro-apoptotic activity of α-bisabolol in preclinical models of primary human acute leukemia cells. *Journal of Translational Medicine* 9: 45. doi: 10.1186/1479-5876-9-45.

Cemek, M., Yilmaz, E., and Büyükokuroğlu, M. E. 2010. Protective effect of *Matricaria chamomilla* on ethanol-induced acute gastric mucosal injury in rats. *Pharmaceutical Biology* 48(7): 757–763.

Chan, M. M., Mattiacci, J. A., Hwang, H. S., Shah, A., and Fong, D. 2000. Synergy between ethanol and grape polyphenols, quercetin, and resveratrol, in the inhibition of the inducible nitric oxide synthase pathway. *Biochemical Pharmacology* 60(10): 1539–1548.

Chandrashekhar, V. M., Halagali, K. S., Nidavani, R. B., Shalavadi, M. H., Biradar, B. S., Biswas, D., and Muchchandi, I. S. 2011. Anti-allergic activity of German chamomile (*Matricaria recutita* L.) in mast cell mediated allergy model. *Journal of Ethnopharmacology* 137(1): 336–340.

Chandrashekhar, V. M., Ranpariya, V. L., Ganapaty, S., Parashar, A., and Muchandi, A. A. 2010. Neuroprotective activity of *Matricaria recutita* Linn against global model of ischemia in rats. *Journal of Ethnopharmacology* 127(3): 645–651.

Cheng, Y. D., Al-Khoury, L., and Zivin, J. A. 2004. Neuroprotection for ischemic stroke: Two decades of success and failure. *NeuroRx* 1(1): 36–45.

Chiu-Yuan, C., Wen-Huang, P., Kuen-Daw, T., and Shih-Lan, H. 2007. Luteolin suppresses inflammation-associated gene expression by blocking NF-κB and AP-1 activation pathway in mouse alveolar macrophages. *Life Sciences* 81(23–24): 1602–1614.

Cho, S., Park, S., Kwon, M., Jeong, T., Bok, S., Choi, W., Jeong, W. et al. 2003. Quercetin suppresses proinflammatory cytokines production through MAP kinases and NF-κB pathway in lipopolysaccharide-stimulated macrophage. *Molecular and Cellular Biochemistry* 243(1–2): 153–160.

Chuang, C. C., Martinez, K., Xie, G., Kennedy, A., Bumrungpert, A., Overman, A., Jia, W., and McIntosh, M. K. 2010. Quercetin is equally or more effective than resveratrol in attenuating tumor necrosis factor-{alpha}-mediated inflammation and insulin resistance in primary human adipocytes. *American Journal of Clinical Nutrition* 92(6): 1511–1521.

Cinco, M., Banfi, E., Tubaro, A., and Della Loggia, R. 1983. A microbiological survey on the activity of a hydroalcoholic extract of chamomile. *International Journal of Crude Drug Research* 21: 145–151.

Comalada, M., Camuesco, D., Sierra, S., Ballester, I., Xaus, J., Gálvez, J., and Zarzuelo, A. 2005. In vivo quercitrin anti-inflammatory effect involves release of quercetin, which inhibits inflammation through down-regulation of the NF-κB pathway. *European Journal of Immunology* 35(2): 584–592.

Courtney, N. 2009. The anti-inflammatory mechanisms of the flavonoid apigenin in vitro and *in vivo*. MSc Dissertation. Ohio State University, Columbus, OH. https://etd.ohiolink.edu /ap/0?0:APPLICATION_PROCESS%3DDOWNLOAD_ETD_SUB_DOC _ACCNUM::: F1501_ID:osu1259783472%2Cinline (Accessed March 13, 2014).

Cui, L. L., Francis, F., Heuskin, S., Lognay, G., Liu, Y., Dong, J., Chen, J., Song, X., and Liu, Y. 2012. The functional significance of E-β-farnesene: Does it influence the populations of aphid natural enemies in the fields? *Biological Control* 60(2): 108–112.

Das, M., Ram, A., and Ghosh, B. 2003. Luteolin alleviates bronchoconstriction and airway hyperreactivity in ovalbumin sensitized mice. *Inflammation Research* 52(3): 101–106.

Datta, K., Sinha, S., and Chattopadhyaya, P. 2000. Reactive oxygen species in health and disease. *National Medical Journal of India* 13: 304–310.

de Siqueira, R. J., Freire, W. B., Vasconcelos-Silva, A. A., Fonseca-Magalhães, P. A., Lima, F. J., Brito, T. S., Mourão, L. T., Ribeiro, R. A., Lahlou, S., and Magalhães, P. J. 2012. In-vitro characterization of the pharmacological effects induced by (−)-α-bisabolol in rat smooth muscle preparations Canadian. *Journal of Physiology and Pharmacology* 90(1): 23–35.

Delarmelina, J. M., Batitucci, C. M. do C. P., Gonçalves, J. L. de O. 2012. The cytotoxic, genotoxic and mutagenic effects of *Matricaria chamomilla* L. tincture *in vivo*. *Revista Cubana de Plantas Medicinales* 17(2): 149–159.

Della Loggia, R., Carle, R., Sosa, S., and Tubaro, A. 1990. Evaluation of the anti-inflammatory activity of chamomile preparations. *Planta Medica* 56: 657–658.

Della Loggia, R., Tubaro, A., Dri, P., Zilli, C., and Del Negro, P. 1986. The role of flavonoids in the anti-inflammatory activity of *Chamomilla recutita*. In: E. Middleton, V. Cody and J. B. Harborne (eds.) *Plant Flavonoids in Biology and Medicine: Biochemical, Pharmacological, and Structure-Activity Relationships*. New York: Alan R. Liss Inc., pp. 481–484.

Della Loggia, R., Traversa, U., Scarcia, V., and Tubaro, A. 1982. Depressive effects of *Chamomilla recutita* (L.) Rausch, tubular flowers, on central nervous system in mice. *Pharmacological Research Communications* 14: 154–162.

Demir, E., Kaya, B., Marcos, R., Cenkci, S. K., and Çetin, H. 2013. Investigation of the genotoxic and antigenotoxic properties of essential oils obtained from two *Origanum* species by *Drosophila* wing SMART assay. *Turkish Journal of Biology* 37: 1–10.

Diegelmann, R. F. and Evans, M. C. 2004. Wound healing: An overview of acute, fibrotic and delayed healing. *Frontiers in Bioscience* 9: 283–289.

Doseff, A., Arango, D., Cardenas, H., Nicholas, C., Nuovo, G., and Grotewold, E. 2011. Apigenin, a potent anti-inflammatory flavonoid: Its efficacy in lung inflammation and target identification provides novel potential alternative therapeutic approaches. *American Journal of Respiratory* and *Critical Care Medicine* 183: A4508. http://www.atsjournals.org/doi/abs/10.1164/ajrccm-conference.2011.183.1_MeetingAbstracts.A4508 (Accessed March 13, 2014).

Drummond, E. M., Harbourne, N., Marete, E., Martyn, D., Jacquier, J. C., O'Riordan, D., and Gibney, E. R. 2013. Inhibition of proinflammatory biomarkers in THP1 macrophages by polyphenols derived from chamomile, meadowsweet and willow bark. *Phytotherapy Research* 27(4): 588–594.

El-Galil, K. A., Mahmoud, H. A., Hassan, A. M., and Morsy, A. S. 2011. Effect of addition chamomile flower meal to laying Japanese quail diets on productive and reproductive performance and some physiological functions. *Egyptian Journal of Nutrition and Feeds* 14(1): 147–158.

El-Garhy, G. M. 2012. Effect of feeding chamomile by-product (*Matricaria chamomilla*) on performance of lactating buffaloes. *Egyptian Journal of Nutrition and Feeds* 15(2): 217–222.

Emam, M. A. 2012. Comparative evaluation of antidiabetic activity of *Rosmarinus officinalis* and *Chamomile recutita* in streptozotocin induced diabetic rats. *Agriculture and Biology Journal of North America* 3(6): 247–252.

Esmaeili, M. H., Honarvaran, F., Kesmati, M., Jahani Hashemi, H., Jaafari, H., and Abbasi, E. 2007. Effects of *Matricaria chamomilla* extract on morphine withdrawal syndrome in mice. *Journal of Qazvin University of Medical Sciences* 11(2): 13–18.

Evans, S., Dizeyi, N., Abrahamsson, P. A., Persson, J. 2009. The effect of a novel botanical agent TBS-101 on invasive prostate cancer in animal models. *Anticancer Research*, 29(10): 3917–3924.

Fabian, D., Juhás, Š., Bukovská, A., Bujňáková, D., Grešáková, Ľ., and Koppel, J. 2011. Anti-inflammatory effects of chamomile essential oil in mice *Slovak Journal of Animal Science* 44(3): 111–116.

Fabri, R.L., Nogueira, M.S., Dutra, L.B., Bouzada, M.L.M., Scio, E.2011. Potencial antioxidante e antimicrobiano de espécies da família Asteraceae (Antioxidant and antimicrobial potential of Asteraceae species). *Revista Brasilera De Plantas Medicinais*, 13 (2):183–189.

Fidler, P., Loprinzi, C. L., O'Fallon, J. R., Leitch, J. M., Lee, J. K., Hayes, D. L., Novotny, P., Clemens-Schutjer, D., Bartel, J., and Michalak, J. C. 1996. Prospective evaluation of a chamomile mouthwash for prevention of 5-FU-induced oral mucositis. *Cancer* 77(3): 522–525.

Flynn, M. E. and Rovee, D. T. 1982. Wound healing mechanisms. *American Journal of Nursing* 82(10): 1544–1549.
Foster, H. B., Niklas, H., and Lutz, S. 1980. Antispasmodic effect of some medicinal plants. *Planta Medica* 40: 309–319.
Friedberg, E. C., Walker, G. C., and Siede, W. 1995. *DNA Repair and Mutagenesis*. Washington, DC: American Society for Microbiology.
Füller, E., Blaschek, W., and Franz, G. 1991. Characterization of water soluble polysaccharides from chamomile flowers. *Planta Medica* 57(Suppl 2): PA40.
Funakoshi-Tago, M., Nakamura, K., Tago, K., Mashino, T., and Kasahara, T. 2011. Anti-inflammatory activity of structurally related flavonoids, Apigenin, Luteolin and Fisetin. *International Immunopharmacology* 11(9): 1150–1159.
Gershbein, L. L. 1977. Regeneration of rat liver in the presence of essential oils and their components. *Food and Cosmetics Toxicology* 15(3): 173–181.
Ghavimi, H., Shayanfar, A., Hamedeyazdan, S., Shiva, A., and Garjani, A. 2012. Chamomile: An ancient pain remedy and a modern gout relief—a hypothesis. *African Journal of Pharmacy and Pharmacology* 6(8): 508–511.
Ghonime, M., Eldomany, R., Abdelaziz, A., and Soliman, H. 2011. Evaluation of immunomodulatory effect of three herbal plants growing in Egypt. *Immunopharmacology and Immunotoxicology* 33(1): 141–145.
Gomaa, A., Hashem, T., Mohamed, M., and Ashry, E. 2003. *Matricaria chamomilla* extract inhibits both development of morphine dependence and expression of abstinence syndrome in rats. *Journal of Pharmacological Sciences* 92(1): 50–55.
González, R., Ballester, I., López-Posadas, R., Suárez, M. D., Zarzuelo, A., Martínez-Augustin, O., and Sánchez De Medina, F. 2011. Effects of flavonoids and other polyphenols on inflammation. *Critical Reviews in Food Science and Nutrition* 51(4): 331–362.
González-Segovia, R., Quintanar, J. L., Salinas, E., Ceballos-Salazar, R., Aviles-Jiménez, F., and Torres-López, J. 2008. Effect of the flavonoid quercetin on inflammation and lipid peroxidation induced by *Helicobacter pylori* in gastric mucosa of guinea pig. *Journal of Gastroenterology* 43(6): 441–447.
Gould, L., Reddy, C. V. R., and Comprecht, R. F. 1973. Cardiac effect of chamomile tea. *Journal of Clinical Pharmacology* 13: 475–479.
Guardia, T., Rotelli, A. E., Juarez, A. O., and Pelzer, L. E. 2001. Anti-inflammatory properties of plant flavonoids. Effects of rutin, quercetin and hesperidin on adjuvant arthritis in rat. *Farmaco* 56(9): 683–687.
Guimarães, R., Barros, L., Dueñas, M., Calhela, R. C., Carvalho, A. M., Santos-Buelga, C., Queiroz, M. J. R. P., and Ferreira, I. C. F. R. 2013. Infusion and decoction of wild German chamomile: Bioactivity and characterization of organic acids and phenolic compounds. *Food Chemistry* 136(2): 947–954.
Gültekin, E. and Yildiz, F. 2006. Introduction to phytoestrogens. In: F. Yildiz (ed.) *Phytoestrogens in Functional Foods*. Boca Raton, FL: CRC Press, pp. 3–4.
Gulzar, P., Kausar, T., Sahaf, K. A., Munshi, N. A., Ahmad, S., and Raja, T. A. 2013. Screening of ethanolic extracts of various botanicals against *Fusarium pallidoroseum*. (Cooke) Sacc.—the causal agent of twig blight of mulberry. *Indian Journal of Sericulture* 52(1): 24–28.
Guo, S. and DiPietro, L. A. 2010. Factors affecting wound healing. *Journal of Dental Research* 89(3): 219–229.
Gupta, A. K. and Misra, N. 2006. Hepatoprotective activity of aqueous ethanolic extract of chamomile capitula in paracetamol intoxicated albino rats. *American Journal of Pharmacology and Toxicology* 1(1): 17–20.
Habersang, S., Leuschner, F., Isaac, O., and Thiemer, K. 1979. Pharmacological studies on toxicity of (−)-α-bisabolol. *Planta Medica* 37: 115–123.
Hahn, B. and Hoelzl, J. 1987. Absorption, distribution and metabolism of carbon 14 levomenol in skin. *Arzneim Forsch* 37(6): 716–720.

Halász, P. and Rásonyi, G. 2004. Neuroprotection and epilepsy. *Advances in Experimental Medicine and Biology* 541: 91–109.

Hamodi, S. J. 2007. The use of chamomile flowers (*Matricaria recutita*) in layer hens diet. *Egyptian Journal of Nutrition and Feeds* 10(2): 289–301.

Hänler, M., Suuronen, T., Salminen, A., Kaarniranta, K., and Kauppinen, A. 2012. Polyphenolic compounds reduce inflammation in ARPE-19 cells. *Acta Ophthalmologica* 90(Suppl s249): 0. http://onlinelibrary.wiley.com/doi/10.1111/j.1755-3768.2012.F079.x/abstract.

Harris, G. K., Qian, Y., Leonard, S. S., Sbarra, D. C., and Shi, X. 2006. Luteolin and chrysin differentially inhibit cyclooxygenase-2 expression and scavenge reactive oxygen species but similarly inhibit prostaglandin-E2 formation in RAW 264.7 cells. *Journal of Nutrition* 136(6): 1517–1521.

Hausen, B. M., Busker, E., and Carle, R. 1984. The sensitizing capacity of compositae plants: 7. Experimental investigations with extracts and compounds of *Chamomilla recutita* and *Anthemis cotula*. *Planta Medica* 50(3): 229–234.

Heidari, M. R., Dadollahi, Z., Mehrabani, M., Mehrabi, H., Pourzadeh-Hosseini, M., Behravan, E., and Etemad, L. 2009. Study of antiseizure effects of *Matricaria recutita* extract in mice. *Annals of the New York Academy of Sciences* 1171: 300–304.

Hernández-Ceruelos, A., Madrigal-Bujaidar, E., and Cruz, C. de la. 2002. Inhibitory effect of chamomile essential oil on the sister chromatid exchanges induced by daunorubicin and methyl methanesulfonate in mouse bone marrow. *Toxicology Letters* 135(1–2): 103–110.

Hernández-Ceruelos, A., Madrigal-Santillán, E., Morales-González, J. A., Chamorro-Cevallos, G., Cassani-Galindo, M., and Madrigal-Bujaidar, E. 2010. Antigenotoxic effect of *Chamomilla recutita* (L.) Rauschert essential oil in mouse spermatogonial cells, and determination of its antioxidant capacity in vitro. *International Journal of Molecular Sciences* 11(10): 3793–3802.

Hirano, T., Higa, S., Arimitsu, J., Naka, T., Shima, Y., Ohshima, S., Fujimoto, M., Yamadori, T., Kawase, I., and Tanaka, T. 2004. Flavonoids such as luteolin, fisetin and apigenin are inhibitors of interleukin-4 and interleukin-13 production by activated human basophils. *International Archives of Allergy and Immunology* 134: 135–140.

Hoyos, J. M. Á., Alves, E., Rozwalka, L. C., de Souza, E. A., and Zeviani, W. M. 2012. Antifungal activity and ultrastructural alterations in *Pseudocercospora griseola* treated with essential oils. *Ciência e Agrotecnologia* 36(3): 270–284.

Imulan BioTherapeutics. 2008. Immune selective anti-inflammatory derivatives. ImSAIDs Technical Summary. http://www.imulan.com/Resources/ImSAIDS_Tech_Summary.pdf (Accessed January 02, 2013).

Isaac, O. 1979. Pharmacological investigations with compounds of chamomile I. On the pharmacology of (−)-α-bisabolol and bisabolol oxides (Review). *Planta Medica* 35: 118–124.

Jacob, S. E. and Hsu, J. W. 2010. Reactions to Aquaphor®: Is bisabolol the culprit? *Pediatric Dermatology* 27(1): 103–104.

Jakovlev, V., Isaac, O., and Flaskamp, E. 1983. Pharmacological investigations with compounds of chamomile. VI. Investigations on the antiphlogistic effect of chamazulene and matricine. *Planta Medica* 49(10): 67–73.

Jakovlev, V., Isaac, O., Thiemer, K., and Kunde, R. 1979. Pharmacological investigations with compounds of chamomile. II. New investigations on the antiphlogistic effects of (−)-α-bisabolol and bisabololoxides. *Planta Medica* 35(2): 125–140.

Jarrahi, M., Vafaei, A. A., Taherian, A. A., Miladi, H., and Rashidi Pour, A. 2010. Evaluation of topical *Matricaria chamomilla* extract activity on linear incisional wound healing in albino rats. *Natural Product Research* 24(8): 697–702.

Javaheri, M. N., Kesmati, M., and Pilevarian, A. A. 2009. Phytoestrogenic effect of *Matricaria recutita* L. and its relationship with endogenous estrogens on locomotor activity in open field test. *Iranian Journal of Medicinal and Aromatic Plants* 24(4): 519–529.

Johari, H., Sharifi, E., Mardan, M., Kafilzadeh, F., Hemayatkhah, V., Kargar, H., and Nikpoor, N. 2011. The effects of a hydroalcoholic extract of *Matricaria chamomilla* flower on the pituitary-gonadal axis and ovaries of rats. *International Journal of Endocrinology and Metabolism* 9(2): 330–334.

Julius, D. and Basbaum, A. I. 2001. Molecular mechanisms of nociception. *Nature* 413: 203–210.

Kakuta, H., Yano-Kakuta, E., and Moriya, K. 2007. Psychological and physiological effects in humans of eating chamomile jelly. *ISHS Acta Horticulturae* 749: 187–192.

Kang, O., Choi, J., Lee, J., and Kwon, D. 2010. Luteolin isolated from the flowers of *Lonicera japonica* suppresses inflammatory mediator release by blocking NF-κB and MAPKs activation pathways in HMC-1 cells. *Molecules* 15(1): 385–398.

Karbalay-Doust, S. and Noorafshan, A. 2009. Antiulcerogenic effects of *Matricaria chamomilla* extract in experimental gastric ulcer in mice. *Iranian Journal of Medical Sciences* 34(3): 198–203.

Kathleen, A. and Kelly, G. S. 2009. Nutrients and botanicals for stress: Adrenal fatigue, neurotransmitter imbalance and restless sleep. *Alternative Medicine Review* 14(2): 114–140.

Kedzia, B. 1991. Antimicrobial activity of oil of chamomilla and its components. *Herba Polonica* 37: 29–38.

Kempuraj, D., Tagen, M., Iliopoulou, B. P., Clemons, A., Vasiadi, M., Boucher, W., House, M., Wolfberg, A., and Theoharides, T. C. 2008. Luteolin inhibits myelin basic protein-induced human mast cell activation and mast cell-dependent stimulation of Jurkat T cells. *British Journal of Pharmacology* 155(7): 1076–1084.

Kesmati, M., Abbasi, Z. Z., and Fathi M. H. 2008. Study of benzodiazepine like effects of *Matricaria recutita* on morphine withdrawal syndrome in adult male rats. *Pakistan Journal of Medical Sciences* 24(5): 735–739.

Kesmati, M., Moghadam, A. Z., Nia, A. H., and Abasizadeh, Z. 2009. Comparison between *Matricaria recutita* L. aqueous and hydroalcoholic extract on morphine withdrawal signs in the presence and absence of Tamoxifen. *Iranian Journal of Medicinal and Aromatic Plants* 25(2): 170–181.

Kholif, S. M. 2010. Effect of chamomile flower or black seed additives on milk yield and composition of lactating goats. *Egyptian Journal of Nutrition and Feeds* 13(3): 403–414.

Khorshed, M. M., El-Galil, E. R. I. A., and El-Shewy, A. A. 2011. Effect of the natural additives on the productive performance of lactating goats. *Egyptian Journal of Nutrition and Feeds*. 14(2): 183–189.

Kim, S., Jung, E., Kim, J. H., Park, Y. H., Lee, J., and Park, D. 2011. Inhibitory effects of (−)-α-bisabolol on LPS-induced inflammatory response in RAW264.7 macrophages. *Food and Chemical Toxicology* 49(10): 2580–2585.

Kim, Y. C. 2010. Neuroprotective phenolics in medicinal plants. *Archives of Pharmacal Research* 33(10): 1611–1632.

Kim, J., Kim, D., Kang, O., Choi, Y., Park, H., Choi, S., Kim, T., Yun, K., Nah, Y., and Lee, Y. 2005. Inhibitory effect of luteolin on TNF-α-induced IL-8 production in human colon epithelial cells. *International Immunopharmacology* 5(1): 209–217.

Kim, S., Shin, K., Kim, D., Kim, Y., Han, M. S., Lee, T. G., Kim, E., Rvu, S. H., and Suh, P. 2003. Luteolin inhibits the nuclear factor-κB transcriptional activity in Rat-1 fibroblasts. *Biochemical Pharmacology* 66(6): 955–963.

Ko, F. N., Huang, T. F., and Teng, C. M. 1991. Vasodilatory action mechanisms of apigenin isolated from *Apium graveolens* in rat thoracic aorta. *Biochimica et Biophysica Acta* 1115(1): 69–74.

Kobayashi, Y., Suzuki, A., Kobayashi, A., Ayaka Kasai, A., Ogata, Y., Kumada, Y., Takahashi, R., and Ogino, F. 2007. Suppression of sensory irritation by chamomile essential oil and its active component bisabololoxide A. *ISHS Acta Horticulturae* 749: 163–174.

Kogiannou, D. A. A., Kalogeropoulos, N., Kefalas, P., Polissiou, M. G., Kaliora, A. C., Kouretas, D., and Tsatsakis, A. 2013. Herbal infusions, their phenolic profile, antioxidant and anti-inflammatory effects in HT29 and PC3 cells. *Food and Chemical Toxicology* 61: 152–159.

Kotanidou, A., Xagorari, A., Bagli, E., Kitsanta, P., Fotsis, T., Papapetropoulos, A., and Roussos, C. 2002. Luteolin reduces lipopolysaccharide-induced lethal toxicity and expression of proinflammatory molecules in mice. *American Journal of Respiratory and Critical Care Medicine* 165(6): 818–823.

Kroenke, K., Spitzer, R. L., Williams, J. B., Monahan, P. O., and Löwe, B. 2007. Anxiety disorders in primary care: Prevalence, impairment, comorbidity, and detection. *Annals of Internal Medicine* 146(5): 317–325.

Kumar, A. 2012. A review on hepatoprotective herbal drugs. *International Journal of Research in Pharmacy and Chemistry* 2(1): 92–102.

Kumar, J. S., Rajvaidya, S., Singh, G. K., Jain, P., and Nagori, B. P. 2012. Investigation of herbal extract as hepatoprotective. *Research Journal of Pharmaceutical Sciences* 1(3): 16–18.

Langhorst, J., Varnhagen, I., Schneider, S. B., Albrecht, U., Rueffer, A., Stange, R., Michalsen, A., and Dobos, G. J. 2013. Randomised clinical trial: A herbal preparation of myrrh, chamomile and coffee charcoal compared with mesalazine in maintaining remission in ulcerative colitis—a double-blind, double-dummy study. *Alimentary Pharmacology & Therapeutics* 38(5): 490–500.

Laskova, I. L. and Uteshev, B. S. 1992. Immunomodulating action of heteropolysaccharides isolated from chamomile flower clusters. *Antibiotiki i Khimioterapia* 37(6): 15–18.

Le Bars, D., Gozariu, M., and Cadden, S. W. 2001. Animal models of nociception. *Pharmacological Reviews* 53(4): 597–652.

Lebish, I. J. and Moraski, R. M. 1987. Mechanisms of immunomodulation by drugs. *Toxicologic Pathology* 15(3): 338–345.

Lee, J. H., Zhou, H. Y., Cho, S. Y., Kim, Y. S., Lee, Y. S., and Jeong, C. S. 2007. Anti-inflammatory mechanisms of apigenin: Inhibition of cyclooxygenase-2 expression, adhesion of monocytes to human umbilical vein endothelial cells, and expression of cellular adhesion molecules. *Archives of Pharmacal Research* 30(10): 1318–1327.

Lee, S., Heo, Y., and Kim, Y. 2010. Effect of German chamomile oil application on alleviating atopic dermatitis-like immune alterations in mice. *Journal of Veterinary Science* 11(1): 35–41.

Lee, Y., Lee, C. H., and Oh, U. 2005. Painful channels in sensory neurons. *Molecules and Cells* 20(3): 315–324.

Lelevich, S. V., Lelevich, V. V., and Novokshonov, A. A. 2009. Neurotransmitter mechanisms of morphine withdrawal syndrome. *Bulletin of Experimental Biology and Medicine* 148(2): 184–187.

Lindahl, M. and Tagesson, C. 1993. Selective inhibition of group II phospholipase A_2 by quercetin. *Inflammation* 17(5): 573–582.

Lins, R., Vasconcelos, F. H. P., Leite, R. B., Coelho-Soares, R. S., and Barbosa, D. N. 2013. Clinical evaluation of mouthwash with extracts from aroeira (*Schinus terebinthifolius*) and chamomile (*Matricaria recutita* L.) on plaque and gingivitis. *Revista Brasileira de Plantas Medicinais* 15(1): 112–120.

Lopez-Lazaro, M. 2009. Distribution and biological activities of the flavonoid luteolin. *Mini Reviews in Medicinal Chemistry* 9(1): 31–59.

Lorenzo, P. S., Rubio, M. C., Medina, J. H., and Adler-Graschinsky, E. 1996. Involvement of monoamine oxidase and noradrenaline uptake in the positive chronotropic effects of apigenin in rat atria. *European Journal of Pharmacology* 312(2): 203–207.

Luqman, S., Kaushik, S., Srivastava, S., Kumar, R., Bawankule, D. U., Pal, A., Darokar, M. P., and Khanuja, S. P. S. 2009. Protective effect of medicinal plant extracts on biomarkers of oxidative stress in erythrocytes. *Pharmaceutical Biology* 47(6): 483–490.

Lydiard, R. B. 2003. The role of GABA in anxiety disorders. *Journal of Clinical Psychiatry* 64(Suppl 3): 21–27.
Mai, E. and Buysse, D. J. 2008. Insomnia: Prevalence, impact, pathogenesis, differential diagnosis, and evaluation. *Sleep Medicine Clinics* 3(2): 167–174.
Mamani-Matsuda, M., Kauss, T., Al-Kharrat, A., Rambert, J., Fawaz, F., Thiolat, D., Moynet, D., Coves, S., Malvy, D., and Mossalayi, M. D. 2006. Therapeutic and preventive properties of quercetin in experimental arthritis correlate with decreased macrophage inflammatory mediators. *Biochemical Pharmacology* 72(10): 1304–1310.
Man, M. Q., Hupe, M., Sun, R., Man, G., Mauro, T. M., and Elias, P. M. 2012. Topical apigenin alleviates cutaneous inflammation in murine models. *Evidence-Based Complementary and Alternative Medicine*, Article ID 912028, 7 pages. http://www.hindawi.com/journals/ecam/2012/912028/ (Accessed November 20).
Matić, I. Z., Juranić, Z., Šavikin, K., Zdunić, G., Nađvinski, N., and Gođevac, D. 2013. Chamomile and marigold tea: Chemical characterization and evaluation of anticancer activity. *Phytotherapy Research* 27(6): 852–858.
Mazokopakis, E. E., Vrentzos, G. E., Papadakis, J. A., Babalis, D. E., and Ganotakis, E. S. 2005. Wild chamomile (*Matricaria recutita* L.) mouthwashes in methotrexate-induced oral mucositis. *Phytomedicine* 12(1–2): 25–27.
Mazokopakis, E. E. and Ganotakis, E. S. 2007. The efficacy of wild chamomile (*Matricaria recutita* L.) as anti-inflammatory and mucosa protective agent. In: J. N. Govil, V. K. Singh, and R. Bhardwaj (eds.) *Phytomedicines*. Houston, TX: Studium Press LLC, pp. 1–8.
Merfort, I., Heilmann, I., Hagedorn-Leeveke, U., and Lippold, B. C. 1994. In vivo skin penetration studies of chamomile flavones. *Pharmazie* 49: 509–511.
Michael, J. B., Fredi, K., Andreana, L. O., Marian, R., Adriane, F. B., O'Connor, B., Maria, R., Patricia, L., and Daniel, A. 2000. Medicinal plants used by Latino healers for women's health conditions in New York City. *Economic Botany* 54(3): 344–357.
Middleton, E., Kandaswami, C., and Theoharides T. C. 2000. The effects of plant flavonoids on mammalian cells: Implications for inflammation, heart disease, and cancer. *Pharmacological Reviews* 52(4): 673–751.
Miller, T., Wittstock, U., Lindequist, U., and Teuscher, E. 1996. Effects of some components of the essential oil of chamomile, *Chamomilla recutita*, on histamine release from rat mast cells. *Planta Medica* 62: 60–61.
Min, Y. D., Choi, C. H., Bark, H., Son, H. Y., Park, H. H., Lee, S., Park, J. W., Park, E. K., Shin, H. I., and Kim, S. H. 2007. Quercetin inhibits expression of inflammatory cytokines through attenuation of NF-κB and p38 MAPK in HMC-1 human mast cell line. *Inflammation Research* 56(5): 210–215.
Min-zhu, C., Wen-zhen, J., Li-min, D., and Shu-yun, X. 1986. Effects of luteolin on inflammation and immune function. *Chinese Journal of Pharmacology and Toxicology* (01). http://en.cnki.com.cn/Article_en/CJFDTOTAL-YLBS198601010.htm (Accessed November 19, 2012).
Morales-Yuste, M., Morillas-Márquez, F., Martín-Sánchez, J., Valero-López, A., and Navarro-Moll, M. C. 2010. Activity of (−)-alpha-bisabolol against *Leishmania infantum* promastigotes. *Phytomedicine* 17(3–4): 279–281.
Morikawa, K., Nonaka, M., Narahara, M., Torii, I., Kawaguchi, K., Yoshikawa, T., Kumazawa, Y., and Morikawa, S. 2003. Inhibitory effect of quercetin on carrageenan-induced inflammation in rats. *Life Sciences* 74(6): 709–721.
Moses, J. M., Ebert, R. H., Graham, R. C., and Brine, K. L. 1964. Pathogenesis of inflammation. I. The production of an inflammatory substance from rabbit granulocytes in vitro and its relationship to leucocyte pyrogen. *Journal of Experimental Medicine* 120(1): 57–82.
Muñoz-Velázquez, E. E., Rivas-Díaz, K., Loarca-Piña, M. G. F., Mendoza-Diaz, S., Reynoso-Camacho, R., and Ramos-Gómez, M. 2012. Comparison of phenolic content, antioxidant capacity and anti-inflammatory activity of commercial herbal infusions. *Revista Mexicana de Ciencias Agrícolas* 3(3): 481–495.

National Cancer Institute. 2013. What is cancer? http://www.cancer.gov/cancertopics/cancerlibrary/what-is-cancer (Accessed January 04, 2013).
National Collaborating Centre for Primary Care (UK). 2007. Post Myocardial Infarction: Secondary Prevention in Primary and Secondary Care for Patients Following a Myocardial Infarction [Internet]. London: Royal College of General Practitioners (UK); (NICE Clinical Guidelines, No. 48.) http://www.ncbi.nlm.nih.gov/books/NBK49343/(Accessed January 04, 2013).
Niederhofer, H. 2009. Observational study: *Matricaria chamomilla* may improve some symptoms of attention-deficit hyperactivity disorder. *Phytomedicine* 16(4): 284–286.
Nimbekar, T., Wanjari, B., and Bais, Y. 2012. Herbosomes—herbal medicinal system for the management of periodontal disease. *International Journal of Biomedical and Advance Research* 3(6): 468–472.
Nouri, M. H. K. and Abad, A. N. A. 2012. Antinociceptive effect of *Matricaria chamomilla* on vincristine-induced peripheral neuropathy in mice. *Journal of Pharmacy and Pharmacology* 6(1): 24–29.
Nwoye, L. O. 2013. Protective and therapeutic effects of *Chamomilla recutita* extract on sub-acute ethanol intoxication in white albino rats. *African Journal of Biotechnology* 12(18): 2378–2385.
Leite, G. de O., Leite, L. H., Sampaio, R. de S., Araruna, M. K., de Menezes, I. R., da Costa, J. G., and Campos, A. R. 2011. (−)-α-Bisabolol attenuates visceral nociception and inflammation in mice. *Fitoterapia* 82(2): 208–211.
O'Connor, T. M., O'Connell, J., O'Brien, D. I., Goode, T., Bredin, C. P., and Shanahan, F. 2004. The role of substance P in inflammatory disease. *Journal of Cell Physiology* 201(2): 167–180.
O'Leary, K. A., de Pascual-Tereasa, S., Needs, P. W., Bao, Y. P., O'Brien, N. M., and Williamson, G. 2004. Effect of flavonoids and vitamin E on cyclooxygenase-2 (COX-2) transcription. *Mutation Research* 551(1–2): 245–254.
Ogata, I., Kawanai, T., Hashimoto, E., Nishimura, Y., Oyama, Y., and Seo, H. 2010. Bisabololoxide A, one of the main constituents in German chamomile extract, induces apoptosis in rat thymocytes. *Archives of Toxicology* 84(1): 45–52.
Ogata-Ikeda, I., Seo, H., Kawanai, T., Hashimoto, E., and Oyama, Y. 2011. Cytotoxic action of bisabololoxide A of German chamomile on human leukemia K562 cells in combination with 5-fluorouracil. *Phytomedicine* 18(5): 362–365.
Ogata, J., Minami, K., Horishita, T., Shiraishi, M., Okamoto, T., Terada, T., and Sata, T. 2005. Gargling with sodium azulene sulfonate reduces the postoperative sore throat after intubation of the trachea. *Anesthesia and Analgesia* 101: 290–293.
Padín, S. B., Fusé, C., Urrutia, M. I., and Dal Bello, G. M. 2013. Toxicity and repellency of nine medicinal plants against *Tribolium castaneum* in stored wheat. *Bulletin of Insectology* 66(1): 45–49.
Pandey, M. K., Singh, A. K., and Singh, R. B. 2002. Mycotoxic potential of some higher plants. *Plant Disease Research* 17(1): 51–56.
Patel, N. B. 2010. Physiology of pain. In: A. Kopf and N. B. Patel (eds.) *Guide to Pain Management in Low-Resource Settings*. Seattle, WA: IASP Press, pp. 13–17.
Patisaul, H. B. and Jefferson, W. 2010. The pros and cons of phytoestrogens. *Front Neuroendocrinol* 31(4): 400–419.
Paulsen, E., Otkjaer, A., and Andersen, K. E. 2010. The coumarin herniarin as a sensitizer in German chamomile [*Chamomilla recutita* (L.) Rauschert, Compositae]. *Contact Dermatitis* 62(6): 338–342.
Paulsen, E., Christensen, L., and Andersen, K. E. 2008. Cosmetics and herbal remedies with compositae plant extracts—are they tolerated by compositae-allergic patients. *Contact Dermatitis* 58(1): 15–23.

Pearlman, D. S. 1999. Pathophysiology of the inflammatory response. *Journal of Allergy and Clinical Immunology* 104(4): S132–S136.
Pereira, N. P., Cunico, M. M., Miguel, O. G., and Miguel, M. D. 2008. Promising new oil derived from the seeds of *Chamomilla recutita* (L.) Rauschert produced in Southern Brazil. *Journal of the American Oil Chemists' Society* 85: 493–494.
Pizarro Espín, A., Valido Díaz, A., Santiesteban Muñoz, D., Valdés Álvarez, M., and Mena Linares, Y. 2012. Evaluation of gastroprotective activity of *Matricaria recutita* in Sprague Dawley. *Revista Electronica de Veterinaria* 13(8): 081206.
Poráčová, J., Blasčákova, M., Zahatňanská, M., Taylorová, B., Sutiaková, I., and Sály, J. 2007a. Effect of chamomile essential oil application on the weight of eggs in laying hens Hisex Braun. *ISHS Acta Horticulturae* 749: 203–206.
Poráčová, J., Zahatňanská, M., Blasčákova, M., Taylorová, B., Sutiaková, I., Tanishima, K., Takabayashi, H., and Cheng, B. J. 2007b. Effect of chamomile essential oil on eggs production and weight of laying hens Hisex Braun. *ISHS Acta Horticulturae* 749: 207–210.
Pourabbas, R., Delazar, A., and Chitsaz, M. T. 2005. The effect of German chamomile mouthwash on dental plaque and gingival inflammation. *Iranian Journal of Pharmaceutical Research* 2: 105–109.
Presibella, M. M., Santos, C. A. M., and Weffort-Santos, A. M. 2007. In vitro antichemotactic activity of *Chamomilla recutita* hydroethanol extract. *Pharmaceutical Biology* 45(2): 124–130.
Presibella, M. M., Villas-Bôas, L. De B., da Silva Belletti, K. M., de Moraes Santos, C. A., and Weffort-Santos, A. M. 2006. Comparison of chemical constituents of *Chamomilla recutita* (L.) Rauschert essential oil and its anti-chemotactic activity. *Brazilian Archives of Biology and Technology* 49(5): 717–724.
Raal, A., Volmer, D., Sõukand, R., Hratkevitš, S., and Kalle, R. 2013. Complementary treatment of the common cold and flu with medicinal plants—results from two samples of pharmacy customers in Estonia. *PLoS ONE* 8(3): e58642.
Radwan, M. A., Abu-Elamayem, M. M., Farrag, S. A. A., and Ahmed, N. S. 2011. Integrated management of *Meloidogyne incognita* infecting tomato using bio-agents mixed with either oxamyl or organic amendments. *Nematologia Mediterranea* 39(2): 151–156.
Radwan, M. A., Farrag, S. A. A., Abu-Elmayen, M. M., and Ahmed, N. S. 2012. Efficacy of dried seed powder of some plant species as soil amendment against *Meloidogyne incognita* (Tylenchida: Meloidogynidae) on tomato. *Archives of Phytopathology and Plant Protection* 45(10): 1246–1251.
Ramadan, M., Goeters, S., Watzer, B., Krause, E., Lohmann, K., Bauer, R., Hempel, B., and Imming, P. 2006. Chamazulene carboxylic acid and matricin: A natural profen and its natural prodrug, identified through similarity to synthetic drug substances. *Journal of Natural Products* 69(7): 1041–1045.
Ranpariya, V. L., Parmar, S. K., Sheth, N. R., Chandrashekhar, V. M. 2011. Neuroprotective activity of *Matricaria recutita* against fluoride-induced stress in rats. *Pharmaceutical Biology* 49(7): 696–701.
Reider, N., Sepp, N., Fritsch, P., Weinlich, G., and Jensen-Jarolim, E. 2000. Anaphylaxis to camomile: Clinical features and allergen cross-reactivity. *Clinical & Experimental Allergy* 30(10): 1436–1443.
Reis, L. S., Pardo, P. E., Oba, E., Kronka, S. N., and Frazzatti-Gallina, N. M. 2006. *Matricaria chamomilla* CH_{12} decreases handling stress in Nelore calves. *Journal of Veterinary Science* 7(2): 189–192.
Rekka, E. A., Kourounakis, A. P., and Kourounakis, P. N. 1996. Investigation of the effect of chamazulene on lipid peroxidation and free radical processes. *Research Communications in Molecular Pathology and Pharmacology* 92(3): 361–364.
Rezaie, A., Mohajeri, D., Zarkhah, A., and Nazeri, M. 2012. Comparative assessment of *Matricaria chamomilla* and zinc oxide on healing of experimental skin wounds on rats. *Annals of Biological Research* 3(1): 550–560.

Roberts, A. and Williams, J. M. G. 1992. The effect of olfactory stimulation on fluency, vividness of imagery and associated mood: A preliminary study. *British Journal of Medical Psychology* 65: 197–199.
Roby, M. H. H., Sarhan, M. A., Selim, K. A. H., and Khalel, K. I. 2013. Antioxidant and antimicrobial activities of essential oil and extracts of fennel (*Foeniculum vulgare* L.) and chamomile (*Matricaria chamomilla* L.). *Industrial Crops and Products* 44: 437–445.
Rocha, M. N. F., Venâncio, E. T., Moura, B. A., Gomes, S., Maria, I., Aquino, N., Manoel R. et al. 2010. Gastroprotection of (−)-α-bisabolol on acute gastric mucosal lesions in mice: The possible involved pharmacological mechanisms. *Fundamental & Clinical Pharmacology* 24(1): 63–71.
Rocha, N. F., Rios, E. R., Carvalho, A. M., Cerqueira, G. S., Lopes, A. A., Leal, L. K., Dias, M. L., de Sousa, D. P., and de Sousa, F. C. 2011. Anti-nociceptive and anti-inflammatory activities of (−)-α-bisabolol in rodents. *Naunyn-Schmiedeberg's Archives of Pharmacology* 384(6): 525–533.
Roder, E. 1982. Secondary effects of medicinal herbs. *Deutsche Apotheker Zeitung* 122(41): 2081–2092.
Rogerio, A. P., Kanashiro, A., Fontanari, C., da Silva, E. V. G., Lucisano-Valim, Y. M., Soares, E. G., and Faccioli, L. H. 2007. Anti-inflammatory activity of quercetin and isoquercitrin in experimental murine allergic asthma. *Inflammation Research* 56(10): 402–408.
Romero, M. del C., Valero, A., Martín-Sánchez, J., and Navarro-Moll, M. C. 2012. Activity of *Matricaria chamomilla* essential oil against anisakiasis. *Phytomedicine* 19(6): 520–523.
Rotelli, A. E., Guardia, T., Juárez, A. O., la Rocha, N. E., and Pelzer, L. E. 2003. Comparative study of flavonoids in experimental models of inflammation. *Pharmacological Research* 48(6): 601–606.
Russell, K. and Jacob, S. E. 2010. Bisabolol. *Dermatitis* 21(1): 57–58.
Safahyi, H., Sabieraj, J., Sailer, E. R., and Ammon H. P. T. 1994. Chamazulene: An antioxidant type inhibitor of leukotriene B4 formation. *Planta Medica* 60(5): 410–413.
Safari, E. 2010. Prevention effect of *Matricaria recutita* L. on withdrawal syndrome in morphine dependent female rats. *Planta Medica* 76: P600.
Safayhi, H., Sabieraj, J., Sailer, E. R., and Ammon, H. P. T. 1993. Chamazulene: An antioxidant-type inhibitor of leukotriene B4 formation *Planta Medica* 60(5): 410–413.
Saklatvala, J. 2002. Glucocorticoids: Do we know how they work? *Arthritis Research* 4(3): 146–150.
Salama, R. H. 2012. *Matricaria chamomilla* attenuates cisplatin nephrotoxicity. *Saudi Journal of Kidney Diseases and Transplantation* 23(4): 765–772.
Saleh, H. M., Saleh, S. A., and El-Bordeny, N. 2009. Effect of adding a mixture of ground medicinal plants to the diets of suckling lambs on their performance. *Egyptian Journal of Nutrition and Feeds* 12(2): 269–277.
Savino, F., Capasso, R., Palumeri, E., Tarasco, V., Locatelli, E., and Capasso, F. 2008. Advances on the effects of the compounds of a phytotherapic agent (ColiMil®) on upper gastrointestinal transit in mice. *Minerva Pediatrica* 60(3): 285–290.
Savino, F., Cresi, F., Castagno, E., Silvestro, L., and Oggero, R. 2005. A randomized double-blind placebo-controlled trial of a standardized extract of *Matricariae recutita*, *Foeniculum vulgare* and *Melissa officinalis* (ColiMil®) in the treatment of breastfed colicky infants. *Phytotherapy Research* 19(4): 335–340.
Sawatzky, D. A., Derek, A. W., Coville-Nash, P. R., and Rossi, A. G. 2006. The involvement of the apoptosis-modulating proteins ERK ½, Bcl-x$_L$ and bax in the resolution of acute inflammation *in vivo*. *American Journal of Pathology* 168(1): 33–41.
Schepetkin, I. A. and Quinn, M. T. 2006. Botanical polysaccharides: Macrophage immunomodulation and therapeutic potential. *International Immunopharmacology* 6(3): 317–333.

Schmidt, P. C. and Vogel, K. 1992. Chamomile—Evaluation of the stabilities of chamomile preparation. *Deutsche Apotheker Zeitung* 132: 462–468.

Schmidt, K. 2003. Complementary and alternative medicine for atopic dermatitis. *Focus on Alternative and Complementary Therapies* 8(2): 173–177.

Schnitzler, A. C., Nolan, L. L., and Labbe, R. 1996. Screening of medicinal plants for anti-leishmanial and antimicrobial activity. *ISHS Acta Horticulturae* 426: 235–242.

Schulz, V. 2009. Chamomile (*Matricaria recutita* L.) and the risk of drug interaction. *Zeitschrift für Phytotherapie* 30(4): 162–168.

Segal, R. and Pilote, L. 2006. Warfarin interaction with chamomilla. *Canadian Medical Association Journal* 174(9): 1281–1282.

Sekiya, Y., Nakagawa, T., Ozawa, T., Minami, M., and Satoh, M. 2004. Facilitation of morphine withdrawal symptoms and morphine-induced conditioned place preference by a glutamate transporter inhibitor DL-threo-β-benzyloxyaspartate in rats. *European Journal of Pharmacology* 485: 201–210.

Selzer, A. 1992. Understanding Heart Disease. University of California Press. UC Press E-Books Collection, 1982–2004. California Digital Library, http://publishing.cdlib.org/ucpressebooks/view?docId=ft9w1009p7&chunk.id=d0e3154&toc.id=d0e3154&brand=ucpress (Accessed January 06, 2013).

Sertoli, A., Francalani, S., Giorgini, S., Brusi, C., and Acciai, M. C. 1994. Prevention of allergic contact dermatitis with alternative products. *Contact Dermatitis* 31(5): 322–321.

Sharma, J. N. and Mohsin, S. S. J. 1990. The role of chemical mediators in the pathogenesis of inflammation with emphasis on the kinin system. *Experimental Pathology* 38(2): 73–96.

Shimelis, N. D., Asticcioli, S., Baraldo, M., Tirillini, B., Lulekal, E., and Murgia, V. 2012. Researching accessible and affordable treatment for common dermatological problems in developing countries. An Ethiopian experience. *International Journal of Dermatology* 51(7): 790–795.

Shimoi, K., Saka, N., Kaji, K., Nozawa, R., and Kinae, N. 2000. Metabolic fate of luteolin and its functional activity at focal site. *BioFactors* 12(1–4): 181–186.

Shinomiya, K. A., Inoue, T. B., Utsu, Y. A., Tokunaga, S. A., Masuoka, T. A., Ohmori, A. A., and Kamei, C. 2005. Hypnotic activities of chamomile and passiflora extracts in sleep-disturbed rats. *Biological & Pharmaceutical Bulletin* 28(5): 808–810

Silvia, A., Rita, Y. C., and María, I. C. 2007. First comprehensive contribution to medical ethnobotany of Western Pyrenees. *Journal of Ethnobiology and Ethnomedicine* 3: 26. doi: 10.1186/1746-4269-3-26.

Simonová, M., Strompfová, V., Marciňáková, M., Haviarová, M., Faix, S., Lauková, A., Vasilková, Z., and Salamon, I. 2007. Chamomile essential oil and its experimental application in rabbits. *ISHS Acta Horticulturae* 749: 197–201.

Smitherman, L. C., Janisse, J., and Mathur, A. 2005. The use of folk remedies among children in an urban black community: Remedies for fever, colic, and teething. *Pediatrics* 115: e297–e304.

Snežana, J., Zorica, P., Marina, M. J., Lola, D., Miroslava, M., Branko, K., Miroslava, M., and Pavle, P. 2007. An ethnobotanical study on the usage of wild medicinal herbs from Kopaonik Mountain (Central Serbia). *Journal of Ethnopharmacology* 11(1): 160–175.

Sofiabadi, M., Esmaeili, M. H., Yazdy, H. H., and Zarmehri, H. A. 2012. The effect of intra-peritoneally injection of *Matricaria chamomilla* ethanolic extract on seizure. *Journal of Medicinal Plants* 11(44): Pe86–Pe92.

Srivastava, J. K. and Gupta, S. 2009. Extraction, characterization, stability and biological activity of flavonoids isolated from chamomile flowers. *Molecular and Cellular Pharmacology* 1(3): 138.

Stingeni, L., Agea, E., Lisi, P., and Spinozzi, F. 1999. T-lymphocyte cytokine profiles in Compositae airborne dermatitis. *British Journal of Dermatology*, 141: 689–693.

Strober, W. and Fuss, I. J. 2011. Proinflammatory cytokines in the pathogenesis of inflammatory bowel diseases. *Gastroenterology* 140(6): 1756–1767.
Subiza, J., Subiza, J. L., Alonso, M., Hinojosa, M., Garcia, R., Jerez, M., and Subiza, E. 1990. Allergic conjunctivitis to chamomile tea. *Annals of Allergy* 65(2): 127–132.
Subiza, J., Subiza, J. L, Hinojosa, M., Garcia, R., Jerez, M., Valdivieso, R., and Subiza, E. 1989. Anaphylactic reaction after the ingestion of chamomile tea: A study of cross-reactivity with other composite pollens. *Journal of Allergy and Clinical Immunology* 84(3): 353–358.
Suganda, A. G., Amoros, M., Girre, L., and Fauconner, B. 1983. Inhibitory effects of some crude and semi-purified extracts of native French plants on the multiplication of human herpes virus I and human polio virus 2 in cell culture. *Journal of Nutrition Products* 46: 626–632.
Szelenyi, I., Isaac, O., and Thiemer, K. 1979. Pharmacological experiments with compounds of chamomile. III. Experimental studies on the ulcer protective effect of chamomile. *Planta Medica* 35: 218–227.
Taliou, A., Zintzaras, E., Lykouras, L., and Francis, K. 2013. An open-label pilot study of a formulation containing the anti-inflammatory flavonoid luteolin and its effects on behavior in children with autism spectrum disorders. *Clinical Therapeutics* 35(5): 592–602.
Tolouee, M., Alinezhad, S., Saberi, R., Eslamifar, A., Zad, S. J., Jaimand, K., Taeb, J. et al. 2010. Effect of *Matricaria chamomilla* L. flower essential oil on the growth and ultrastructure of *Aspergillus niger* van Tieghem. *International Journal of Food Microbiology* 139(3): 127–133.
Tubaro, A., Zilli, C., Redaelli, C., and Della Loggia, R. 1984. Evaluation of the anti-inflammatory activity of a chamomile extract after topical application. *Planta Medica* 4(3): 59.
Tudor, F. and Georgescu, M. F. 2011. Ethnobotanical Studies in Dobrogea. Scientific Papers, UASVM Bucharest, Series A, Vol. LIV, ISSN 1222-5339. http://www.agro-bucuresti.ro/fisiere/file/Cercetare/LS_2011/75_%20Ethnobotanical%20studies%20in%20Dobrogea.pdf (Accessed January 10, 2013).
Türi, E., Türi, M., Annuk, H., and Arak, E. 1999. Action of aqueous extracts of bearberry and cowberry leaves and wild camomile and pineapple-weed flowers on *Escherichia coli* surface structures. *Pharmaceutical Biology* 37(2): 127–133.
Tzianabos, A. O. 2000. Polysaccharide immunomodulators as therapeutic agents: Structural aspects and biologic function. *Clinical Microbiology Reviews* 13(4): 523–533.
Umezu, T. 2013. Evaluation of central nervous system acting effects of plant-derived essential oils using ambulatory activity in mice. *Pharmacology and Pharmacy* 4(2): 160–170.
UMMS. 2012. Insomnia—Medications. University of Maryland Medical Center. http://www.umm.edu/patiented/articles/what_drug_treatments_insomnia_000027_8.htm?iframe=true&width=100%25&height=100%25/#ixzz2FwbzmQvj (Accessed January 12, 2013).
Varoni, E.M., Lodi, G., Sardella, A., Carrassi, A., Iriti, M. 2012. Plant polyphenols and oral health: old phytochemicals for new fields. *Current Medicinal Chemistry*, 19(11):1706–20.
Viola, H., Wasowski, G., Levi de Stein, M., Wolfman, C., Silveira, R., Dajas, F., Medina, J. H., and Paladini A. C. 1995. Apigenin, a component of *Matricaria recutita* flowers, is a central benzodiazepine receptors-ligand with anxiolytic effects. *Planta Medica* 61: 213–216.
Wen-hui, F. and Mao, H. 2006. Effects of luteolin on airway remodeling in asthmatic mice. *Anhui Medical and Pharmaceutical Journal* (09). http://en.cnki.com.cn/Article_en/CJFDTOTAL-AHYY200609002.htm (Accessed January 12, 2013).
Weng, Z., Zhang, B., Asadi, S., Sismanopoulos, N., Butcher, A., Fu, X., Katsarou-Katsari, A., and Antoniou, C. 2012. Quercetin is more effective than cromolyn in blocking human mast cell cytokine release and inhibits contact dermatitis and photosensitivity in humans. *PLoS ONE* 7(3): e33805. doi: 10.1371/journal.pone.0033805.

WHO. 2013. Leishmaniasis. http://www.who.int/leishmaniasis/en/ (Accessed August 19, 2013).

Wilens, T. E. 2006. Mechanism of action of agents used in attention deficit hyperactivity disorder. *Journal of Clinical Psychiatry* 67(Suppl 8): 32–37.

Wood, J. N. 2004. Recent advances in understanding molecular mechanisms of primary afferent activation. *Gut* 53(Suppl II): ii9–ii12.

Xagorari, A., Papapetropoulos, A., Mauromatis, A., Economou, M., Fotsis, T., and Roussos, C. 2001. Luteolin inhibits an endotoxin-stimulated phosphorylation cascade and pro-inflammatory cytokine production in macrophages. *Journal of Pharmacology and Experimental Therapeutics* 296(1): 181–187.

Yamada, K., Miura, T., Mimaki, Y., and Sashida, Y. 1996. Effect of inhalation of chamomile oil vapour on plasma ACTH level in ovariectomized rat under restriction stress. *Biological & Pharmaceutical Bulletin* 19(9): 1244–1246.

Ying, B., Yang, T., Song, X., Hu, X., Fan, H., Lu, X., Chen, L. et al. 2009. Quercetin inhibits IL-1 beta-induced ICAM-1 expression in pulmonary epithelial cell line A549 through the MAPK pathways. *Molecular Biology Reports* 36: 1825–1832.

Zaidi, S. F., Muhammad, J. S., Saeeda, S., Khan, U., Anwarul-Hassan, G., Wasim, J., and Sugiyama, T. 2012. Anti-inflammatory and cytoprotective effects of selected Pakistani medicinal plants in *Helicobacter pylori*-infected gastric epithelial cells. *Journal of Ethnopharmacology* 141(1): 403–410.

Zick, S. M., Wright, B. D., Sen, A., and Arnedt, J. T. 2011. Preliminary examination of the efficacy and safety of a standardized chamomile extract for chronic primary insomnia: A randomized placebo-controlled pilot study. *Complementary and Alternative Medicine* 11(78). doi: 10.1186/1472-6882-11-78.

Zu, Y., Yu, H., Liang, L., Fu, Y., Efferth, T., Liu, X., and Wu, N. 2010. Activities of ten essential oils towards *Propionibacterium acnes* and PC-3, A-549 and MCF-7 Cancer Cells. *Molecules* 15: 3200–3210.

4 Chamomile Oil and Extract

Localization, Chemical Composition, Extraction, and Identification

4.1 INTRODUCTION

The various kinds of medicinal preparations, nutraceuticals, cosmeceuticals, and other health-care products, which are based on chamomile, use the essential oil or extract of chamomile. The essential oil is obtained from the flowers through hydrodistillation while the extract through the use of solvents such as ethanol. In addition, many of the formulations and preparations use individual chemicals fractionated from either the essential oil or the extract. For this, the extract used is prepared from the flowers, from the whole plant, or from the roots, which gives a higher yield of the compounds.

The essential oil is chiefly extracted from the flowers, but it is also found in the stems, leaves, and roots. The oil content varies in these different organs of the plant. The flowers contain more amount of oil as compared to the herb or the root. The oil content also varies from one place to another. Several studies have been carried out on the essential oil collected from plants growing in different locations in a country, and it has been found that the essential oil content (Franz et al. 1978a, 1986) and composition (Hoelzl and Demuth 1975) indeed vary in the plants growing in different locations. Further, the oil content varies depending on the flowering stage when it was harvested, with a fully opened flower containing more oil (Franz et al. 1978b; Franz 1980). Also the oil content may vary depending on the nutrition and irrigation regime (Franz 1983). These aspects are discussed in detail in this chapter.

Chamomile is known to contain more than 120 compounds in the essential oil and extract (Salamon et al. 2010). These compounds belong to the various groups of terpenes, phenols, flavonoids, flavones, coumarins, and polysaccharides. The major compounds used in the health-care preparations, cosmetics, and foodstuffs are chamazulene, α-bisabolol, bisabolol oxide A, en-yn-dicycloethers, apigenin, quercetin, and luteolin. Some of these compounds are lipophilic and some are hydrophilic in nature. The terpenes and coumarins are lipophilic whereas the flavonoids and phenols are hydrophilic. Many of these compounds are volatile in nature and many others decompose easily in extreme conditions.

4.2 LOCALIZATION AND CONTENT OF THE CHAMOMILE ESSENTIAL OIL

The essential oil is synthesized and accumulated predominantly from the flowers. The glandular trichomes present on tubular florets of the capitula have been shown to contain essential oil. When the florets are exposed to chlorine or bromine (10% chloroform solution), appearance of blue color in the trichomes confirms the presence of chamazulene, which is a characteristic constituent of the essential oil of chamomile (Vaverkova and Herichova 1980). The essential oil is also present in the herb (leaves and stem) and the root. Anatomical investigations had revealed that the oil is present in the modified parenchyma: oil cells, idioblasts, and schizogenous oil passages (Jackson and Snowdon 1968). Glandular hairs have been found in the stems, which secrete essential oil. In addition, schizogenous ducts are also found in the stem and roots, which secrete essential oil (Andreucci et al. 2008).

The essential oil has also been found in callus cultures. The oil has been distilled from callus (Magiatis et al. 2001) and from genetically transformed hairy roots (Szöke et al. 2004a; Szöke et al. 2004c) and the oil content and chemical composition determined.

4.2.1 Essential Oil Content in Flowers

The essential oil content of the flowers has been reported to be 0.46%–0.67% by Debska in 1958 (Lawrence 1987). Piccaglia et al. (1993) reported the oil content of seven chamomile varieties grown in Ozzano (Bologna). These varieties were sourced from Egypt, Argentina, Hungary, Germany, Holland, Czechoslovakia, and Yugoslavia. The oil content in these varieties was reported as follows:

- Egypt—0.22%
- Argentina—0.36%
- Hungary—0.31%
- Germany—0.22%
- Holland—0.22%
- Czechoslovakia—0.29%
- Yugoslavia—0.39%

In 1994, Salamon carried out a breeding experiment on the diploid Bona and tetraploid Goral varieties of chamomile. He reported that the essential oil content of Bona was 0.93% and of Goral was 1.3% (Salamon 1994). The ploidy levels determine the oil content in the plant, with the tetraploids yielding more than the diploids. Das (1999) also found that the essential oil content varied widely among different genotypes in India. In the 1995–1996 field experiment, she found that the oil content of 45 chamomile selections grown in Lucknow, India, ranged from 0.15% to 0.2%. In the next year, after selecting 26 genotypes and estimating the oil content in them, she found the oil content to range from 0.42% to 0.76%. The effect of the increase in oil content in the selections in the second year should be attributed to selection for high-yielding genotypes. The genetic aspects of the essential oil content will be discussed in detail in Chapter 5.

4.2.1.1 Variation in Essential Oil Content in Different Locations

The oil content varies with place or location. If the same variety is grown under different climatic conditions, the yield or content of the essential oil gets affected. Many researchers, such as Bazek and Starey, Juracec, Tyihak et al., von Schanz and Salonen, and Peneva, have reported that the essential oil content in the flowers varies depending on the place of origin and cultivars or varieties (Lawrence 1987). Franz et al. (1978a) reported a difference in the essential oil contents obtained from 14 chamomile varieties cultivated in two different regions of Germany. Gašić et al. (1983) reported that in the flowers of the chamomile plants taken from different localities of Yugoslavia, the essential oil content ranged from 0.12% to 0.64%. Franz et al. (1986) reported that the yield of essential oil was found to be extremely variable in the samples collected from different places, such as India, eastern Europe, Egypt, and Argentina, which points out to the effect of ecological factors on the oil content. Salamon et al. (2010) reported that the naturally occurring chamomile variety collected from 20 localities in Iran had a significant variation in the essential oil content, which ranged from 0.1% to 0.95%. The essential oil content was reported to be 0.3%–0.4% in south India (Nidagundi and Hegde 2007) and 0.4%–0.5% in northeast India (Tandonet al. 2013).

4.2.1.2 Variation in Essential Oil Content in Different Stages of Flowering

The oil content and oil quality of the capitulum vary according to the stage of capitulum development (Motl et al. 1977; 1982; Marczal and Petri 1979; Gasic et al. 1986). Franz (1980) studied the essential oil content and composition in the flowers of four chamomile cultivars during their ontogenetical development. They divided the flower development into four stages: I, buds; II, ligulate flowers developed; III, tubular flowers opened; and IV, decaying flower heads. On commencement of flowering, the flowers were harvested five times in intervals of a week. On steam distilling the oil, they found that the flowers of stages I (0.85% oil) and III (0.9% oil), that is, the buds and the tubular flowers opened, had more essential oil content than stages II (0.75% oil) and IV (0.6% oil).

4.2.1.3 Variation in Essential Oil Content Due to Farming Practices

A variation in the essential oil content is attributed to farming practices such as growing season, plant spacing, irrigation schedule, fertilizers, and harvesting time. Franz et al. (1985) studied in detail the effect of sowing time on the essential oil content using two varieties of chamomile H-29 and BK-2. The seeds of these two varieties were sown in August, September, October, April, May, and June in the years 1979–1980 and 1981–1982. They found that there was a variation in the essential oil content in the flowers. The essential oil content of the H-29 variety was found to range from 0.7% plants grown in April (1980) to 1.04% in those grown in June (1980). The essential oil content of the BK-2 variety was found to range from 0.60% in plants grown in April (1980) to 1.00% in those grown in October (1979). They reported that sowing in spring led to less yield than that in autumn. Highest oil contents were obtained when the sowing was done in late spring. This variation clearly shows that the sowing time has an impact on the essential oil content and has to be taken into consideration during cultivation to make optimum use of the climatic conditions.

Franz (1980) had reported, as mentioned earlier, that the flowering stages at the time of harvest determine the essential oil content. They further reported that the harvesting dates may be optimized according to the flowering stage. The flowers may be harvested during stage III, and the most favorable time would be 1 week after the beginning of the flowering (Franz et al. 1985).

The irrigation and fertilization regime also seems to have an indirect effect on the essential oil content through an increase in the biomass (Franz 1983). Mirshekari (2011) studied the effect of irrigation and nitrogen fertilizer on the Bodegold variety of chamomile. The irrigation was designed to be given after 60, 120, and 180 mm evaporation. Urea was applied at the rate of 50, 100, and 150 kg/ha. It was applied during planting, stem elongation, and early flowering stage. Irrigating after 180 and 60 mm evaporation resulted in the emergence of buds after 70 and 78 days, respectively. This finding indicates the possibility that less water is required for bud development. Further, it was found that application of urea during planting and stem elongation increased flower yield and consequently the essential oil yield (2.82 L/ha).

4.2.1.4 Variation in Composition of Organ-Specific Essential Oil

Oil composition also differs depending on the plant parts used. The essential oils of shoot, root, and flowers, show an organ-specific differential distribution of certain chemical constituents. Lawrence (1987) compiled the data from different studies and presented the essential oil content in different parts of chamomile plant, such as the flower, tubular florets, ligulate florets, receptacle, seeds, the plant without flowers, leaves, and stem. The essential oil content in these parts was reported as follows: whole flower (0.91%), tubular florets (0.8%), ligulate florets (0.3%), receptacle (0.9%), seeds (0.3%), whole plant without flowers (0.6%), and leaves (0.2%). No oil content was reported from the stems or roots. Pöthke and Bulin (1969) reported that a high amount of oil is present in the receptacle and the tubular florets. Reichling et al. (1979) carried out a comparative assessment of the essential oil content and composition of the various tissues such as the stem tumors, roots, and herb. They reported that the stem tumors and the herb (stem and leaves) contained essential oil. The oil content in the stem tumor and the herb was found to be similar. Reichling and Beiderbeck (1991) reported that there was 0.6% essential oil in the whole plant without flowers and 0.03% in the roots. Das (1999) studied the oil content and composition of different parts of the plant, that is, the whole flower, tubular (disc) floret, ligulate (ray) floret, leaves, stem, herb (leaves and stem), and root. The fresh parts were used to extract the oil. The oil content was highest in the tubular florets. On the basis of the oil content, the various parts could be arranged in the following order:

Tubular (disc) floret > whole flower > ligulate (ray) floret > leaves > root > herb > stem.

The average values of oil content and color from the different parts of the chamomile plant are given in Table 4.1.

TABLE 4.1
Oil Content and Oil Color from Different Parts of Chamomile Plant

S. No.	Plant Part	Oil Content (% Fresh Weight)	Oil Color
1.	Whole flower	0.08	Ink blue
2.	Tubular (disc) floret	0.1	Ink blue
3.	Ligulate (ray) floret	0.062	Light blue
4.	Leaves	0.025	Pale yellow
5.	Stem	0.006	Pale yellow
6.	Herb (stems and leaves)	0.01	Green
7.	Root	0.018	Yellow

4.3 DISCOVERY OF THE COMPOUNDS IN THE ESSENTIAL OIL

The chemical composition of chamomile oil, obtained through hydrodistillation of the flowers, was not known much before the 1950s, other than the fact that it contained some sesquiterpenes. In the subsequent years, researchers isolated and identified the compounds that are now well known in chamomile oil. The major constituents of chamomile oil, such as α-bisabolol and chamazulene, were isolated, studied, and reported by Šorm et al. in 1951 and 1953, and by Čekan et al. in 1954, respectively (Hitziger et al. 2003). The en-yn-dicycloethers were isolated by Bohlmann et al. in 1960 (Hitziger et al. 2003).

Lawrence (1987) in a review of the chamomile essential oil mentioned several researchers, such as Šorm, who characterized cadenine in 1951; Piquet et al., who found 12 components in lab-distilled essential oil and characterized β-farnesene, α-bisabolol, and chamazulene in 1970; Flaskamp et al., who determined the stereochemistry of bisabolol and its oxides in 1971; Glasl, who characterized *cis*- and *trans*-en-yn-dicycloether in 1975; and Motl, who characterized spathulenol in 1978. In 1989, Retamar et al. reported myrcene, 1,8-cineole, linalool, pulegone, α-terpineol, and β-caryophyllene in chamomile essential oil (Lawrence 1996).

In 1954, Čekan et al. isolated matricin, a precursor of chamazulene, and in the same year Stahl isolated chamazulene carboxylic acid (CCA), another precursor of chamazulene.

Ness et al. (1996) reported the presence of a new compound 8-desacetylmatricin, a degradation product of matricin.

Kumar et al. (2001) identified eight compounds, which were hitherto unknown to chamomile essential oil namely n-hexanol, α-gurjunene, α-humulene, globulol, camphor, β-bourbonene, camphene, and geranyl acetate. n-Hexanol was identified in the essential oil of the leaves and stem. α-Gurjunene, camphor, and β-bourbonene were identified in the essential oil of the flowers, leaves and stem, and root. α-Humulene and globulol were identified in the essential oil of the root. Camphene was identified in the essential oil of the flowers. Geranyl acetate was identified in the essential oil of the leaves and stem, and roots.

Das et al. (2002) identified three more compounds new to chamomile essential oil: phytol (diterpene), isophytol (diterpene), and hexadec-11-yn-13,15-diene. Hexadec-11-yn-13,15-diene was identified in the root oil. Isophytol was found in the essential oils of the whole flower, disc florets, ray florets, leaves, stem, and roots. Phytol was identified in the essential oil of the leaves and the roots.

With the use of advanced separation instruments such as gas chromatography (GC), high-performance liquid chromatography–gas chromatography, gas chromatography-mass spectrometry (GC-MS), and vacuum headspace analysis, more compounds new to chamomile essential oil were identified. Similarly, with the new extraction methods with efficient solvent systems under controlled conditions, such as supercritical carbon dioxide fluid extraction (SCFE), hitherto unknown compounds in chamomile extracts were reported.

Raal et al. (2003) reported 3-octanol, p-cymen-8-ol, (E)-anethole, decanol, decanoic acid, α-bergamotene, and allo-aromadendrene in the essential oil of chamomile.

Gosztola et al. (2010) identified 1-epi-cubenol and epi-bisabolol in the essential oil of a chamomile population named K/28.

Nurzyńska-Weirdak (2011) reported yomogi alcohol, artemisyl acetate, dehydrosesquicineole, α-acoradiene, helofolen-12 al, and apiole in chamomile essential oil.

Petronilho et al. (2011) identified seven sesquiterpenoids for the first time in chamomile. These are ar-curcumene, calamenene, isoledene, α-calacorene, dendrasaline, cuparene, and dihydrocurcumene.

Tschiggerl and Bucar (2012) identified three sesquiterpenes new to chamomile in volatile fraction of chamomile infusion or tea using the hydrodistillation method and the solid-phase extraction (SPE) method. They reported for the first time the presence of leucodin and acetoxyachillin. They also confirmed the presence of achillin, which was previously identified only by mass spectral data library search.

Tandon et al. (2013) identified trans-limonene oxide and 1,6-dioxaspironon-3-ene for the first time in chamomile oil.

New biotechnological methods, such as tissue culture and hairy root culture, were used to develop artificial tissues and callus in an attempt to extract the essential oil. These oils show a different composition than the natural essential oil.

Magiatis et al. (2001) identified gossonorol, cubenol, α-cadinol, 1-azulenethanol acetate, and (–)-α-bisabolol acetate in the essential oil from callus cultures of chamomile.

Szöke et al. (2004c) identified berkheyaradulene, 4-(2′,4′,4′-trimethylbicyclo[4.1.0] hept-2′-en-3′-yl)-3-buten-2-one, geranyl isovalerate, and cedrol as new components from the hairy root culture of chamomile plants. Szöke et al. (2004a) also identified a new sesquiterpene compound β-eudesmol in the intact and in vitro organized root of chamomile. They also identified β-selenine, a new compound, in the genetically transformed hairy root cultures.

4.4 CHEMICAL COMPOSITION OF THE FLOWER ESSENTIAL OIL

Chamomile oil mainly contains terpenes. The main terpenes are the sesquiterpenes chamazulene, farnesene, bisabolol, bisabolol oxides A and B, and bisabolone oxide A. Some of these compounds have been found to be pharmaceutically active.

Chamomile Oil and Extract

Other compounds found in small amounts are the spiroethers, aliphatic esters, aliphatic alcohols, benzenoid compounds, and so on. The identification process, characterization, and use of these compounds are presented in detail in Sections 4.4, 4.5, 4.10, and 4.11.

Lawrence (1987) reported that as early as 1958, Debska had found that the essential oil contained 26.1%–46.2% azulenes. Schilcher in 1973 reported that the chamazulene content in the flower essential oils was found in the range of 1.45%–17.69%, α-bisabolol in the range of 4.37%–77.21%, bisabolol oxide A in the range of 2.13%–52.25%, bisabolol oxide B in the range of 3.17%–58.85%, and spiroether in the range of 1.92%–12.00%. In 1977, Motl et al. provided a more detailed report on the composition of the essential oil of some commercial chamomile samples. They reported that the essential oil contained chamazulene in the range of 2.16%–35.59%, α-bisabolol in the range of 1.72%–67.25%, bisabolol oxide A in the range of 0%–55%, bisabolol oxide B in the range of 4.35%–18.93%, and bisabolone oxide A in the range of 0%–63.85% (Lawrence 1987).

The chemical composition of the chamomile oil has been found to differ depending on the various cultivars studied or when plants growing under different environment conditions have been used. Franz (1992) reported variation in the chemical constituents in four cultivars, Badakalaszi-2, Degumille, Csomori, and Bona.

- Badakalaszi-2 flower essential oil contained chamazulene (12%–17.6%), α-bisabolol (1.3%–3.3%), bisabolol oxide A (33.4%–39.0%), bisabolol oxide B (2.4%–5.1%), and farnesene (17.4%–40.7%).
- Degumille flower essential oil contained chamazulene (17.8%–20.8%), α-bisabolol (34.6%–40.9%), bisabolol oxide A (2.2%–6.2%), bisabolol oxide B (2.1%–7.5%), and farnesene (15.4%–34.8%).
- Csomori flower essential oil contained chamazulene (10.8%–18.3%), α-bisabolol (32.4%–57.6%), bisabolol oxide A (0.0%–1.7%), bisabolol oxide B (0.2%–2.4%), and farnesene (11.6%–43.8%).
- Bona flower essential oil contained chamazulene (11.6%–21.8%), α-bisabolol (33.4%–60.1%), bisabolol oxide A (0.4%–1.4%), bisabolol oxide B (0.0%–2.2%), and farnesene (14.0%–37.8%).

Piccaglia et al. (1993) reported that the amounts of the compounds differed in the different chamomile samples collected from seven countries.

- The chamazulene content was found to be as follows: Egypt, 2.7%; Argentina, 10.3%; Hungary, 19.7%; Germany, 7.6%; Holland, 3.5%; Czechoslovakia, 17.5%; and Yugoslavia, 15.8%.
- The α-bisabolol content was found to be as follows: Egypt, 2.9%; Argentina, 6.8%; Hungary, 5.9%; Germany, 1.6%; Holland, 2.8%; Czechoslovakia, 58.5%; and Yugoslavia, 1.7%.
- The bisabolol oxide A content was found to be as follows: Egypt, 53.6%; Argentina, 11.5%; Hungary, 35.0%; Germany, 44.7%; Holland, 51.6%; Czechoslovakia, 2.5%; and Yugoslavia, 28.6%.

- The bisabolol oxide B content was found to be as follows: Egypt, 9.5%; Argentina, 50.5%; Hungary, 8.4%; Germany, 13.5%; Holland, 10.7%; Czechoslovakia, 3.9%; and Yugoslavia, 21.3%.
- The farnesene content was found to be as follows: Egypt, 8.3%; Argentina, 5.8%; Hungary, 8.1%; Germany, 7.7%; Holland, 8.9%; Czechoslovakia, 6.5%; and Yugoslavia, 5.2%.
- The spiroether content was found to be as follows: Egypt, 5.9%; Argentina, 4.4%; Hungary, 6.4%; Germany, 5.9%; Holland, 7.0%; Czechoslovakia, 5.8%; and Yugoslavia, 6.1%.

Salamon (1994) reported the chemical composition of the diploid Bona and tetraploid Goral varieties as follows:

- Bona contained 16.4% chamazulene, 57.3% α-bisabolol, 1.7% bisabolol oxide A, 1.2% bisabolol oxide B, 17.3% farnesene, and 4.7% cis-en-yn-dicycloether.
- Goral contained 14.8% chamazulene, 31.7% α-bisabolol, 13.3% bisabolol oxide A, 9.5% bisabolol oxide B, 21.5% farnesene, and 1.5% cis-en-yn-dicycloether.

Das et al. (1999) reported the composition of the essential oils of two varieties (Vallary and CR-3A) and a selection (CR-SPL) as follows:

- Vallary was found to contain 2.4% chamazulene, 14.8% α-bisabolol, 38.0% bisabolol oxide A, 6.4% bisabolol oxide B, 12.6% β-farnesene, and 4.4% cis-en-yn-dicycloether.
- CR-3A was found to contain 18.2% chamazulene, 2.2% α-bisabolol, 13.7% bisabolol oxide A, 16.1% bisabolol oxide B, 10.1% β-farnesene, and 7.7% cis-en-yn-dicycloether.
- The selection CR-SPL was found to contain 3.5% chamazulene, 2.2% α-bisabolol, 15.5% bisabolol oxide A, 6.3% bisabolol oxide B, 8.8% β-farnesene, and 4.2% cis-en-yn-dicycloether.

In the same year, Kumar et al. (1999) reported the composition of the flower essential oil of a variety Prashant. It was found to contain 7.5% chamazulene, 4.8% α-bisabolol, 25.5% bisabolol oxide A, 42.2% bisabolol oxide B, 4.9% β-farnesene, and 4.8% cis-en-yn-dicycloether.

The chamomile plants of European origin are known to be devoid of chamazulene (Mann and Staba 1986). Das (1999, p. 232) found that the flowers of several plants yielded oils whose colors ranged from white to brown. These oils were found to be devoid of chamazulene.

Salamon et al. (2010) reported variations in the essential oil composition of the 20 chamomile populations naturally growing in Iran. In these populations, chamazulene was found to range from 0.0% to 13.0%. The plants collected from Khoozestan-Hendijan (Khuzestan-Hendijan) did not contain any chamazulene. These populations had α-bisabolol ranging from 4.0% to 61%, bisabolol oxide A ranging from 0.0% to 62%, and β-farnesene ranging from 0.2% to 5.0%.

Chamazulene content is high in the oil extracts from florets and the seeds. Interestingly, the receptacle of the capitulum, which is highly rich in the essential oil, is more or less devoid of chamazulene. The compound en-yn-dicycloether is in high amounts in the essential oil of the receptacle. The oil of the tubular florets has high amounts of chamazulene, bisabolol, and its oxides (Marczal and Petri 1988; Menziani et al. 1990; Repcak et al. 1993).

4.4.1 Major Compounds

Some of the major compounds found in chamomile oil are discussed in Sections 4.4.1.1–4.4.1.4. These compounds, besides being found in high amounts, are used extensively in different medicinal formulations, flavoring and perfumery, and cosmetics.

4.4.1.1 Chamazulene and Its Precursors

Chamazulene is a blue-colored compound present in the hydrodistilled oil of chamomile flowers. It is chamazulene that renders the characteristic ink blue color to the oil. Chamazulene is not found naturally in the oil. It is derived from its precursor matricin in the presence of water during hydrodistillation. The conversion of chamazulene from matricin may also occur if the pH is altered to more acidic conditions (Schmidt et al. 1991; Schmidt and Ness 1993; Ness and Schmidt 1995).

The blue color of is due to the five conjugated double bonds in chamazulene. The color of the oil may range from blue to green due to azulene. The chemical structure of chamazulene was elucidated in 1953 independently by the team of Sŏrm, Novák and Herout, and the team of Meisels and Weizmann (Hitziger et al. 2003) (Figure 4.1).

Chamazulene is a sesquiterpene and the common name assigned to it is 1,4-dimethyl-7-ethylazulene. The IUPAC name is 7-ethyl-1,4-dimethylazulene. It has the chemical formula $C_{14}H_{16}$ with a molar mass of 184.28 g/mol. It has a density of 1.0 ± 0.1 g/cm^3 and boiling point of 299.1 ± 20.0 °C at 760 mmHg (The Merck CSID 10268). Its light absorption peak at UV_{max} is 370 mm (Mann and Staba 1986). It is extensively used in medicinal preparations owing to its pharmacological properties.

Matricin is a pale yellow compound found to be present in the tubular and ligulate flowers of chamomile. These are not present in the receptacle (Hitziger et al. 2003). It was isolated by Čekan in 1954, who assigned its structure to be (3S,3aR,4S,9R,9aS,9bS)-4 acetyoxy-3a,4, 5,9,9a,9b-hexahydro-9-hydroxy-3,6,9-trimethylazuleno[4,5-b]furan-2(3H)-one (Schilcher et al. 2005). The relative structure of matricin was elucidated by Flaskamp in 1982 using nuclear magnetic resonance (NMR) spectroscopy and also

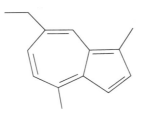

FIGURE 4.1 Chamazulene

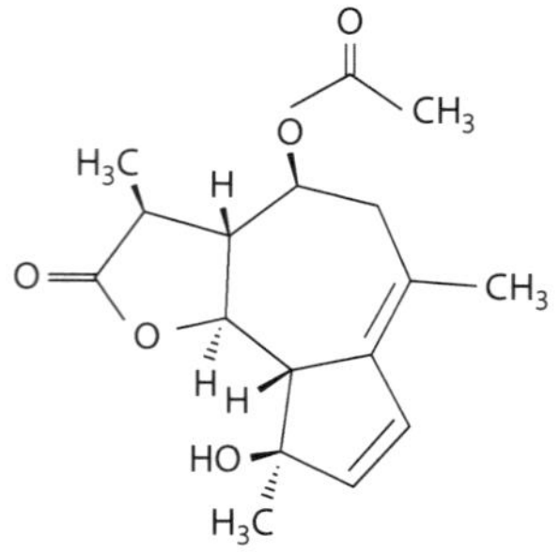

FIGURE 4.2 Matricin

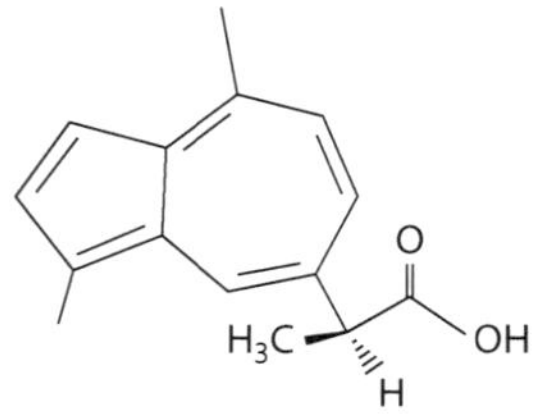

FIGURE 4.3 *S*-Chamazulene carboxylic acid

by Appendino in 1985, who used x-ray analysis of four epimatrins to further elucidate the chemical structure of matricin (Hitziger et al. 2003). In 2001, Goeters, Imming, Pawlitzki, and Hempel established the 3*S*,3a*R*,4*S*,9*R*,9a*S*,9b*S* configuration of matricin using synthetic, NMR, and circular dichroism spectroscopy studies (Goeters et al. 2001).

The IUPAC name of matricin is (3*S*,3a*R*,4*S*,9*R*,9a*S*,9b*S*)-9-hydroxy-3,6,9-trimethyl-2-oxo-2*H*,3*H*,3a*H*,4*H*,5*H*,9*H*,9a*H*,9b*H*-azuleno[4,5-*b*]furan-4-yl acetate. The chemical formula of matricin is $C_{17}H_{22}O_5$ and its molar mass is 306.35 g/mol (Matricin 2013). The melting point of matricin is 159°C and its optical rotation in chloroform is +30 (Mann and Staba 1986) (Figure 4.2).

Matricin is extremely unstable and decomposes readily on exposure to water into chamazulene. The first degradation product of matricin is 8-desacetylmatricin (Ness et al. 1996), which further degrades into CCA.

CCA was isolated and characterized chemically and stereochemically; its pharmacological activities were investigated by Goeters et al. in 2001. It was found to be a natural profen with anti-inflammatory properties (Hitziger et al. 2003). The chemical structure of CCA was elucidated by Cuong et al. using mass spectrometry and NMR (Schilcher et al. 2005). Its chemical properties are yet to be studied (Figure 4.3).

4.4.1.2 Bisabolols and Bisabolol Oxides

Bisabolol, commonly known as (−)-α-bisabolol, is also called levomenol. It is a monoterpene, which is a pale yellow liquid with a mild woody and floral odor. It is speculated that bisabolol occurs as a mixture of four stereoisomers (optical isomers),

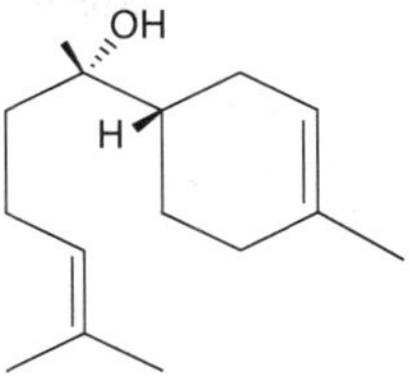

FIGURE 4.4 α-Bisabolol

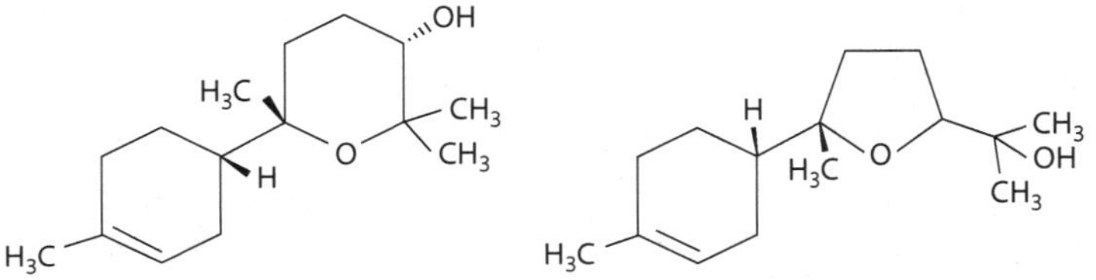

FIGURE 4.5 (a) (–)-α-Bisabolol oxide A and (b) (–)-α-Bisabolol oxide B

namely (–)-α-bisabolol, (–)-*epi*-α-bisabolol, (+)-α-bisabolol, and (+)-*epi*-α-bisabolol. However, in the nature, (–)-α-bisabolol is mostly found (Kamatou and Viljoen 2010).

The IUPAC name of bisabolol is 6-methyl-2-(4-methylcyclohex-3-en-1-yl)hept-5-en-2-ol and its molecular formula is $C_{15}H_{26}O$. Bisabolol has a low density (0.93 g/cm^3) and a boiling point of 154°C–156°C at 13 mmHg. It has a refractive index of 1.4939. It is insoluble in water but soluble in alcohol (Mann and Staba 1986; Kamatou and Viljoen 2010) (Figure 4.4).

Bisabolol tends to oxidize into (–)-α-bisabolol oxide A, (–)-α-bisabolol oxide B, (–)-α-bisabolol oxide C, and (–)-α-bisabolone oxide A (Franz et al. 2005) (Figure 4.5).

Bisabolol is extensively used in the perfume and flavor industry, especially in body lotions and creams. The U.S. Food and Drug Administration has given bisabolol the generally recognized as safe status (Kamatou and Viljoen 2010).

4.4.1.3 (*E*)-β-Farnesene

(*E*)-β-Farnesene or *trans*-β-farnesene is a sesquiterpene. It is found in nature. Its IUPAC name is (6*E*)-7,11-dimethyl-3-methylene-1,6,10-dodecatriene and molecular formula is $C_{15}H_{24}$. It has an average mass of 204.351105 Da (CSID: 4444850).

It is considered an aphid alarm pheromone (Pickett and Griffiths 1980), which essentially repels the aphids from feeding on the plant (Figure 4.6).

4.4.1.4 Spiroethers

Spiroethers in the chamomile oil exist in two isomeric forms, *cis*-en-yn-dicycloether and *trans*-en-yn-dicycloether. The IUPAC name of *cis*-en-yn-dicycloether is (2*Z*)-2-hexa-2,4-diynylidene-1,6-dioxaspiro[4.4]non-3-ene (CID 5352494) and that of *trans*-en-yn-dicycloether is (2*E*)-2-hexa-2,4-diynylidene-1,6-dioxaspiro[4.4]non-3-ene (CID 5352496). The molecular formula is $C_{13}H_{12}O_2$ and the molecular weight is 200.23318 g/mol.

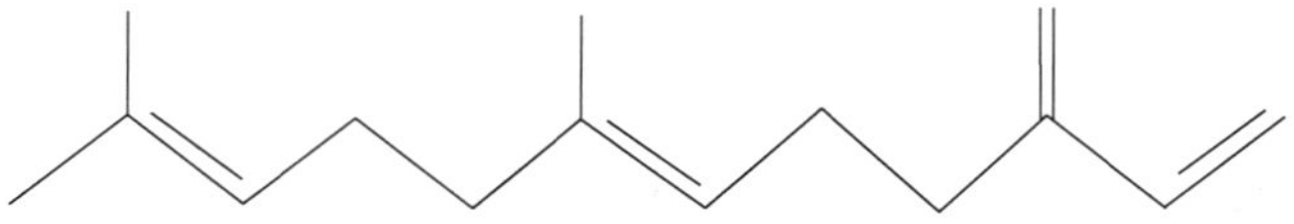

FIGURE 4.6 (*E*)-β-Farnesene

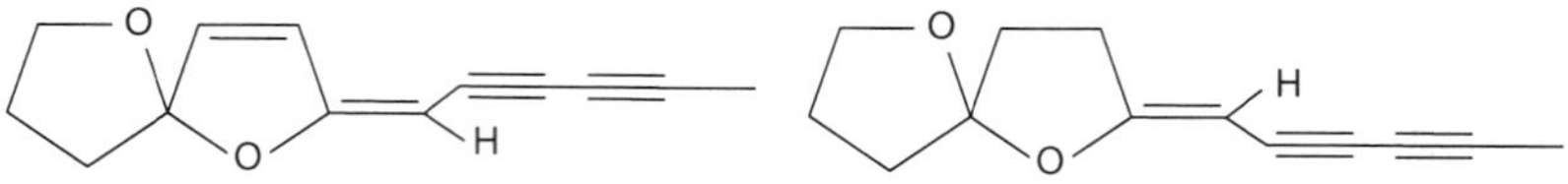

FIGURE 4.7 (a) *cis*-Dicycloether and (b) *trans*-dicycloether

The spiroethers are used as whitening agents in skin creams (CID 5352496). The *cis* form is more pharmacologically active than the *trans* form (Mann and Staba 1986) (Figure 4.7).

4.4.2 Minor Compounds

The minor compounds found in chamomile oil are described briefly in Sections 4.4.2.1–4.4.2.17. The uses of these compounds are also mentioned.

4.4.2.1 Germacrene D

Germacrene D is a sesquiterpene with the IUPAC name (1*E*,6*E*,8*S*)-1-methyl-5-methylidene-8-(propan-2-yl)cyclodeca-1,6-diene (CID 5317570). It has a molecular formula of $C_{15}H_{24}$. It has a molecular weight of 204.35106 g/mol and a boiling point of 268.014°C at 760 mmHg (CSID: 16787784). It has a density of 0.793 g/cm³.

Germacrene D is a precursor to many sesquiterpenes and is known to possess antimicrobial properties (Rivero-Cruz et al. 2006; Setzer 2008) (Figure 4.8).

4.4.2.2 (*E*)-Nerolidol

(*E*)-Nerolidol is a sesquiterpene, which is also known as peruviol. Its IUPAC name is (3*S*,6*Z*)-3,7,11-trimethyldodeca-1,6,10-trien-3-ol (CID 5356544). Its molecular formula is $C_{15}H_{26}O$. It has a molecular weight of 222.366302 g/mol and a boiling point of 275.999°C. It has a density of 0.87 g/cm³, and a refractive index of 1.48 (CSID: 4447568).

Nerolidol is a straw colored aromatic compound with a floral rose-like and woody aroma. It is extensively used in perfume and flavor industry. It is also known to possess antimicrobial properties (Pículo et al. 2011) (Figure 4.9).

4.4.2.3 Farnesol

(*E*,*E*)-Farnesol is a sesquiterpene alcohol found in chamomile essential oil. The (*Z*,*E*)-farnesol is also found in chamomile oil. The IUPAC name of farnesol is (2*E*,6*E*)-3,7,11-trimethyldodeca-2,6,10-trien-1-ol (CID 445070). Its molecular formula is $C_{15}H_{26}O$. It has a refractive index of 1.485 and a density of 0.876 g/cm³. Its boiling point is 283.37°C (CSID: 392816).

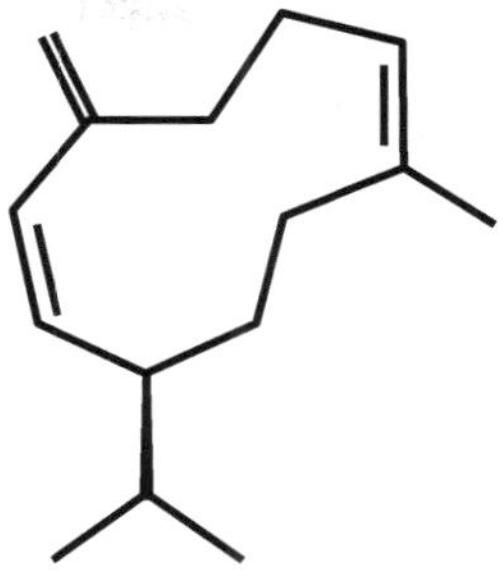

FIGURE 4.8 Germacrene D

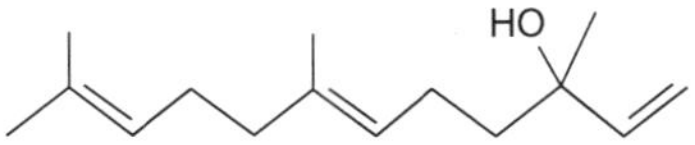

FIGURE 4.9 (*E*)-Nerolidol

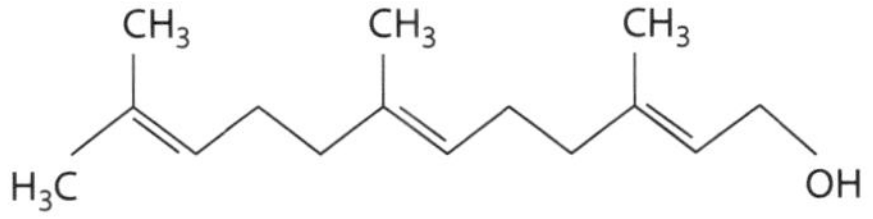

FIGURE 4.10 Farnesol

Farnesol is a colorless oily liquid, which is used in perfumery (Farnesol 2013). It is also used in flavoring cigarettes, gelatins, puddings, and candies (Burdock 2005; Martin 2010). It is used in cosmetic aerosols and sprays, deodorants, skin care products, and antidandruff shampoos (Ash and Ash 2004). In addition, pharmacological tests on several animal models suggest that it functions as a chemopreventative and antitumor agent in vivo (Joo and Jetten 2009) (Figure 4.10).

4.4.2.4 Spathulenol

Spathulenol is a terpene alcohol having the molecular formula $C_{15}H_{24}O$. Its IUPAC name is (7S)-1,1,7-trimethyl-4-methylidene-1a,2,3,4a,5,6,7a,7b-octahydrocyclopropa[*h*]azulen-7-ol. It has a molecular weight of 220.35046 g/mol (CID 6432640).

Spathulenol has been found to possess properties of an adjuvant in chemotherapy (Martins et al. 2010) (Figure 4.11).

4.4.2.5 n-Hexanol

n-Hexanol, also known as 1-hexyl alcohol, is a colorless alcohol. It has herbaceous, woody, fruity odor and aromatic flavor (Burdock 2005). Its IUPAC name is hexan-1-ol and molecular formula is $C_6H_{14}O$. It has an average mass of 102.174797 g/mol and a boiling point of 157°C. It has a specific gravity of 0.816–0.821 and a refractive index of 1.415–1.420 (CID 8103; CSID: 7812).

n-Hexanol is used to flavor baked goods, frozen dairy, gelatins, puddings, nonalcoholic beverages, and soft candies (Burdock 2005) (Figure 4.12).

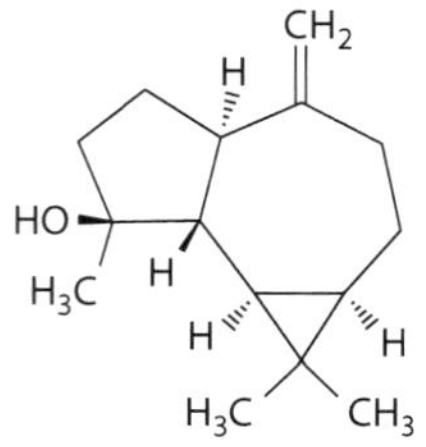

168   Chamomile: Medicinal, Biochemical, and Agricultural Aspects

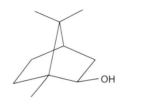

FIGURE 4.11 Spathulenol

FIGURE 4.12 n-Hexanol

FIGURE 4.13 Isoborneol

4.4.2.6 Isoborneol

Isoborneol or exoborneol is a monoterpene, which appears as whitish crystals when isolated in pure form. It has a piney and camphoraceous odor (CSID: 4882019; Burdock 2005). The IUPAC name of isoborneol is (1R,3R,4R)-4,7,7-trimethylbicyclo[2.2.1]heptan-3-ol (CID 6321405). The molecular formula is $C_{10}H_{18}O$. It has a boiling point of 214°C. It has a refractive index of 1.05 and a density of 0.993 g/cm³.

Isoborneol is used for flavoring baked goods, chewing gum, frozen dairy, gelatins, puddings, nonalcoholic beverages, and soft candies (Burdock 2005) (Figure 4.13).

4.4.2.7 Nerol

Nerol is a colorless liquid with a fresh sweet rose-like aroma and a bitter flavor. Its IUPAC name is (2Z)-3,7-dimethyl-2,6-octadien-1-ol and has the molecular formula $C_{10}H_{18}O$. It has an average mass of 154.249298 g/mol. Its boiling point is 226°C and has a refractive index of 1.467–1.478 (CSID: 558917).

Nerol is used to flavor food and beverages (Burdock 2005) (Figure 4.14).

4.4.2.8 Phytol

Phytol is a diterpene with the IUPAC name (E,7R,11R)-3,7,11,15-tetramethyl-2-hexadecen-1-ol. Its molecular formula is $C_{20}H_{40}O$ and has a molecular weight of 296.531006 g/mol (CID 5280435).

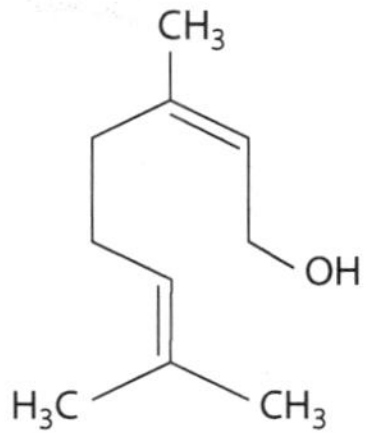

FIGURE 4.14 Nerol

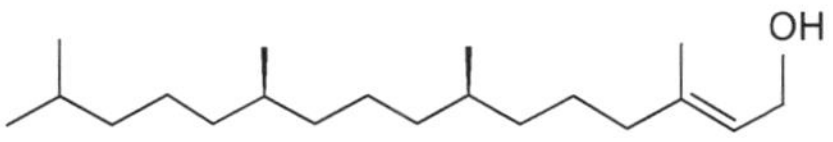

FIGURE 4.15 Phytol

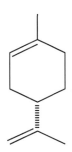

FIGURE 4.16 Limonene

According to some researchers, phytol may have antitrypanosomal and larvicidal activities (Bero et al. 2013; Senthilkumar et al. 2013) (Figure 4.15).

4.4.2.9 Limonene

Limonene, a terpene, exists as L- and D-limonene. D-Limonene is the predominant form, has a pleasant orange-like odor, and is used in perfumery. The IUPAC name of limonene is (4S)-1-methyl-4-prop-1-en-2-ylcyclohexene and the molecular formula is $C_{10}H_{16}$. Its molecular weight is 136.234 g/mol (CID 439250).

Limonene is used to flavor food items such as fruit juices, soft drinks, baked goods, ice cream, and pudding. It also has notable pharmacological properties. It is used to dissolve cholesterol-containing gallstones. It has also been used for relief of heartburn and gastroesophageal reflux disease because it can neutralize gastric acid and can also support normal peristalsis. It also shows chemopreventive activity against many types of cancer (Sun 2007) (Figure 4.16).

4.4.2.10 Camphene

Camphene is a bicyclic monoterpene and a colorless crystalline solid. It has a mild camphoraceous aroma. Its IUPAC name is 3,3-dimethyl-2-methylidenebicyclo[2.2.1]

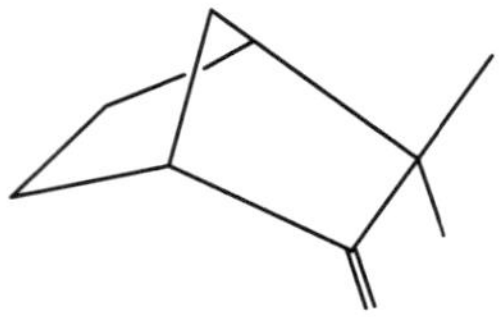

FIGURE 4.17 Camphene

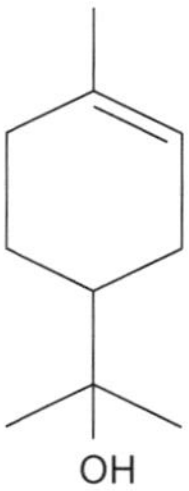

FIGURE 4.18 α-Terpineol

heptanes with the molecular formula $C_{10}H_{16}$. Its average mass is 136.233994 Da and boiling point is 159°C. It has a specific gravity of 0.850. It is soluble in alcohol and most acids, but insoluble in water (CID 6616; CSID: 6364; Burdock 2005) (Figure 4.17).

4.4.2.11   α-Terpineol

α-Terpineol is an oxygenated monoterpene. In its pure form, it is a white crystalline powder and has the odor of a lilac flower. It has a sweet lime taste. It has the molecular formula $C_{10}H_{18}O$. Its IUPAC name is 2-(4-methylcyclohex-3-en-1-yl)propan-2-ol and molecular weight is 154.24932 g/mol. Its boiling point is 218–221°C. Its specific gravity is 0.935 and has a refractive index of 1.4820. It is soluble in benzene, acetone, and propylene glycol. It is insoluble in water (CID 17100).

α-Terpineol is used in perfumery and flavoring alcoholic drinks. It has also been shown to possess antibacterial properties (Trinh et al. 2011; Park et al. 2012) (Figure 4.18).

4.4.2.12   Chamomillol

Chamomillol is a sesquiterpene alcohol with the molecular formula $C_{15}H_{26}O$. Its molecular weight is 222.3663 g/mol (Chamomillol 2013). The properties and uses of chamomillol are yet to be fully elucidated (Figure 4.19).

4.4.2.13   β-Elemene

β-Elemene is a sesquiterpene. Its IUPAC name is (1S,2S,4R)-1-ethenyl-1-methyl-2,4-bis(prop-1-en-2-yl)cyclohexane. It has the molecular formula of $C_{15}H_{24}$ and the molecular weight of 204.35106 g/mol (CID 6918391).

β-Elemene has been reported to have anticancer properties (Liu et al. 2011; Zou et al. 2013) and has the potential to be used in cancer therapy (Figure 4.20).

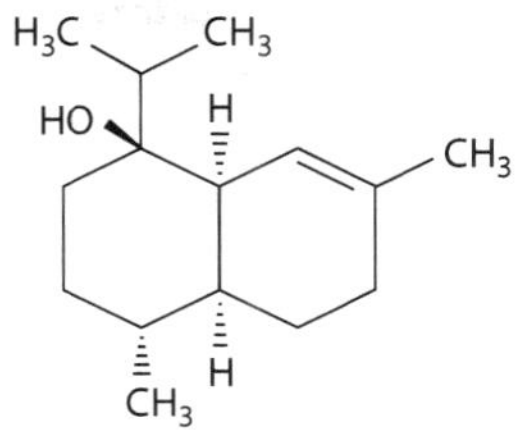

Chamomile Oil and Extract

FIGURE 4.19 Chamomillol

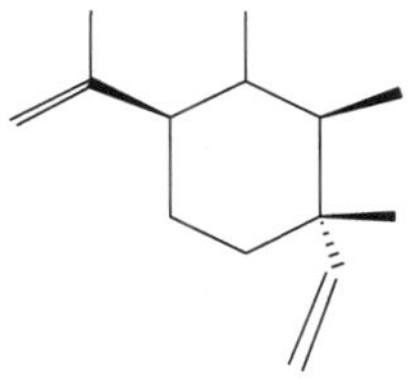

FIGURE 4.20 β-Elemene

4.4.2.14 α-Humulene

α-Humulene is also known as α-caryophyllene and is a sesquiterpene compound. Its IUPAC name is (1*E*,4*E*,8*E*)-2,6,6,9-tetramethylcycloundeca-1,4,8-triene and its molecular formula is $C_{15}H_{24}$. It has a molecular weight of 204.35106 g/mol (CID 5281520). It has a specific gravity of 0.89 and a boiling point of 123°C (CSID: 4444853).

α-Humulene has been found to possess anti-inflammatory properties (Fernandes et al. 2007; Rogerio et al. 2009). It could be a potential candidate for developing drug interventions for asthma (Figure 4.21).

4.4.2.15 Hexadec-11-yn-13,15-diene

Hexadec-11-yn-13,15-diene or (3*E*)-1,3-hexadecadien-5-yne has the molecular formula $C_{16}H_{26}$. It has a molecular weight of 218.377594 g/mol. The predicted density is 0.825 g/cm³ and the boiling point is 309.441°C (CSID: 4509368; Hexadec-11-yn-13,15-diene 2013). The other chemical and physical properties and pharmacological or other uses of hexadec-11-yn-13,15-diene are yet to be elucidated (Figure 4.22).

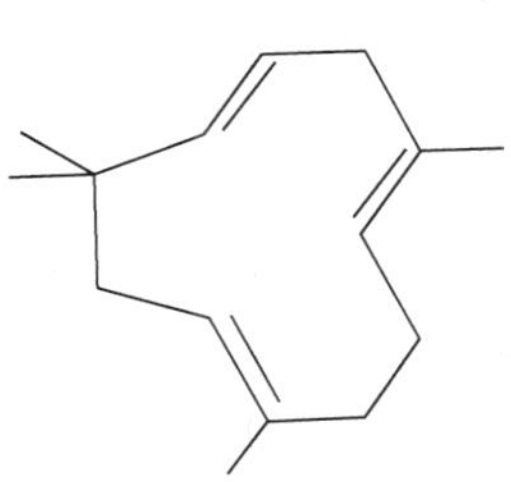

FIGURE 4.21 α-Humulene

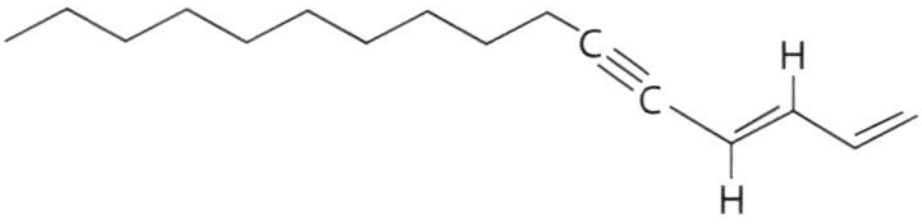

FIGURE 4.22 Hexadec-11-yn-13,15-diene

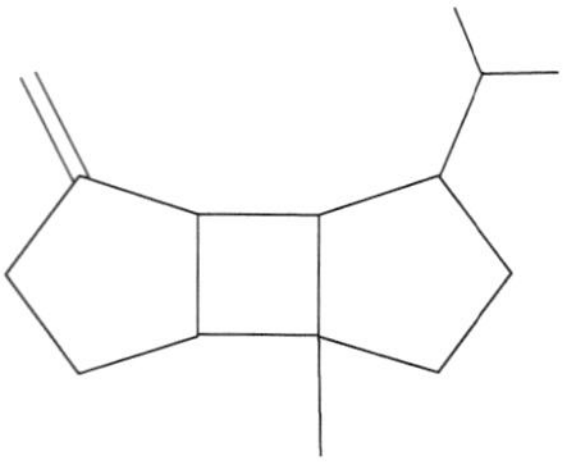

FIGURE 4.23 β-Bourbonene

4.4.2.16 β-Bourbonene

β-Bourbonene is also known as (1*S*,3a*S*,3b*R*,6a*S*,6b*R*)-1-isopropyl-3a-methyl-6 methylenedecahydrocyclopenta[3,4]cyclobuta[1,2]cyclopentene. Its molecular formula is $C_{15}H_{24}$ and the molecular weight is 204.351105 g/mol (CSID: 56330).

β-Bourbonene might have antimicrobial activities (Liu et al. 2009) or may have potential pesticidal properties (Beck et al. 2009). However, these findings are yet to be studied in detail (Figure 4.23).

4.4.2.17 τ-Cadinol

τ-Cadinol is a sesquiterpene alcohol, which is also known as cedrelanol. Its systematic name is (1*S*,4*S*,4a*R*,8a*R*)-4-isopropyl-1,6-dimethyl-1,2,3,4,4a,7,8,8a-octahydro-1-naphthalenol. Its molecular formula is $C_{15}H_{26}O$ and the molecular weight is 222.366302 g/mol. The predicted density is 0.937 g/cm³, the refractive index is 1.49, and the boiling point is estimated to be 303.393°C at 760 mmHg (CSID: 141284).

τ-Cadinol has been reported to possess muscle-relaxing properties (Claeson et al. 1991; Pattiram et al. 2011). It has also been reported that it could induce dendritic cells from monocytes, and hence may be used in dendritic cell-based immunotherapy for cancer (Takei et al. 2006) (Figure 4.24).

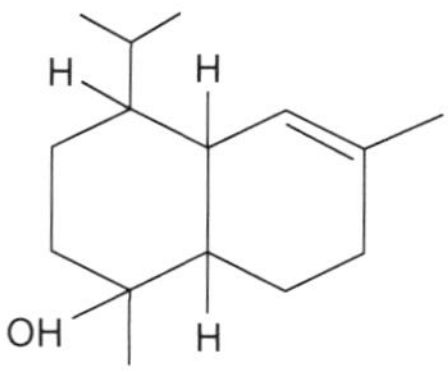

FIGURE 4.24 τ-Cadinol

Chamomile Oil and Extract

4.5 PHYSICAL PROPERTIES OF THE FLOWER ESSENTIAL OIL

The hydrodistilled chamomile oil is normally deep ink blue in color with a specific gravity ranging from 0.92 to 0.96. It has an acid value of 5–50 and ester value of 3–93. The oil is soluble in polar solvents (The Wealth of India 1962). Because of the intense blue color, the refractive index and optical rotation have not been determined. However, using colorimetry, a relation has been established between chamazulene content and the blue color of the oil (Karawya et al 1968).

Chamomile oil undergoes degradation on prolonged storage because of polymerization and the color of the oil changes from blue to greenish blue to brown. The degradation of chamomile oil can be monitored using color indexes.

4.6 CHEMICAL COMPOSITION OF THE HERB ESSENTIAL OIL (STEM AND LEAVES)

Chamazulene has not been found to be present in the oil extracted from the green parts of the shoot. The color of the oil derived from the stem and leaves separately is light blue to greenish whereas that of the oil from the herb (stem and leaves taken together) is greenish to brown. The oil of the herb was found to contain predominantly (E)-β-farnesene (37.7% in Prashant), cis-en-yn-dicycloether (15% in Prashant), germacrene D (14.6% in Prashant), (E)-nerolidol (3.5% in Vallary), farnesol (3.4% in Vallary), spathulenol (3.3% in Vallary), and n-hexanol (3.1% in Vallary) (Kumar et al. 2001).

The essential oil of the leaves, when separately distilled and analyzed, was found to have a preponderance of (E,E)-α-farnesene (9.3%), isoborneol (4.4%), α-bisabolol oxide A (4%), α-bisabolol oxide B (3.9%), nerol (3.7%), and phytol (2.6%). The essential oil of the stem without leaves, when distilled separately, showed a preponderance of limonene (23.9%), (Z)-en-yn-dicycloether (15.7%), camphene (3.1%), (E)-β-farnesene (2.9%), α-terpineol (2.7%), and (Z)-en-yn-dicycloether (2.6%) (Das et al. 2002).

The presence of the volatile constituents in the leaves as well as roots provides a high possibility of using these for medicinal, cosmetic, and aromatic purposes (Kobayashi et al. 2013).

4.7 CHEMICAL COMPOSITION OF THE ROOT ESSENTIAL OIL

The root oil lacks chamazulene and is pale yellow. Reichling et al. (1979, 1983) had reported the presence of (E)-β-farnesene, (E,E)-α-farnesene, β-caryophyllene, caryophyllene oxide, chamomillol, cis- and trans-dicycloethers, and chamomilla easters.

The root oil has as a predominace of (E)-β-farnesene (44.4%), chamomillol (5.8%), (Z)-en-yn-dicycloether (10.7%), β-elemene (5.1%), α-humulene (2.8%), and hexadec-11-yn-13,15-diene (2.2%). α-Humulene and hexadec-11-yn-13,15-diene are unique to root oil. In addition, the root oil contains α-bisabolol oxide A, α-bisabolol oxide B, β-bourbonene, and τ-cadinol (Das et al. 2002).

4.8 COMPARATIVE ACCOUNT OF THE ESSENTIAL OILS OF VARIOUS PARTS OF CHAMOMILE PLANT

Das (1999, pp. 223–224), Kumar et al. (2001), and Das et al. (2002) studied the composition of the flower oil, herb (stem and leaves) oil, and root oil. Table 4.2 provides the comparative account of the oils obtained from these various parts of the plant. It was observed that there were 61 compounds in the essential oils of the flower, herb, and root taken together. Of these 61 compounds, the flower oil contained 53 compounds and lacked 5 compounds, the herb oil contained 49 compounds and lacked 9 compounds, and the root oil contained 31 compounds and lacked 27 compounds. Thus, the flower oil was found to be comparatively richer in compounds than the herb oil or root oil. As expected, chamazulene was not found in the herb oil and the root oil. The other pharmacologically important compounds such as α-bisabolol, bisabolol oxide A, bisabolol oxide B, and α-bisabolone oxide were present in the essential oils of flower and herb. Bisabolol oxides A and B were found in the root.

The ratio of the main components in chamomile essential oil was estimated by Franz et al. (2005) as follows:

- Chamazulene and matricin: up to 15%
- Bisabolols and bisabolol oxides: up to 50%
- (E)-β-Farnesene: up to 45%
- *cis-* and *trans-*Isomers of the spiroethers: up to 25%

The compounds so far identified in the hydrodistilled oil are listed in Table 4.3.

4.9 CHEMICAL COMPOSITION OF CHAMOMILE FLOWER EXTRACT

The chamomile plant is known to possess several other compounds in addition to those mentioned in Table 4.3. These compounds include flavonoids and other organic compounds. Most of these compounds are found in the flowers (Mann and Staba 1986). These compounds are extracted using solvents because they are nonvolatile and cannot be hydrodistilled. Owing to their specific nonvolatile feature, these compounds cannot be detected by techniques such as GC-MS.

However, these compounds are soluble in various solvents and can be extracted. The extracts have been obtained using various solvents such as ethanol and water, glycerine and propylene glycol, toluene, pentane, sunflower oil, and n-hexane. In addition, SCFE of chamomile flowers has been carried out. Thereafter, these compounds are detected using techniques such as chromatography and high-performance thin-layer chromatography (HPTLC). The extract is almost always yellow, which indicates the presence of the flavonoids.

As early as 1979, the flavonoids found in the chamomile flower extract were found to be 18 in number and were listed by Kunde and Isaac (1979a) as follows:

1. Apigenin and its glucosides
2. Luteolin and its glucosides

TABLE 4.2
Comparative Account of the Essential Oil Compounds of Various Organs of the Chamomile Plant

S. No.	Category	Compound	Whole Flower	Tubular (Disc) Florets	Ligulate (Ray) Florets	Leaves	Stems	Roots
1.	Aliphatic alcohol	(E)-2-Hexenol	+	+	+	+	–	–
2.		(Z)-3-Hexenol	+	tr	tr	–	–	–
3.		n-Hexanol	–	–	–	+	+	–
4.	Monoterpene	Tricyclene	+	+	+	–	–	–
5.		α-Pinene	+	+	+	+	+	–
6.		Camphene	+	+	+	–	+	–
7.	Aliphatic ketone	6-Methylhept-5-en-2-one	+	+	+	+	–	–
8.	Monoterpene	Myrcene	+	+	+	+	+	–
9.	Aliphatic ester	(Z)-3-Hexenyl acetate	+	+	+	–	–	+
10.	Monoterpene	α-Phellandrene	tr	–	+	+	+	–
11.		p-Cymene	+	+	+	+	+	–
12.		Limonene	+	+	+	+	+	–
13.		(Z)-β-Ocimene	+	+	–	–	–	–
14.		(E)-β-Ocimene	+	+	+	+	–	+
15.		Artemisia ketone	+	+	+	–	–	+
16.		γ-Terpinene	+	+	+	+	+	–
17.	Oxygenated monoterpene	Artemisia alcohol	+	+	+	–	–	–
18.		Linalool	+	+	+	+	+	+
19.		Camphor	+	+	+	+	+	–

(Continued)

TABLE 4.2 (Continued)
Comparative Account of the Essential Oil Compounds of Various Organs of the Chamomile Plant

S. No.	Category	Compound	Whole Flower	Tubular (Disc) Florets	Ligulate (Ray) Florets	Leaves	Stems	Roots
20.		Isoborneol	−	+	+	+	+	−
21.		Borneol	+	+	−	−	+	+
22.		Terpinen-4-ol	+	+	+	+	+	−
23.		α-Terpineol	+	+	+	+	+	−
24.		Nerol	−	+	+	+	−	−
25.		Pulegone	+	+	−	−	−	−
26.		Geraniol	−	+	+	+	+	+
27.		Geranyl acetate	tr	+	+	+	−	−
28.	Sesquiterpene	α-Copaene	+	+	−	−	−	+
29.		β-Bourbonene	+	+	+	+	+	+
30.		β-Elemene	−	−	−	+	−	+
31.		α-Gurjunene	+	+	+	+	+	+
32.		(Z)-Caryophyllene	−	−	−	−	−	+
33.		(E)-Caryophyllene	+	+	−	+	−	+
34.		(E)-β-Farnesene	+	+	+	+	+	+
35.		α-Humulene	−	−	−	−	+	+
36.		γ-Muurolene	−	+	−	−	−	+
37.		Germacrene D	+	+	−	+	−	−
38.		α-Patchoulene	+	+	+	+	−	+
39.		(E,E)-α-Farnesene	+	+	+	+	−	+
40.		γ-Cadenine	+	+	+	+	−	−
41.		δ-Cadenine	+	+	+	−	−	+

Chamomile Oil and Extract

42.	Oxygenated sesquiterpene	(E)-Nerolidol	+	+	+	+	+	+	+
43.		Spathulenol	−	+	+	+	+	+	+
44.		Caryophyllene oxide	+	+	+	+	−	−	+
45.	Sesquiterpene alcohol	Globulol	−	−	−	−	−	−	+
46.	Oxygenated sesquiterpene	Chamomillol	−	−	−	+	+	−	+
47.		τ-Cadinol	+	+	+	+	+	+	+
48.		α-Muurolol	+	+	+	+	+	+	−
49.		Hexadec-11-yn-13,15-diene	−	−	−	−	−	−	+
50.		β-Eudesmol	+	+	+	+	+	−	+
51.		α-Bisabolol oxide B	+	+	+	+	+	+	+
52.		α-Bisabolone oxide	+	+	+	+	+	+	−
53.		α-Bisabolol	+	+	+	+	+	+	−
54.		(E,E)-Farnesol	−	−	−	−	+	−	−
55.	Sesquiterpene	Chamazulene	+	+	+	+	+	−	+
56.	Oxygenated sesquiterpene	α-Bisabolol oxide A	+	+	+	+	+	−	+
57.		(E,E)-Farnesyl acetate	−	−	−	−	−	−	+
58.	Polyene	(Z)-en-yn-Dicycloether	+	+	+	+	+	+	+
59.		(E)-en-yn-Dicycloether	+	+	+	+	+	+	+
60.	Diterpene	Isophytol	+	+	+	+	+	+	+
61.		Phytol	−	−	−	+	−	−	+

+, Present; −, absent; tr, traces.

TABLE 4.3
Chemical Compounds Identified as Constituents of Chamomile Oil Distilled from Flowers and Callus Cultures

S. No.	Type of Compound	Compound	Reference
1.	Aliphatic ester	Butyl 3-methyl butyrate	Brunke et al. (1993), Lawrence (1996)
2.		Ethyl 3-methyl butyrate	Brunke et al. (1993), Lawrence (1996)
3.		Butyl 2-methyl butyrate	Brunke et al. (1993)
4.		Ethyl 2-methyl butyrate	Shaath and Azzo (1993), Lawrence (1996)
5.		Hexyl acetate	Brunke et al. (1993), Lawrence (1996)
6.		(Z)-3-Hexenyl acetate	Surburg et al. (1996), Lawrence (1996)
7.		Hexyl propionate	Brunke et al. (1993), Surburg et al. (1996)
8.		(Z)-3-Hexenyl propionate	Surburg et al. (1996), Lawrence (1996)
9.		2-Methylbutyl 2-methylbutyrate	Surburg et al. (1996), Lawrence (1996)
10.		Methyl tiglate	Brunke et al. (1993), Lawrence (1996)
11.		2-Methylbutyl propionate	Surburg et al. (1996), Lawrence (1996)
12.		Propyl 2-methylbutyrate	Shaath and Azzo (1993), Lawrence (1996)
13.		Methyl alkyl angelate	Shaath and Azzo (1993), Lawrence (1996)
14.	Aliphatic alcohol	(Z)-3-Hexenol	Brunke et al. (1993), Lawrence (1996)
15.		(E)-2-Hexenol	Brunke et al. (1993), Lawrence (1996)
16.		n-Hexenol	Kumar et al. (2001)
17.		Phytol	Das et al. (2002)
18.		Isophytol	Das et al. (2002)
19.	Aliphatic aldehylde	Nonanal	Shaath and Azzo (1993), Lawrence (1996)
20.	Aliphatic ketone	6-Methylhept-5-en-2-one	Reverchon and Senatore (1994)
21.	Aromatic (benzenoid) compound	Benzaldehyde	Brunke et al. (1993), Lawrence (1996)
22.		Methyl benzoate	Brunke et al. (1993), Lawrence (1996)
23.		Methyl salicylate	Surburg et al. (1996), Lawrence (1996)
24.		2-Methoxy-4-methyl phenol	Surburg et al. (1996)
25.		Xanthoxylin (2-hydroxy-4, 6-dimethoxyacetophenone)	Motl and Repcak (1979)
26.	Monoterpene	δ-3-Carene	Reichling and Biederbeck (1991)
27.		p-Cymene	Shaath and Azzo (1993), Lawrence (1996)
28.		Myrcene	Retamar et al. (1989), Lawrence (1996)
29.		Limonene	Brunke et al. (1993), Lawrence (1996)
30.		(Z)-β-Ocimene	Surburg et al. (1996), Lawrence (1996)
31.		(E)-β-Ocimene	Surburg et al. (1996), Lawrence (1996)
32.		Sabinene	Brunke et al. (1993), Lawrence (1996)
33.		α-Terpinene	Tsutsulova and Antonova (1984)
34.		γ-Terpinene	Shaath and Azzo (1993), Lawrence (1996)
35.		α-Pinene	Shaath and Azzo (1993), Lawrence (1996)

TABLE 4.3 (*Continued*)
Chemical Compounds Identified as Constituents of Chamomile Oil Distilled from Flowers and Callus Cultures

S. No.	Type of Compound	Compound	Reference
36.		β-Pinene	Brunke et al. (1993), Lawrence (1996)
37.		2,6-Dimethylocta-1,3(*E*),5(*Z*),7-tetraene	Brunke et al. (1993), Lawrence (1996)
38.		2,6-Dimethylocta-1,3(*E*),5(*E*),7-tetraene	Brunke et al. (1993), Lawrence (1996)
39.		Camphor	Kumar et al. (2001)
40.		Camphene	Kumar et al. (2001)
41.		Geranyl acetate	Kumar et al. (2001)
42.	Oxygenated monoterpene	Borneol	Retamar et al. (1989), Lawrence (1996)
43.		1,8-Cineol	Retamar et al. (1989), Lawrence (1996)
44.		Artemisia ketone	Brunke et al. (1993), Lawrence (1996)
45.		Artemisia alcohol	Brunke et al. (1993), Lawrence (1996)
46.		2,6-Dimethyl octa-3(*E*),5(2),7-trien-2-ol	Brunke et al. (1993), Lawrence (1996)
47.		2,6-dimethyl octa-35 (*E*), 7-trien-2-ol	Brunke et al. (1993), Lawrence (1996)
48.		Isoborneol	Reverchon and Senatore (1994)
49.		Linalool	Retamar et al. (1989), Lawrence (1996)
50.		Pulegone	Retamar et al. (1989), Lawrence (1996)
51.		Lavandulyl acetate	Brunke et al. (1993), Lawrence (1996)
52.		α-Terpineol	Retamar et al. (1989), Lawrence (1996)
53.		Terpinen-4-ol	Reverchon and Senatore (1994)
54.		Lavandulol	Brunke et al. (1993), Lawrence (1996)
55.		Menthol	Reverchon and Senatore (1994)
56.		Menthyl acetate	Reverchon and Senatore (1994)
57.		Nerol	Reverchon and Senatore (1994)
58.		Geraniol	Reverchon and Senatore (1994)
59.		Thujone	DePasquale and Silvestri (1976), Mann and Staba (1986)
60.		2-Methyl-5,7-decadienal	Retamar et al. (1989), Lawrence (1996)
61.		Bornyl acetate	DePasquale and Silvestri (1976)
62.	Sesquiterpene	Azulene	Bohlman and Skuballa (1970), Reichling et al. (1983)
63.		β-Caryophyllene	Retamar et al. (1989), Lawrence (1996)
64.		*cis*-Caryophyllene	Reichling et al. (1983)
65.		Caryophyllene epoxide	Reichling et al. (1983)
66.		Chamazulene	Šorm et al. (1953), Cekan (1954), Lawrence (1987)
67.		Calamene	Reichling and Biederbeck (1991)
68.		α-Cubebene	Motl and Repcak (1979), Mann and Staba (1986)

(*Continued*)

TABLE 4.3 (*Continued*)
Chemical Compounds Identified as Constituents of Chamomile Oil Distilled from Flowers and Callus Cultures

S. No.	Type of Compound	Compound	Reference
69.		Bisabolene	Thune and Solberg (1980)
70.		*trans*-α-Farnesene	Lawrence (1987), Ramaswami et al. (1988), Lawrence (1996)
71.		*trans*-β-Farnesene	Lawrence (1987), Ramaswami et al. (1988), Marczal and Petri (1989), Lawrence (1996)
72.		β-Elemene	Brunke et al. (1993), Reverchon and Senatore (1994), Lawrence (1996)
73.		Germacrene D	Takahashi et al. (1981), Lawrence (1996)
74.		Cadinene	Reichling and Biederbeck (1991)
75.		γ-Cadinene	Matos et al. (1993), Lawrence (1996)
76.		α-Copaene	Tsutsulova and Antonova (1984)
77.		Bicyclogermacrene	Surburg et al. (1996), Lawrence (1996)
78.		α-Humulene	Brunke et al. (1993), Lawrence (1996)
79.		β-Selenine	Brunke et al. (1993), Lawrence (1996)
80.		δ-Cadinene	Shaath and Azzo (1993), Lawrence (1996)
81.		Guaiazulene	DePasquale and Silvestri (1976), Mann and Staba (1986)
82.		α-Patchoulene	Retamar et al. (1989)
83.		δ-Muurolene	Tsutsulova and Antonova (1984)
84.		δ-Muurolene	Motl and Repcak (1979), Mann and Staba (1986)
85.		α-Gurjunene	Kumar et al. (2001)
86.		Globulol	Kumar et al. (2001)
87.		β-Bourbonene	Kumar et al. (2001)
88.	Oxygenated sesquiterpene	α-Bisabolol	Šorm et al. (1953), Cekan (1954), Flaskamp et al. (1981), Hitziger et al. (2003)
89.		α-Bisabolol oxide A	Pöthke and Bulin (1969), Lawrence (1987)
90.		α-Bisabolol oxide B	Pöthke and Bulin (1969), Lawrence (1987)
91.		α-Bisabolol oxide C	Schilcher (1977), Mann and Staba (1986)
92.		α-Bisabolone oxide A	Marczal and Petri (1989), Mann and Staba (1986)
93.		Matricin	Flaskamp et al. (1981), Mann and Staba (1986)
94.		τ-Cadinol	Reverchon and Senatore (1994)
95.		Caryophyllene oxide	Reverchon and Senatore (1994)
96.		Chamavioline	Reichling and Biederbeck (1991)
97.		Matricarin	Cekan et al. (1959), Mann and Staba (1986)
98.		Chamomillol	Reichling and Biederbeck (1991)
99.		(*E*)-Nerolidol	Retamar et al. (1989), Lawrence (1996)
100.		(*E,E*)-Farnesol	Pöthke and Bulin (1969), Lawrence (1987)
101.		(*Z,E*)-Farnesol	Pöthke and Bulin (1969), Lawrence (1987)

TABLE 4.3 (Continued)
Chemical Compounds Identified as Constituents of Chamomile Oil Distilled from Flowers and Callus Cultures

S. No.	Type of Compound	Compound	Reference
102.	Others	Spathulenol	Motl et al.(1978), Lawrence (1987)
103.		Linear sesquiterpene lactone	Yamazaki et al. (1982), Mann and Staba (1986)
104.		Desacetylmatricin	Ness and Schmidt (1995), Ness et al. (1996)
105.		(cis-en-yn-dicycloether, MW 200)	Reichling et al. (1983)
106.		(trans-en-yn-dicycloether MW 200)	Reichling et al. (1983)
107.		cis-Spiroether (MW 214)	Lawrence (1987), Marczal and Petri (1989)
108.		trans-Spiroether (MW 214)	Lawrence (1987), Marczal and Petri (1989)
109.		Indole	Surburg et al. (1996), Lawrence (1996)
110.		Capric acid	Shaath and Azzo (1993), Lawrence (1996)
111.		Mint sulfide	Takahashi et al. (1981), Lawrence (1996)
112.		Caryophyllene epoxide	Reichling and Biederbeck (1991)
113.		Chamomilla ester I	Reichling et al. (1983), Mann and Staba (1986)
114.		Chamomilla ester II	Reichling et al. (1983), Mann and Staba (1986)
115.		Hexadec-11-yn-13,15-diene	Das et al. (2002)
116.		Oleanolic acid	Ahmad and Mishra (1997)
117.		β-Sitosterol	Ahmad and Mishra (1997)
118.		β-Sitosterol glucoside	Ahmad and Mishra (1997)
119.		3-Octanol	Raal et al. (2003)
120.		p-Cymen-8-ol	Raal et al. (2003)
121.		(E)-Anethole	Raal et al. (2003)
122.		Decanol	Raal et al. (2003)
123.		Decanoic acid	Raal et al. (2003)
124.		α-Bergamotene	Raal et al. (2003)
125.		Allo-aromadendrene	Raal et al. (2003)
126.		1-epi-Cubenol	Gosztola et al. (2010)
127.		epi-Bisabolol	Gosztola et al. (2010)
128.		n-Octanal	Raal et al. (2011)
129.		cis-3-Hexenyl isovalerate	Raal et al. (2011)
130.		α-Copaene	Raal et al. (2011)
131.		Isofaurinone	Raal et al. (2011)
132.		Dendrolasin	Raal et al. (2011)
133.		Dihydronerolidol	Raal et al. (2011)
134.		Viridoflorol	Raal et al. (2011)
135.		γ-Eudesmol	Raal et al. (2011)
136.		Cubenol	Raal et al. (2011)
137.		γ-Cadinol	Raal et al. (2011)

(Continued)

TABLE 4.3 (*Continued*)
Chemical Compounds Identified as Constituents of Chamomile Oil Distilled from Flowers and Callus Cultures

S. No.	Type of Compound	Compound	Reference
138.		α-Eudesmol	Raal et al. (2011)
139.		Geranyl tiglate	Raal et al. (2011)
140.		n-Octadecane	Raal et al. (2011)
141.		Hexahydrofarnesyl acetone	Raal et al. (2011)
142.		n-Nonadecane	Raal et al. (2011)
143.		Palmitic acid	Raal et al. (2011)
144.		n-Eicosane	Raal et al. (2011)
145.		γ-Palmitolactone	Raal et al. (2011)
146.		*cis*-Linoleneic acid	Raal et al. (2011)
147.		n-Tricosane	Raal et al. (2011)
148.		n-Pentacosane	Raal et al. (2011)
149.		Yomogi alcohol	Nurzyńska-Weirdak (2011)
150.		Artemisyl acetate	Nurzyńska-Weirdak (2011)
151.		Dehydro-sesquicineole	Nurzyńska-Weirdak (2011)
152.		α-Acoradiene	Nurzyńska-Weirdak (2011)
153.		Helofolen-12 al	Nurzyńska-Weirdak (2011)
154.		Apiole	Nurzyńska-Weirdak (2011)
155.		Achillin	Tschiggerl and Bucar (2012)
156.		Acetoxyachillin	Tschiggerl and Bucar (2012)
157.		Leucodin	Tschiggerl and Bucar (2012)
158.		*trans*-Limonene oxide	Tandon et al. (2013)
159.		1,6-Dioxaspironon-3-ene	Tandon et al. (2013)
160.		Gossonorol	Magiatis et al. (2001)
161.		Cubenol	Magiatis et al. (2001)
162.		α-Cadinol	Magiatis et al. (2001)
163.		1-Azulenethanol acetate	Magiatis et al. (2001)
164.		(−)-α-Bisabolol acetate	Magiatis et al. (2001)
165.		Berkheyaradulene	Szoke et al. (2004c)
166.		Geranyl isovalerate	Szoke et al. (2004c)
167.		Cedrol	Szoke et al. (2004c)
168.		β-Eudesmol	Szoke et al. (2004c)
169.		β-Selenine	Szoke et al. (2004c)
170.		*ar*-Curcumene	Petronilho et al. (2011)
171.		Calamenene	Petronilho et al. (2011)
172.		Isoledene	Petronilho et al. (2011)
173.		α-Calacorene	Petronilho et al. (2011)
174.		Dendrasaline	Petronilho et al. (2011)
175.		Cuparene	Petronilho et al. (2011)
176.		Dihydrocurcumene	Petronilho et al. (2011)

3. Quercetin and its glucosides
4. Patuletin and its glucoside
5. Isorhamnetin and its glucoside
6. Rutin
7. Chrysoeriol-7-glucoside

The flavonoids in chamomile flower extract were classified into five flavones according to the increasing polarity of the compounds by Kunde and Isaac (1979a). These were arranged as follows:

1. Methoxylated flavone aglyca
2. Hydroxylated flavone aglyca
3. Acetylated flavone monoglycoside
4. Flavone monoglycoside
5. Flavone diglycoside

The first flavonoid to be isolated from chamomile flowers was apigenin. This was followed by the isolation of quercimetrin by Lang and Schwand in 1957; apiin by Wagner and Kirmayer in 1957; apigenin glycosides by Tyihak, Sarkany-Kiss, and Varzár Petri in 1962; luteolin-7-glucoside, quercetin-7-glucoside, apigenin-7-glucoside, and patulitrin by Hörhammer, Wagner, and Salfner in 1963; quercetin-3-rutinoside and quercetin-3-galactoside by Elkiey, Darwish, and Mustafa in 1963; chrysoplenitin by Hänsel, Rimpler, and Walther in 1966; and aglyca of isorhamnetin and chrysoeriol by Reichling, Becker, Exner, and Dräger in 1979 (see Schilcher et al. 2005).

Kunde and Isaac (1979a) identified a new compound—a flavone derivative—and named it apigenin-7-glucoside acetate. Redaelli et al. (1979) reported the presence of phytosterols in the dried chamomile flowers. Two more new compounds were reported by Redaelli et al. (1980), which were apigenin-7-O-β-glucoside and its 2″- and 6″-monoacetates in the ligulate flowers of chamomile. Redaelli et al. (1982) further reported the presence of (2″,3″)- and (3″,4″)-diacetates of apigenin-7-O-β-glucoside. These diacetates were found in the ligulate or ray florets of chamomile.

Exner et al. (1981) found for the first time the presence of six methylated aglycones in the chamomile flowers growing in Egypt. These compounds were jaceidin, chrysoplenol, eupatoletin, spinacetin, axillarin, and eupalitin. Exner et al. (1980, 1981) also identified 6-methoxy kaempferol in chamomile flower extracts.

Ahmad and Mishra (1997) reported the presence of oleanolic acid, β-sitosterol, and β-sitosterol glucoside for the first time from the n-hexane extract of chamomile flowers.

Mulinacci et al. in 2000 reported the presence of caffeic acid and ferulic acid derivatives in the methanolic chamomile extracts. They identified 5-caffeoylquinic acid, 3-caffeoylquinic acid, 4-caffeoylquinic acid, quinic acid, ferulic acid-1-O-glycoside, caffeoylquinic acid derivative, ferulic acid-7-O-glycoside, dicaffeoylquinic acid derivative, 1,3-dicaffeoylquinic acid, quinic acid derivative, ferulic acid derivative, and quercetin derivative.

Zenkevich et al. (2002) reported the presence of posthumulone in the nonvolatile fraction of the oil extracted from chamomile flowers.

Švehlíková et al. (2004) identified the malonyl and caffeoyl derivatives of apigenin-7-O-glucoside, namely apigenin-7-(6″-malonylglucoside) and apigenin-7-(6″-caffeoylglucoside). These are unstable and rapidly degrade to form apigenin-7-glucoside.

Schilcher et al. (2005) reported that a Ukranian group claimed that they had isolated four coumarins, namely coumarin, isoscopoletin, esculentin, and scopoletin, from chamomile flowers.

The flavanone naringenin was reported in the chamomile flower extracts by Fonseca et al. (2007). They also reported the presence of the phenylpropanoids caffeic acid and chlorogenic acid.

Nováková et al. (2010) found the presence of rutin in chamomile tea extracts.

Lin and Harnly (2012) identified several flavonoids and hydroxycinnamates new to chamomile extract.

Petruľová-Poracká et al. (2013) identified skimmin (umbelliferone-7-O-β-D-glucoside), daphnin (daphnetin-7-O-β-D-glucoside), and daphnetin (7,8-dihydroxycoumarin) for the first time in diploid and tetraploid leaves and flowers of chamomile.

The distribution of these compounds has been found to be specific to different parts of the flowers. For example, Redaelli et al. (1980) reported that the dried ligulate florets contained 0.3%–0.5% apigenin and 7%–9% apigenin glucosides, but the tubular florets contained 0.2%–0.3% apigenin and its glucosides. Schreiber et al. (1990) and Pekic et al. (1994) also reported that the ligulate (disc) florets contain higher amounts of apigenin compared to the tubular (ray) florets. Redaelli et al. (1981b) found that the ligulate florets contained more umbelliferone (53.7 mg/100 g) and herniarin (129.8 mg/100 g) than the tubular florets, which contained 5.5 mg/100 g umbelliferone and 34/9 mg/100 g herniarin. Greger in 1975 reported the presence of quercetin 7-glucoside, isorhmnetin-7-glucoside, luteolin-7-glucoside, chrysoeriol-7-glucoside, 6-hydroxyluteolin-glucoside, luteolin-7-rhamnoglucoside, and chrysoeriol-7-rhamnoglucoside in the leaves (see Schilcher et al. 2005).

Reichling and Beidebeck (1991) reported that the flavonoids are not found in the shoot (stem and leaves) and the root of the chamomile plant. The coumarins and the phenolic acids are also not found in the shoot and root. However, Schilcher et al. (2005) found that there were some reports that documented the presence of isorhamnetin and luteolin in all parts of the plant except the root.

Schilcher et al. (2005) reported that the total flavonoid content differed according to the place of origin. They cited a study where the flavonoid content was determined in flowers obtained from 12 different locations but cultivated under the same conditions. It was found that the flavonoid content ranged from 1.0% to 2.57%.

The effect of heat on the flavonoids apigenin, luteolin, and chrysoeriol was studied by Hostetler et al. (2013). They found that these flavonoids were most stable at 100°C and pH 3 but progressively degraded at pH 5 or 7.

The flavonoids and their derivatives, coumarins, and other compounds identified in the solvent extracts of chamomile flowers are presented in Table 4.4.

4.9.1 MAJOR COMPOUNDS

The major compounds found in the chamomile extracts are described in Sections 4.9.1.1–4.9.1.5. Information about their uses are also presented.

4.9.1.1 Apigenin

Apigenin is a yellow-colored flavone. Its IUPAC name is 5,7-dihydroxy-2-(4-hydroxyphenyl)chromen-4-one. Its molecular formula is $C_{15}H_{10}O_5$ and the molecular weight is 270.2369 g/mol. It has a melting point of 345°C–350°C. Purified apigenin appears as yellow crystallized needles (CID 5280443).

The predicted density of luteolin is 1.548 g/cm^3 and its refractive index is 1.732 (CSID: 4444100).

Apigenin has been used as a dye or colorant since ancient times (Melo 2009). It is an anti-inflammatory and spasmolytic compound. It has also been found to have anticancerous properties (Figure 4.25). The pharmacological uses are described in Chapter 3.

4.9.1.2 Luteolin

Luteolin is a yellow-colored flavone. Its IUPAC name is 2-(3,4-dihydroxyphenyl)-5,7-dihydroxychromen-4-one. Its molecular formula is $C_{15}H_{10}O_6$ and the molecular weight is 286.2363 g/mol (CID 5280445). The predicted density of luteolin is 1.655 g/cm^3 and its refractive index is 1.768 (CSID: 4444102).

Luteolin has been used as a dye (Melo 2009). It is known to possess anti-inflammatory and antiasthmatic properties (Figure 4.26). The pharmacological properties of luteolin are described in detail in Chapter 3.

4.9.1.3 Quercetin

Quercetin is a flavone, which appears like yellow needles when purified. Its IUPAC name is 2-(3,4-dihydroxyphenyl)-3,5,7-trihydroxychromen-4-one. Its molecular formula is $C_{15}H_{10}O_7$ and the molecular weight is 302.2357 g/mol. Its predicted density is 1.799 g/cm^3 and the refractive index is 1.823 (CSID: 4444051).

Quercetin is a compound with bitter taste. It is insoluble in water; however, it is extremely soluble in ether, methanol, ethanol, acetone, pyridine, and acetic acid (CID 5280343). It is used in coloring, especially foods (Tanaka et al. 1993) and dyes (Melo 2009). It also has antioxidant properties (Figure 4.27). The pharmacological properties of quercetin are discussed in Chapter 3.

4.9.1.4 Herniarin

Herniarin is a coumarin, which appears as a white crystalline powder when purified. Its IUPAC name is 7-methoxychromen-2-one. Its molecular formula is $C_{10}H_8O_3$ and the molecular weight is 176.16872 g/mol (CID 10748). Its melting point is 118°C. Its predicted density is 1.249 g/cm^3 and the refractive index is 1.572 (CSID: 10295).

TABLE 4.4
Flavonoids and Coumarins Identified in Solvent Extracts of Chamomile Capitula

S. No.	Type of Compound	Compound	Reference
1.	Methoxylated flavones and flavonol	Jaceidin	Exner et al. (1981)
2.		Chrysoplenol	Exner et al. (1981)
3.		Eupatoletin	Exner et al. (1981)
4.		Spinacetin	Exner et al. (1981)
5.		Axillarin	Exner et al. (1981)
6.		Eupalitin	Exner et al. (1981)
7.		6-Methoxykaempferol	Exner et al. (1980, 1981)
8.		Chrysoeriol	Reichling and Beiderbeck (1991)
9.		Isorhamnetin	Reichling and Beiderbeck (1991)
10.		Patuletin	Reichling and Beiderbeck (1991)
11.		Chrysoplenetin	Reichling and Beiderbeck (1991)
12.		6,7-Dimethoxy quercetin	Exner et al. (1980)
13.		Apiin	Schilcher et al. (2005)
14.		Patulitrin	Schilcher et al. (2005)
15.	Hydroxylated flavones and flavonol	Luteolin	Reichling and Beiderbeck (1991)
16.		Quercetin	Reichling and Beiderbeck (1991)
17.		Apigenin	Reichling and Beiderbeck (1991)
18.		Quercimetrin	Schilcher et al. (2005)
19.		Naringenin	Fonseca et al. (2007)
20.	Acetylated Flavone monoglycoside	Apigenin-7-O-(6-O-acetyl) glycoside	Redaelli et al. (1980), Reichling and Beiderbeck (1991)
21.		Apigenin-7-O-(2-O-acetyl) glycoside	Redaelli et al. (1980), Reichling and Beiderbeck (1991)
22.		(2″,3″)-Diacetate of apigenin-7-O-β-glucoside	Redaelli et al. (1982)
23.		(3″,4″)-Diacetate of apigenin-7-O-β-glucoside	Redaelli et al. (1982)
24.	Acylated glucosides	Apigenin-7-(6″-malonyl-glucoside)	Švehlíková et al. (2004)
25.		Apigenin-7-(6″-acetyl-glucoside)	Švehlíková et al. (2004)
26.		Apigenin-7-(6″-coffeoyl-glucoside)	Švehlíková et al. (2004)
27.		Apigenin-7 (4″-acetyl-glucoside)	Švehlíková et al. (2004)
28.		Apigenin-7-(4″-acetyl, 6″-malonyl-Glucoside)	Švehlíková et al. (2004)

Chamomile Oil and Extract

TABLE 4.4 (*Continued*)
Flavonoids and Coumarins Identified in Solvent Extracts of Chamomile Capitula

S. No.	Type of Compound	Compound	Reference
29.		Apigenin-7-(monoacetyl/monomalonyl-glucoside)	Švehlíková et al. (2004)
30.	Flavone and flavonol monoglycoside	Apigenin-7-*O*-glucoside	Reichling and Beiderbeck (1991)
31.		Luteolin-7-*O*-glucoside	Reichling and Beiderbeck (1991)
32.		Luteolin-4-glucoside	Reichling and Beiderbeck (1991)
33.		Quercetin-7-*O*-glucoside	Reichling and Beiderbeck (1991)
34.		Quercetin-3-*O*-galactoside	Reichling and Beiderbeck (1991)
35.		Patuletin-7-*O*-glucoside	Reichling and Beiderbeck (1991)
36.		Isorhamnetin-7-*O*-glucoside	Reichling and Beiderbeck (1991)
37.	Flavone and flavonol diglycoside	Apigenin-7-*O*-apiosyl-glycoside	Reichling and Beiderbeck (1991)
38.		Apigenin-7-*O*-rutinoside	Reichling and Beiderbeck (1991)
39.		Apigenin-7-*O*-neohesperidoside	Reichling and Beiderbeck (1991)
40.		Luteotin-7-*O*-rutinoside	Reichling and Beiderbeck (1991)
41.		Quercetin-3-*O*-rutinoside	Reichling and Beiderbeck (1991)
42.	Coumarin	Herniarin	Reichling and Beiderbeck (1991)
43.		Umbelliferone	Reichling and Beiderbeck (1991)
44.		Coumarin	Schilcher et al. (2005)
45.		Isoscopoletin	Schilcher et al. (2005)
46.		Esculentin	Schilcher et al. (2005)
47.		Scopoletin	Schilcher et al. (2005)
48.	Others	Phytosterols	Redaelli et al. (1980)
49.		Oleanolic acid	Ahmad and Mishra (1997)
50.		β-Sitosterol	Ahmad and Mishra (1997)
51.		β-Sitosterol glucoside	Ahmad and Mishra (1997)
52.		Posthumulone	Zenkevich et al. (2002)
53.		5-Caffeoylquinic acid	Mulinacci et al. (2000)
54.		3-Caffeoylquinic acid	Mulinacci et al. (2000)
55.		4-Caffeoylquinic acid	Mulinacci et al. (2000)
56.		Quinic acid	Mulinacci et al. (2000)
57.		Ferulic acid-1-*O*-glycoside	Mulinacci et al. (2000)
58.		Caffeoylquinic acid derivative	Mulinacci et al. (2000)
59.		Ferulic acid-7-*O*-glycoside	Mulinacci et al. (2000)
60.		Dicaffeoylquinic acid derivative	Mulinacci et al. (2000)
61.		1,3-Dicaffeoylquinic acid	Mulinacci et al. (2000)
62.		Quinic acid derivative	Mulinacci et al. (2000)
63.		Ferulic acid derivative	Mulinacci et al. (2000)

(*Continued*)

TABLE 4.4 (Continued)
Flavonoids and Coumarins Identified in Solvent Extracts of Chamomile Capitula

S. No.	Type of Compound	Compound	Reference
64.		Quercetin derivative	Mulinacci et al. (2000)
65.		Caffeic acid	Fonseca et al. (2007)
66.		Chlorogenic acid	Fonseca et al. (2007)
67.		Ferulic acid glucose	Lin and Harnly (2012)
68.		3,4-Dicaffeoylquinic acid	Lin and Harnly (2012)
69.		1,5-Dicaffeoylquinic acid	Lin and Harnly (2012)
70.		4,5-Dicaffeoylquinic acid	Lin and Harnly (2012)
71.		Hexahydroxyflavone 3-O-hexoside	Lin and Harnly (2012)
72.		Pentahydroxyflavone 7-O-hexoside	Lin and Harnly (2012)
73.		Hexahydroxyflavone 3-O-dihexoside	Lin and Harnly (2012)
74.		Apigenin 7-O-malonylglucoside	Lin and Harnly (2012)
75.		Apigenin 7-O-caffeoylglucoside	Lin and Harnly (2012)
76.		Apigenin 7-O-malonylacetylglucoside	Lin and Harnly (2012)
77.		Dihydroxy-tetramethoxyflavone	Lin and Harnly (2012)
78.		Hexahydroxyflavone 3-O-dihexoside	Lin and Harnly (2012)
79.		Pentahydroxymethoxyflavone 7-O-glucoside	Lin and Harnly (2012)
80.		Pentahydroxymethoxyflavone glucoside	Lin and Harnly (2012)
81.		Isorhamonetin-3-O-rhamnosylgalactoside	Lin and Harnly (2012)
82.		Isorhamnetin 3-O-glucoside	Lin and Harnly (2012)
83.		Petuletin 7-O-glucoside	Lin and Harnly (2012)
84.		Tetrahydroxy-dimethoxyflavone 7-O-glucoside	Lin and Harnly (2012)
85.		Pentahydroxymethoxy flavone caffeoylglucoside	Lin and Harnly (2012)
86.		Pentahydroxymethoxyflavone caffeoylglucoside	Lin and Harnly (2012)
87.		Skimmin (umbelliferone-7-O-β-D-glucoside)	Petruľová-Poracká et al. (2013)
88.		Daphnin (daphnetin-7-O-β-D-glucoside)	Petruľová-Poracká et al. (2013)
89.		Daphnetin (7,8-dihydroxycoumarin)	Petruľová-Poracká et al. (2013)

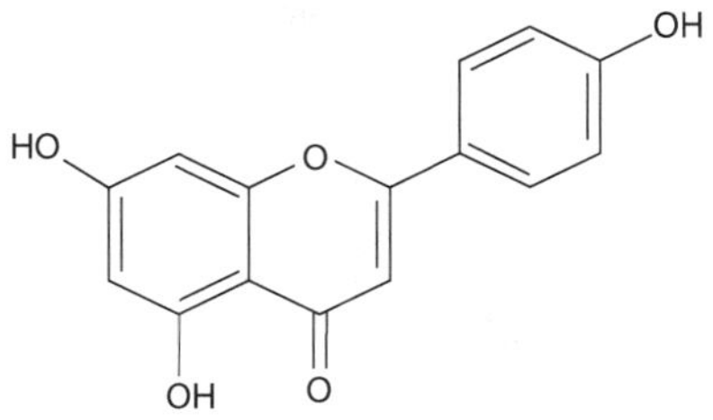

Chamomile Oil and Extract

FIGURE 4.25 Apigenin

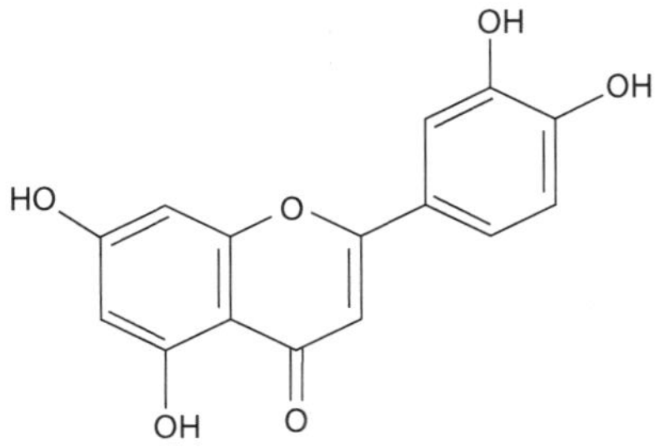

FIGURE 4.26 Luteolin

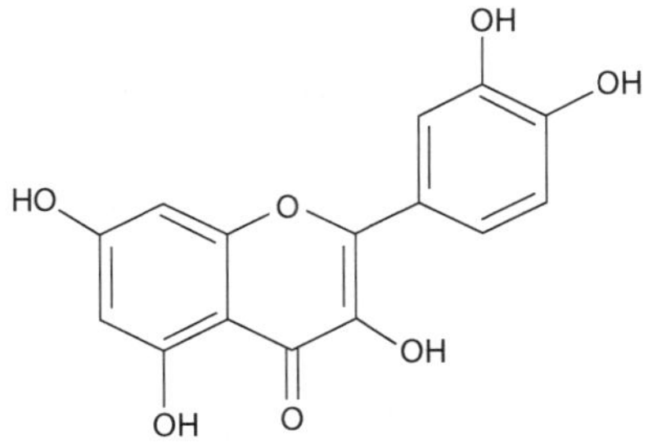

FIGURE 4.27 Quercetin

Herniarin is known to irritate the lungs, skin, and eyes (CSID: 10295; Paulsen et al. 2010) (Figure 4.28).

4.9.1.5 Umbelliferone

Umbelliferone is a hydroxycoumarin, which appears as a white powder when purified. Its IUPAC name is 7-hydroxychromen-2-one. Its molecular formula is $C_9H_6O_3$ and the molecular weight is 162.14214 g/mol (CID 5281426). Its melting point is 235°C. Its predicted density is 1.403 g/cm³ and the refractive index is 1.64 (CSID: 4444774).

Umbelliferone is known to have UV adsorbing properties and, therefore, it is used in sunscreen lotions (Hoffman 2003). It is also known to have antioxidant properties, but more work needs to be done to fully elucidate this property (Figure 4.29).

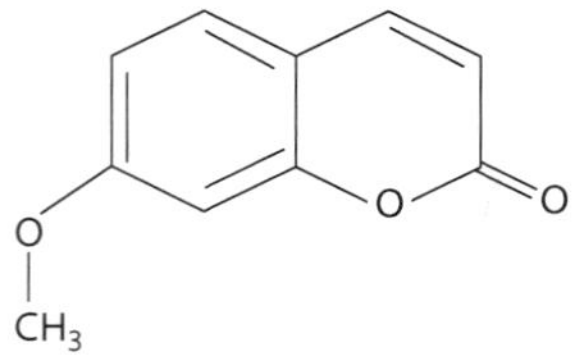

FIGURE 4.28 Herniarin

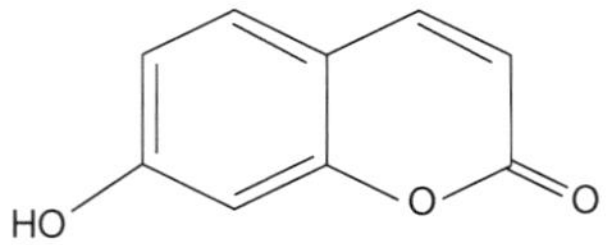

FIGURE 4.29 Umbelliferone

4.9.2 OTHER COMPOUNDS

In addition to the major compounds described earlier, several other useful compounds, such as polysaccharides, lipidic and ceraceous substances, and phenyl carboxylic acids, have been identified in chamomile.

4.9.2.1 Polysaccharides

Chamomile flowers are known to contain polysaccharides. These polysaccharides are located on the cells of the cypsela or the fruit wall. On coming in contact with water, the polysaccharides swell. The polysaccharides have been reported to have immunostimulating properties (Laskova and Uteshev 1992; Uteshev et al. 1999).

4.9.2.2 Lipidic and Ceraceous Substances

The chamomile flowers have been reported to possess lipidic and ceraceous substances such as hydrocarbons, aliphatic esters, sterol esters or phytosterols, triterpenol esters, triglycerides, ketoesters, and esters containing acetylene (Schilcher et al. 2005).

4.9.2.3 Phenyl Carboxylic Acids

The phenyl carboxylic acids identified in chamomile are syringic acid, vanillic acid, anisic acid, and caffeic acid (Schilcher et al. 2005).

4.10 EXTRACTION METHODS

The useful compounds of chamomile are extracted industrially by various methods. There is no single recommended method of extraction, which is because the different methods of extraction yield different types of compound. For example, hydrodistillation of the flowers yield mostly terpenes, and solvent extraction of the flowers yield mostly flavonoids. Therefore, a judicious selection of the extraction method is necessary depending on the nature of the chemical formulation in which the chamomile compounds are going to be used.

Chamomile Oil and Extract

The most common methods of extraction are infusion, steam distillation, and solvent extraction. Other methods being optimized are SCFE and vacuum headspace.

4.10.1 Infusion

Chamomile flowers are commonly taken as tea or as baths in the form of infusion. To prepare the infusion, the flowers are steeped in boiling water, strained, and used. The infusion contains mostly the hydrophilic compounds such as the flavonoids. Several methods have been devised to optimize the process of infusion so that both the hydrophilic and the lipophilic compounds could be extracted to get the maximum benefits.

Harbourne et al. (2009) studied the optimum extracting and processing conditions of infusion and concluded that water extraction at 90°C for 20 minutes yields an extract rich in phenols and low in turbidity. This extract could be used in beverages as an anti-inflammatory agent.

4.10.2 Steam Distillation

The quality of the essential oil extracted from different parts of chamomile plant depends on the method of extraction to a considerable extent. Presently, the most valued oil is the steam-distilled oil of chamomile flowers. The principle used in steam distillation is the collection of the volatile compounds in steam and separating them from the water by vaporization according to their boiling points followed by condensation.

Chamomile flowers are put in a distillation unit retort and steam is applied from below. As the steam passes through the flowers, it ruptures the cells of the flowers that contain the volatile compounds. These compounds are released and mixed with the hot water vapors. As more heat is applied, the constituents evaporate. The condensers of the distillation unit collect the volatile vapors and help them to condense. The oil floats in water and is collected over a column of water in a separator. The oil is collected from the top of the column and sometimes a suitable solvent such as xylene or hexane is used for dissolving any leftover oil for optimum collection. The solvent is evaporated and the oil is stored in dark glass bottles or suitable containers that protect the oil from light and decomposition.

The quantity and quality of the oil depends on the duration of distillation. At different geographical locations, the period of distillation has varied by tradition or as specified by their respective pharmacopoeias. The recommended time or duration of distillation is varied and depends on the prescription of the pharmacopoeias (Mechler 1979). For example, the German pharmacopoeia recommends 2 hours of distillation, which yields 0.29% oil, but the European pharmacopoeia recommends 4 hours of distillation, which yields 0.33% oil.

The fresh chamomile flowers, on steam distillation for a period of 2–13 hours, yield about 0.2%–1.9% essential oil. Clearly, the amount of oil yield increases with the increase in the duration of distillation (Mann and Staba 1986).

During the process of distillation, loss of oil quantity and quality occurs and efforts should be made to minimize these losses. The losses occur due to several factors. Oil loss occurs when the oil sticks to the walls of the condenser due to crystallization. To minimize this loss, proper rinsing with solvents may be carried out wherever possible. There are several components in the oil that are thermolabile. Excessive heat may degrade these products. There are several compounds that are soluble in water. These compounds get dissolved in the steam and are lost when the water is collected and discarded. Therefore, distillation of these collected waters is recommended to obtain these compounds (Mann and Staba 1986).

4.10.3 SOLVENT EXTRACTION

Solvent extraction involves dissolving the compounds present in the flower in solvents of appropriate polarity followed by separating or filtering and collecting the solvent containing the compounds. The residue is discarded. The flowers are macerated or softly broken by soaking in the solvent for several hours. This results in the breaking up of the cells containing the essential oil. The oil containing the compounds is released into the solvent.

The products made from the extracts are sometimes taken orally and sometimes applied on the body. Therefore, the solvents used for extraction should be appropriate in terms of nontoxicity. The most commonly used solvents are ethyl alcohol, water, glycerol, propylene glycol, sunflower oil, and n-hexane. The solvents are evaporated and the residue is kept to the minimum levels so as to make the extract nontoxic.

The oil extracted using sunflower oil is usually a dark blue, viscous, opaque liquid. High levels of en-yn-dicycloethers are found in the extracted oil. Herniarin and the flavonoids are also found in the extracted oil, but these are not found in the distilled oil (Mann and Staba 1986). The oils extracted under alkaline conditions are yellow due to the absence of azulenes (Ness et al. 1996).

Ahmadi-Golsefidi and Soleimani (2007) reported that the chamomile flowers could be extracted by maceration with propylene glycol and water as solvent.

In addition to the flavonoids and coumarins, mucilage and waxes are also extracted by this process. Amino acids, amines, sugars, and oligosaccharides have been detected in dry extracts (Garcia Peña et al. 2009). Pereira et al. (2008) obtained oil from the seeds of chamomile by grinding them and extracting them using a Soxhlet extraction system. They tested the antimicrobial activity of the seed oil and concluded that chamomile seed oil can be obtained using a simple extraction method.

The optimum extraction of the compounds depends not only on the type of the solvents but also on the extraction procedures. The duration of maceration, temperature, and pressure play a very important role. Several extraction procedures have been devised (Carle et al. 1989; Vogel and Schmidt 1993; Ness and Schmidt 1995) and patented (Snuparek et al. 1988).

The medicinal preparations that involve oral intake need to be free of toxic substances and many of the solvents used in extraction may be toxic. Only a few nontoxic solvents are available such as ethyl alcohol, glycerol, propylene glycol, and water.

Carle et al. (1989) devised a new method of producing chamomile extracts using ethyl alcohol and water in the ratio of 25:75. They took both the dried and fresh chamomile flowers and macerated them in the solution at 20°C–30°C for 90 minutes. The extracted material was put through a Wilmes press. After 14 days, the resulting solution was filtered and stored in amber glass bottles at room temperature. The extracts from fresh flowers were found to be stable after 10 months. The extracts from the dried flowers had lost 70% of their matricin content within 2 years.

Vogel and Schmidt (1993) devised a method to obtain an alcohol-free liquid extract of chamomile flowers. They took dried chamomile flowers. The flowers were harvested and dried for 18 hours at 45°C. The dried chamomile flowers were macerated in methyl alcohol/propylene glycol (70:30) for 1 hour with a magnetic stirrer. The flower/solvent ratios were kept at 1:10 and 1:25. The mixture was allowed to percolate. The percolate was then vacuum distilled to remove the methyl alcohol. Water was added during this process. The extract was collected and the color of the extract was described to be the same as that of ethyl alcohol extract of chamomile flowers. The methyl alcohol was quantified to be <100 ppm. Thereafter, propylene glycol was added to adjust to the initial volume. On analysis of the extract, it was found that both the lipophilic and the hydrophilic compounds could be extracted. (−)-α-Bisabolol, bisabolol oxides A and B, apigenin-7-glucoside, apigenin, and *cis*- and *trans*-dicycloethers were obtained in high amounts in the extract. The compounds were assessed for their stability, and they were found to be stable and the amount of content was constant. The stability of the compounds was compared with the ethyl alcohol/water extract and was found to be equivalent.

Meneses-Reyes et al. (2008) reported an optimized process of flavonoid extraction from chamomile flowers. The optimum yield of flavonoids was obtained when the flower particles used were large (>2.00 mm) and subjected to extraction with methanol/water (80:20) for 1 hour at 70°C.

4.10.4 Supercritical Carbon Dioxide Fluid Extraction

Carbon dioxide as a solvent is being increasingly advocated because it can extract the lipophilic compounds, is cheap, and is easy to use. Carbon dioxide can be used at low temperatures, which prevents the degradation of the compounds. The extracts are devoid of toxic solvents or thermally degraded products. The process of extraction using carbon dioxide is called supercritical carbon dioxide fluid extraction (SCFE. In this process, carbon dioxide at high temperature and pressure is used to extract the oil from chamomile flowers. In 1978, Stahl and Schulz explored the possibility of extraction of chamomile essential oil using SCFE. They tested pressures from 72 to 500 bar and temperatures between 31°C and 80°C (Reverchon and Senatore 1994).

Vuorela et al. (1990) attempted to extract the essential oil using a supercritical carbon dioxide at 40°C and 200 bar. They obtained (−)-α-bisabolol, bisabolol oxides A and B, β-farnesene, and traces of matricin. They analyzed the extract using headspace gas chromatography and MS. They found that the most volatile components of the essential oils could be obtained using carbon dioxide extraction.

Reverchon and Senatore (1994) used SCFE to extract the essential oil of chamomile flowers (110 g). They used carbon dioxide at a high pressure of 300 bar at a flow rate of 0.8 kg/h. The temperatures were kept at a range from −10°C to 60°C. The duration of extraction was 150 minutes. The compounds were separated in fractions in a series of separators by reducing the temperature and pressure. The extraction yielded a yellow-colored oil. The oil when assessed by capillary GC and GC-MS was found to contain bisabolol oxide B (16.9%) and bisabolol oxide A (50.4%). Matricin content was evaluated as chamazulene (3.5%). The *cis*- and *trans*-dicycloethers constituted 12.9% of oil. Two more dicycloethers (*cis*- and *trans*-isomers) (MW 214) were also found in the oil in trace amounts. These compounds were identified as *cis*- and *trans*-2[hexadi-yn-(2,4)-ylidene]-16-dioxaspiro-[4,5]-decene.

Scalia et al. (1999) compared the extracts obtained using SCFE with Soxhlet extraction, steam distillation, and maceration. The SCFE using carbon dioxide at 90 atm and 40°C for 30 minutes yielded higher levels of essential oil compared with steam distillation. The SCFE yielded higher levels of apigenin as compared to Soxhlet extraction or maceration. It has been found that the SCFE method is not too efficient for extracting flavonoids, such as apigenin-7-glucoside (Scalia et al. 1999; Kaiser et al. 2004).

Kotnik et al. (2007) extracted the essential oil of chamomile flowers using the SCFE method and found that the highest yields of matricin, (−)-α-bisabolol, and chamazulene were obtained using carbon dioxide at 250 bar and 40°C. A two-step separation process ensured that unwanted components were separated from the essential oil. The extraction process was successfully transferred to a pilot scale.

4.10.5 Vacuum Headspace

Vacuum headspace equipment comprises an airtight dome-shaped apparatus that surrounds the sample and the odor or aroma compounds from the sample are extracted by creating a vacuum. These compounds are adsorbed to a suitable adsorbent and later analyzed through different methods such as GC-MS or NMR spectrometry. The compositions of essential oil in the chamomile flowers extracted by SCFE and hydrodistillation were studied using vacuum headspace method (Lawrence 1996). Difference in area percentages of several constituents was noticed in both the samples; for example, artemisia ketone, which was found as a minor constituent of the hydrodistilled oil (0.44%), was found as a major constituent (10.0%) of the headspace analysis. These results substantiate the fact that the presence of a chemical constituent and its amount in the oil depends on the method of extraction.

4.10.6 Solid-Phase Extraction

SPE is a method by which the sample is enriched before the actual analysis. The mechanism of SPE uses the column chromatography technique. It uses a small column packed with silica that forms the stationary phase. Sometimes the silica is bonded to specific functional groups that enable specific separation. The sample is dissolved in a suitable solvent and passed through the stationary phase. The compounds in the sample get adsorbed in the stationary phase. These are later eluted out with suitable solvents. Tschiggerl and Bucar (2012) isolated the volatile fraction of chamomile infusion or tea by two different methods, the hydrodistillation method and the SPE method. They found a remarkable difference in the amounts of compounds present in the hydrodistilled oil and the volatile portion of the infusion.

Krüger (2010) proposed a method for complete separation of the volatile compounds based on a combination of hydrodistillation and SPE methods. It was named simultaneous distillation SPE. An RP-18 solid phase was used to adsorb and fix the volatile compounds. It helps to characterize the volatile constituents.

4.10.7 Solid-Phase Microextraction

Solid-phase microextraction (SPME) involves a fiber column coated on the outside with an adsorbent. The adsorbent could be a polymer or sorbent such as fused silica. The compounds in the sample get adsorbed to the adsorbent layer. The fiber is then introduced into a hot injector of a gas chromatogram that desorbs the compounds. These compounds are eluted and detected by a mass spectrometer. Several variants of this method in combination with other techniques have been proposed by researchers to effectively analyze chamomile compounds. Rubiolo et al. (2006) proposed headspace–solid-phase microextraction gas chromatography–principal component analysis (HS-SPME GC-PCA) as an alternative to GC-PCA as it was found a rapid technique by which chamomile chemotypes could be distinguished. Rafieiolhossaini et al. (2012) compared the methods of HS-SPME with steam distillation-solvent extraction (SDSE). They found that the HS-SPME contained (E)-β-farnesene (49%), artemisia ketone (10%), and germacrene D (9%) whereas the SDSE contained α-bisabolol oxide A (42%), chamazulene (21%), and (Z)-spiroether (8%). On the basis of their observations, they suggested that HS-SPME was a sensitive technique that could be used for rapid screening and quality assessment of chamomile oil. Merib et al. (2013) found that cold-fiber headspace–solid-phase microextraction (CF-HS-SPME) was effective for the extraction of the more volatile compounds. The extraction time of CF-HS-SPME was also less at 15 minutes as compared to the commonly used HS-SPME technique that took 40 minutes. The researchers were able to determine the volatile fraction using GC-MS.

4.11 IDENTIFICATION METHODS OF CHEMICAL COMPOUNDS IN CHAMOMILE OIL AND EXTRACT

The compounds present in the essential oil or extracts of chamomile flowers are detected by a variety of methods. The most common method is the thin-layer chromatography (TLC) (Pachaly 1982; Wagner and Bladt 1996; Medić-Šarić et al. 1997). The other methods used are column chromatography, pressurized ultra-microchamber combined with TLC (Mincsovics et al. 1979) and HPLC (Redaelli et al. 1981a,b), HPTLC (Zeković et al. 1994), GC (Padula et al. 1977; Cartony et al. 1990; Lakszner et al. 1990), HPLC–high-resolution gas chromatography (Hyvönen et al. 1991), GC-MS (Orav et al. 2001), HSGC (Vuorela et al. 1989), NMR spectroscopy (Motl et al. 1978; Kunde and Isaac 1979b; Carle et al. 1992), and infrared mass spectrometry (Carle et al. 1990). The other techniques include liquid sampling mass spectrometry and ultraviolet and visible spectroscopy (UV-Vis) (Ristić et al. 2007).

4.11.1 Thin-Layer Chromatography

TLC is used to identify the main constituents of chamomile essential oil and extract. TLC is frequently used because of its low cost and flexibility in terms of selection of the reagents used. Various mobile phases are used for the TLC of chamomile compounds such as toluene, ethyl acetate, acetic acid, dichloromethane, and chloroform in various combinations. This is a rapid and simple technique but the resolution of the compounds is low as compared to other advanced separation techniques. Medić-Šarić et al. (1997)

tested 11 mobile phases for separating the essential oil components of chamomile to determine the best phase. The isolated compounds were identified by spraying with vanillin sulfuric acid reagent. Based on their study, they reported that the mobile phases of chloroform/toluene (75:25 v/v) and chloroform/toluene/ethyl acetate (65:30:5) were the most favorable mobile phases that ensured optimum separation of the compounds.

4.11.2 THIN-LAYER CHROMATOGRAPHY–ULTRA VIOLET SPECTROMETRY

TLC could be combined with the detectors such as UV spectrometers to identify the separated compounds. Pakic et al. (1989) reported a method for determining apigenin and apigenin 7-O-β-glucoside from the ligulate flowers of chamomile using TLC-UV spectrometry. They used chloroform/methanol (9:1) as the solvent system for separating apigenin and methylene chloride/methanol (8:2) with 0.1% water for separating apigenin 7-O-β-glucoside. UV spectrophotometry was used to determine the presence of flavonoids in alcohol and water solutions. They reported that the ligulate florets contained 0.33% apigenin and 2.39% apigenin 7-O-β-glucoside.

4.11.3 SPECTROCOLORIMETER

A spectrocolorimeter is a spectrometer that can measure the colors as seen by humans (tristimulus values). Khanina et al. (1995) used a spectrocolorimeter for the quantitative determination of chamazulene. This method can be used to estimate the chamazulene content in chamomile oil.

4.11.4 HIGH-PERFORMANCE THIN-LAYER CHROMATOGRAPHY

HPTLC is an automated form of TLC where the stationary phase of silica is used at a low temperature to separate terpenoids that are sensitive to high temperatures. It is considered the fastest method to determine the quality of the compounds in the biological extracts. Zeković et al. (1994) used HPTLC for qualitative and quantitative analysis of nonvolatile and volatile compounds of chamomile.

4.11.5 HIGH-PERFORMANCE LIQUID CHROMATOGRAPHY

HPLC is a chromatographic technique that uses a column packed with a stationary phase such as silica gel for normal phase and C4-, C8-, or C18-bonded silica for reversed-phase HPLC. The sample is injected along with a mobile solvent such as water or acetonitrile or sometimes methanol under high pressure of 6000 psi. The elution time is reduced compared to TLC, and the separation of the compounds is better than TLC. These compounds are detected with a detector attached to the HPLC apparatus. Redaelli et al. (1981a) used reversed-phase HPLC to analyze apigenin and its glucosides in chamomile flowers. They carried out HPLC using a reverse phase column and eluting with acetonitrile/water, acetic acid system. They could detect the flavonoids apigenin, apigenin-7-glucoside, and apigenin-7-acetyl glucoside in the ligulate florets of chamomile. Using the same method they could not detect apigenin and its glucosides in the tubular florets. They also applied this method to chamomile extracts. HPLC was used by Redaelli et al. (1981b) to determine the amounts of coumarins of chamomile. Pekic et al. (1994) developed a method to determine the

Chamomile Oil and Extract

apigenin 7-O-β-glucoside and its mono- and diacetylated derivatives by the HPLC method. They screened chamomile flowers from different locations. They found that the ligulate flowers contained more flavonoids in comparison to the tubular ones.

The HPLC apparatus could be combined with various detectors such as mass spectrometers or diode array detectors (DAD). Mass spectrometers measure the masses of the molecules, whereas DAD detect the response of the eluting molecules to UV and visible spectrum. Mulinacci et al. (2000) analyzed the polyphenolic compounds using HPLC-MS and HPLC-DAD. They extracted the flowers in methanol and compared the UV-Vis and MS spectra. They were able to identify all the main polyphenolic compounds in different parts of the chamomile flowers.

4.11.6 Overpressured-Layer Chromatography

Today the terms HPLC and HPTLC have been replaced by the term planar chromatography (Reich 2000). At present, a separation technique called the overpressured-layer chromatography (OPLC) that combines the advantages of conventional TLC/HPTLC with those of HPLC is in use. It exploits the advantages of planar-layer system for detection, isolation, and identification of new antimicrobials, antineoplastics, biopesticides, and other biologically active substances. OPLC is also used for studying fundamental biochemical reactions and mechanisms (Tyihák 1987; Tyihák et al. 2012).

4.11.7 Ultra-Performance Liquid Chromatography

HPLC and HPTLC take typically 45–50 minutes for separating the compounds of chamomile (Haghi et al. 2014). A technique called the ultra-performance liquid chromatography (UPLC) uses higher pressure than the HPLC to separate the compounds. The typical pressure used in UPLC is 15,000 psi. Haghi et al. (2014) used UPLC coupled with photodiode array detector for the separation and detection of phenolic compounds in the chamomile extracts. They found that the UPLC method gave a higher method sensitivity, speed, and resolution compared with HPLC.

4.11.8 Capillary Electrochromatography

Capillary electrochromatography (CEC) is a technique that combines capillary electrophoresis and HPLC. In capillary electrophoresis, the sample is separated by the use of an electrically driven flow through the separation column. The chromatographic capillary column is packed with the stationary phase of HPLC and the sample mixed with solvents is passed at a high voltage through the column. Separation occurs due to electroosmosis. Fonseca et al. (2007) used CEC to analyze the phenolic content of chamomile extracts. They used a Hypersil SCX/C18 column with 50% acetonitrile in 2.8 pH phosphate buffer. The compounds were separated in less than 7.5 minutes. They could identify 11 bioactive phenolic compounds.

4.11.9 Gas Chromatography-Mass Spectrometry

GC is a technique where a liquid sample is injected into a column packed or coated with an inert material such as silica and containing a liquid stationary phase. The sample is immediately vaporized on injecting and carried by a gas that forms the mobile phase. The compounds are separated in the stationary phase and eluted, which could be detected using various detectors such as mass spectrometer. The column GC-MS has been used by many researchers to detect the compounds of chamomile. Das et al. (1999), Kumar et al. (2001), and Das et al. (2002) analyzed chamomile essential oil using GC-MS and identified several compounds. Orav et al. (2001) studied the volatile constituents of chamomile oil by GC-MS and identified 37 constituents.

4.11.10 Nuclear Magnetic Resonance Spectroscopy

NMR spectroscopy determines the chemical environment of the nuclei. It is generally used to deduce the molecular structure of the compounds. In this technique, deuterium-labeled solvents are commonly used. Deuterium NMR spectroscopy has been used to detect bisabolol adulterants (Carle et al. 1992). There is a difference in the deuterium isotope distribution between two plants in nature. This difference is attributed to the isotope effects in the primary and secondary metabolism. Using deuterium NMR spectroscopy, it is possible to detect adulterant compounds. The oil of *Vanillosmopsis erythropappa* is used as an adulterant as it has large amount of α-bisabolol. However, the oil is different from chamomile oil where the former oil accumulates more deuterium in the α-bisabolol when subjected to quantitative deuterium NMR spectroscopy. Deuterium gets uniformly distributed in the bisabolol of *V. erythropappa*, whereas chamomile oil has sparse and scanty distribution of deuterium in the α-bisabolol molecule, which is readily detected in the H-NMR spectra.

REFERENCES

Ahmad, A. and Mishra, L. N. 1997. Isolation of herniarin and other constituents from *Matricaria chamomile* flowers. *International Journal of Pharmacognosy* 35: 121–125.

Ahmadi-Golsefidi, M. and Soleimani, M. H. 2007. A new method for determination of effective constituents of chamomile extracts. *ISHS Acta Horticulturae* 749: 193–196.

Andreucci, A. C., Ciccarelli, C., Desideri, I., and Pagni, A. M. 2008. Glandular hairs and secretory ducts of *Matricaria chamomilla* (Asteraceae): Morphology and histochemistry. *Annales Botanici Fennici* 45: 11–18.

Ash, M. and Ash, I. 2004. *A Handbook of Preservatives*. New York: Synapse Information Resources, p. 96.

Beck, J. J., Merrill, G. B., Higbee, B. S., Light, D. M., and Gee, W. S. 2009. In situ seasonal study of the volatile production of almonds (*Prunus dulcis*) var. 'Nonpareil' and relationship to navel orangeworm. *Journal of Agricultural and Food Chemistry* 57(9): 3749–3753.

Bero, J., Beaufay, C., Hannaert, V., Hérent, M. F., Michels, P. A., and Quetin-Leclercq, J. 2013. Antitrypanosomal compounds from the essential oil and extracts of *Keetia leucantha* leaves with inhibitor activity on *Trypanosoma brucei* glyceraldehyde-3-phosphate dehydrogenase. *Phytomedicine* 20(3–4): 270–274.

Bohlman, F. and Skuballa, W. 1970. Polyacetylenic compounds 175. Synthesis and biogenesis of Anthemis thioethers. *Chemische Berichte* 103: 1886–1893.

Brunke, E. J., Hammerschmidt, F. J., and Schmaus, G. 1993. Flower scent of some traditional medicinal plants. In *Bioactive Volatile Compounds from Plants*; Teranishi, R., Buttery, R. G., and Sugisawa, H. (eds.). ACS Symposium Series 525; American Chemical Society: Washington, DC, pp. 282–296.

Burdock, G. A. 2005. *Fenaroli's Handbook of Flavor Ingredients.* Fifth Edition. Boca Raton, FL: CRC Press.
Carle, R., Beyer, J., Cheminat, A., and Krempp, E. 1992. 2H NMR determination of site-specific natural isotope fractionation in (−)-α-bisabolols *Phytochemistry* 31: 171–174.
Carle, R., Dölle, B., and Reinhard E. 1989. A new approach to the production of chamomile extracts. *Planta Medica* 55: 540–543.
Carle, R., Fleischhauer, I., Beyer, J., and Reinhard, E. 1990. Studies on the origin of levo-a-bisabolol and chamazulene in chamomile preparations: Part-I. Investigations by isotope radio mass spectroscopy (IRMS). *Planta Medica* 56(5): 456–460.
Cartony, G., Goretti, G., Russo, M. V., and Zacchei, P. 1990. Microcapillary gas chromatographic analysis of chamomile. *Annali di Chimica* 80: 523–536.
Čekan, Z., Herout, V., and Šorm, F. 1954. On terpenes. 62. Isolation and properties of the pro-chamazulene from *Matricaria chamomilla* L., a further compound of the guaianolide group. *Collection of Czechoslovak Chemical Communications* 19: 798–804.
Čekan, Z., Prochazka, V., Herout, V., Šorm, F. 1959. On terpens. CI. Isolation and constitution of matricarin, another guaianolide from Chamomile (*Matricaria chamomilla* L.). *Collection of Czechoslovak Chemical Communications.* 24, 1554–1557. http://cccc.uochb.cas.cz/24/5/1554/ (Accessed March 18, 2014).
Chamomillol. 2013. http://webbook.nist.gov/cgi/inchi/InChI%3D1S/C15H26O/c1-10%282%2915%2816%298-7-12%284%2913-6-5-11%283%299-14%2813%2915/h9-10,12-14,16H,5-8H2,1-4H3/t12%3F,13%3F,14%3F,15-/m1/s1 (Accessed March 18, 2013).
CID 10748. http://pubchem.ncbi.nlm.nih.gov/summary/summary.cgi?cid=10748 (Accessed March 22, 2013).
CID 17100. http://pubchem.ncbi.nlm.nih.gov/summary/summary.cgi?cid=17100&loc=ec_rcs#x27 (Accessed March 18, 2013).
CID 439250. http://pubchem.ncbi.nlm.nih.gov/summary/summary.cgi?cid=439250 (Accessed March 18, 2013).
CID 445070. http://pubchem.ncbi.nlm.nih.gov/summary/summary.cgi?cid=445070&loc=ec_rcs (Accessed March 18, 2013).
CID 5280343. http://pubchem.ncbi.nlm.nih.gov/summary/summary.cgi?cid=5280343&loc=ec_rcs#x27 (Accessed March 22, 2013).
CID 5280435. http://pubchem.ncbi.nlm.nih.gov/summary/summary.cgi?cid=5280435&loc=ec_rcs (Accessed March 18, 2013).
CID 5280443. http://pubchem.ncbi.nlm.nih.gov/summary/summary.cgi?cid=5280443&loc=ec_rcs#x27 (Accessed March 22, 2013).
CID 5280445. http://pubchem.ncbi.nlm.nih.gov/summary/summary.cgi?cid=5280445&loc=ec_rcs#x27 (Accessed March 22, 2013).
CID 5281426. http://pubchem.ncbi.nlm.nih.gov/summary/summary.cgi?cid=5281426&loc=ec_rcs (Accessed March 22, 2013).
CID 5281520. http://pubchem.ncbi.nlm.nih.gov/summary/summary.cgi?cid=5281520#x27 (Accessed March 18, 2013).
CID 5317570. http://pubchem.ncbi.nlm.nih.gov/summary/summary.cgi?cid=5317570&loc=ec_rcs (Accessed March 18, 2013).
CID 5352494. http://pubchem.ncbi.nlm.nih.gov/summary/summary.cgi?cid=5352494#x400 (Accessed March 13, 2013).
CID 5352496. http://pubchem.ncbi.nlm.nih.gov/summary/summary.cgi?cid=5352496 (Accessed March 13, 2013).
CID 5356544. http://pubchem.ncbi.nlm.nih.gov/summary/summary.cgi?cid=5356544&loc=ec_rcs (Accessed March 18, 2013).
CID 6321405. http://pubchem.ncbi.nlm.nih.gov/summary/summary.cgi?cid=6321405&loc=ec_rcs (Accessed March 18, 2013).

CID 6432640. http://pubchem.ncbi.nlm.nih.gov/summary/summary.cgi?sid=43026438&view opt=PubChem#x332 (Accessed March 18, 2013).
CID 6616. http://pubchem.ncbi.nlm.nih.gov/summary/summary.cgi?cid=6616&loc=ec_rcs (Accessed March 18, 2013).
CID 6918391. http://pubchem.ncbi.nlm.nih.gov/summary/summary.cgi?sid=26527443&view opt=PubChem (Accessed March 18, 2013).
CID 8103. http://pubchem.ncbi.nlm.nih.gov/summary/summary.cgi?cid=8103&loc=ec_rcs (Accessed March 18, 2013).
Claeson, P., Andersson, R., and Samuelsson, G. 1991. T-cadinol: A pharmacologically active constituent of scented myrrh: Introductory pharmacological characterization and high field 1H- and 13C-NMR data. *Planta Medica* 57(4): 352–356.
CSID: 10295. http://www.chemspider.com/Chemical-Structure.10295.html (Accessed March 22, 2013).
CSID: 141284. http://www.chemspider.com/Chemical-Structure.141284.html (Accessed March 18, 2013).
CSID: 16787784. http://www.chemspider.com/Chemical-Structure.16787784.html (Accessed March 15, 2013).
CSID: 392816. http://www.chemspider.com/Chemical-Structure.392816.html (Accessed March 15, 2013).
CSID: 4444051. http://www.chemspider.com/Chemical-Structure.4444051.html (Accessed March 22, 2013).
CSID: 4444100. http://www.chemspider.com/Chemical-Structure.4444100.html (Accessed March 22, 2013).
CSID: 4444102. http://www.chemspider.com/Chemical-Structure.4444102.html (Accessed March 22, 2013).
CSID: 4444774. http://www.chemspider.com/Chemical-Structure.4444774.html (Accessed March 22, 2013).
CSID:4444850. http://www.chemspider.com/Chemical-Structure.4444850.html (Accessed March 13, 2013).
CSID: 4444853. http://www.chemspider.com/Chemical-Structure.4444853.html (Accessed March 18, 2013).
CSID:4447568. http://www.chemspider.com/Chemical-Structure.4447568.html (Accessed March 15, 2013).
CSID: 4509368. http://www.chemspider.com/Chemical-Structure.4509368.html (Accessed March 18, 2013).
CSID: 4882019. http://www.chemspider.com/Chemical-Structure.4882019.html (Accessed March 18, 2013).
CSID: 558917. http://www.chemspider.com/Chemical-Structure.558917.html (Accessed March 18, 2013).
CSID: 56330. http://www.chemspider.com/Chemical-Structure.56330.html (Accessed March 18, 2013).
CSID: 6364. http://www.chemspider.com/Chemical-Structure.6364.html (Accessed March 18, 2013).
CSID: 7812. http://www.chemspider.com/Chemical-Structure.7812.html (Accessed March 18, 2013).
CSID 10268.http://www.chemspider.com/Chemical-Structure.10268.html (Accessed March 9, 2014).
Das, M. 1999. Analysis of the genetic variation in Chamomile (*Matricaria recutita*) for the selection of improved breeding types. PhD Thesis submitted for the award of Doctor of Philosophy in Botany to the Lucknow University, Lucknow, India. 296 pp.

Das, M., Kumar, S., Mallavarapu, G. R., and Ramesh, S. 1999. Composition of the essential oils of the flowers of three accessions of *Chamomilla recutita* (L.) Rausch. *Journal of Essential Oil Research* 11: 615–618.

Das, M., Ram, G., Singh, A., Mallavarapu, G. R., Ramesh, S., Ram, M., and Kumar, S. 2002. Volatile constituents of different plant parts of *Chamomilla recutita* L. Rausch grown in the Indo-Gangetic plains. *Flavour and Fragrance Journal* 17: 9–12.

DePasquale, A. and Silvestri, R. 1976. Content of active substances in the various parts of *Matricaria chamomilla* L. *Essenze e Derivati Agrumari* 45: 292–298.

Exner, J., Reichling, J., and Becker, H.1980. Flavonoide in *Matricaria chamomilla*. *Planta Medica* 39: 219.

Exner, J., Reichling, J., Cole T. C. H., and Becker, H. 1981. Methylated flavonoid-aglycones from "Matricariae flos." *Planta Medica* 41: 198–200.

Farnesol. 2013. http://www.cosmeticsinfo.org/ingredient/farnesol (Accessed March 15, 2013).

Fernandes, E. S., Passos, G. F., Medeiros, R., da Cunha, F. M., Ferreira, J., Campos, M. M., Pianowski, L. F., and Calixto, J. B. 2007. Anti-inflammatory effects of compounds alpha-humulene and (−)-*trans*-caryophyllene isolated from the essential oil of *Cordia verbenacea*. *European Journal of Pharmacology* 569(3): 228–236.

Flaskamp, E., Nonnenmacher, G., Zimmermann, G., and Isaac, O.1981. Zur Stereochemie der Bisaboloide aus Matricaria chamomilla L. *Zeitschrift für Naturforschung* 86B: 1023–1030.

Fonseca, F. N., Tavares, M. F., and Horváth, C. 2007. Capillary electrochromatography of selected phenolic compounds of *Chamomilla recutita*. *Journal of Chromatography A* 1154(1–2): 390–399.

Franz, C. 1982. Genetic, ontogenetic and environmental variability of the constituents of chamomile oil from *Chamomilla recutita* (L.) Rauschert (Syn *Matricaria chamomilla* L.). In: K. H. Kubeczka (ed.) *Aetherische Oele Ergeb Int Arbeitstag*. Stuttgart, Germany: Thieme, pp. 214–224.

Franz, C. 1992. Genetica biochemica e coltivazione della camomilla (*Chamomilla recutita* (L.) Rausch.). *Agricoltura Ricerca* 131: 87–96.

Franz, C., Bauer, R., Carle, R., Tedesco, D., Tubaro, A., and Zitterl-Eglseer, K. 2005. Matricaria recutita. In: *Study on the Assessment of Plant/Herbs, Plant/Herb Extracts and Their Naturally or Synthetically Produced Components as Additives for Use in Animal Production*. pp. 155–169. http://www.efsa.europa.eu/en/scdocs/doc/070828.pdf (Accessed March 07, 2013).

Franz, C., Holzl, J., Mathe, A., and Winklhofer, A. 1985. Recent results on cultivation: Harvest time and breeding of chamomile [*Chamomilla recutita* (L.) Rauschert, syn. *Matricaria chamomilla* L.]. *Chamomile in Industrial and Pharmaceutical Use*. Trieste, Italy, pp. 6–17.

Franz, Ch. 1980. Content and composition of the essential oil in flower heads of *Matricaria chamomilla* L. during its ontogenetical development. *ISHS Acta Horticulturae* 96: 317–322.

Franz, Ch. 1983. Nutrient and water management for medicinal and aromatic plants. *ISHS Acta Horticulturae* 132: 203–216. http://www.actahort.org/books/132/132_22.htm (Accessed March 19, 2014).

Franz, Ch., Müller, E., Pelzmann, H., Hårdh, K., Hälvä, S., and Ceylan, A. 1986. Influence of ecological factors on yield and essential oil of camomile (*Chamomilla recutita* (L.) Rauschert syn, *Matricaria chamomilla* L.). *ISHS Acta Horticulturae* 188: 157–162.

Franz, Ch., Vömel, A., and Hölzl, J. 1978a. Preliminary morphological and chemical characterization of some populations and varieties of *Matricaria chamomilla* L. *ISHS Acta Horticulturae* 73: 109–114.

Franz, Ch., Vömel, A., and Hölzl, J. 1978b. Variation in the essential oil of *Matricaria chamomilla* L. depending on plant age and stage of development. *ISHS Acta Horticulturae* 73: 229–238.
Garcia Peña, C. M., Nguyen, K. B., Nguyen, B. T., Tillan Capo, J., Romero Díaz, J. A., López, O. D., and Fuste Moreno, V. 2009. Secondary metabolites in *Passiflora incarnata* L., *Matricaria recutita* L., and *Morinda latifolia* L. dry extracts. *Revista Cubana de Plantas Medicinales* 14(2). http://scielo.sld.cu/scielo.php?script=sci_arttext&pid=S1028 -47962009000200004&lng=en&nrm=iso&tlng=es (Accessed March 18, 2014)..
Gasic, O., Lukic, V., Adamovic, D., and Canak, N. 1986. Variation in the content and composition of the essential oil in the flower heads of *Matricaria chamomilla* L., during its ontogenetical development. *Acta Pharmaceutica Hungarica* 56(6): 283–288.
Gašić, O., Lukić, V., and Nikolić, A. 1983. Chemical study of *Matricaria chamomilla* L. *Fitoterapia* 54(2): 51–55.
Goeters, S., Imming, P., Pawlitzki, G., and Hempel, B. 2001. On the absolute configuration of matricin. *Planta Medica* 67(3): 292–294.
Gosztola, B., Sárosi, S., and Németh, E. 2010. Variability of the essential oil content and composition of chamomile (*Matricaria chamomilla* L.) affected by weather conditions. *Natural Product Communications* 5(3): 465–470.
Haghi, G., Hatami, A., Safaei, A., and Mehran, M. 2014. Analysis of phenolic compounds in *Matricaria chamomilla* and its extracts by UPLC-UV. *Research in Pharmaceutical Sciences* 9(1): 31–37.
Harbourne, N., Jaquier, J. C., and O'Riordian, D. 2009. Optimisation of the extraction and processing conditions of chamomile (*Matricaria chamomilla* L.) for incorporation into a beverage. *Food Chemistry* 115: 15–19.
Hexadec-11-yn-13,15-diene. 2013. http://webbook.nist.gov/cgi/cbook.cgi?ID=R203166& Units=SI&Mask=2000#ref-1 (Accessed March 18, 2013).
Hitziger, von, T., Höll, P., Ramadan, M., Dettmering, D., Imming, P., and Hempel, B. 2003. Die alte junge Kamille. Phytopharmazie. http://www.pharmazeutische-zeitung. de/index.php?id = titel_05_2003 (Accessed March 18, 2013).
Hoelzl, J. and Demuth, G.1975. Influence of ecological factors on the formation of essential oils and flavors in *Matricaria chamomilla*. *Phytochemistry* 41: 1275–1279.
Hoffman, D. 2003. *Medical Herbalism: The Science and Practice of Herbal Medicine*. Rochester, VT: Healing Arts Press, p. 96.
Hostetler, G. L., Riedl, K. M., and Schwartz, S. J. 2013. Effects of food formulation and thermal processing on flavones in celery and chamomile. *Food Chemistry* 141(2): 1406–1411.
Hyvönen, H., Torkkeli, H., Hakkinen, V. M. A., Rickkola, M. L., and Lethonen, P. J. 1991. Two dimensional separation of chamomile by on-line HPLC-HRGC. *Acta Pharmaceutica Fennica* 100: 269–273.
Jackson, B. P. and Snowdon, D. W. 1968. *Powdered Vegetable Drugs*. London, United Kingdom: Churchill, pp. 82–83.
Joo, J. H. and Jetten, A. M. 2009. Molecular mechanisms involved in farnesol-induced apoptosis. *Cancer Letters* 287(2): 123.
Kaiser, C. S., Römpp, H., and Schmidt, P. C. 2004. Supercritical carbon dioxide extraction of chamomile flowers: Extraction efficiency, stability, and in-line inclusion of chamomile-carbon dioxide extract in beta-cyclodextrin. *Phytochemical Analysis* 15(4): 249–256.
Kamatou, G. P. P. and Viljoen, A. M. 2010. A review of the application and pharmacological properties of α-bisabolol and α-bisabolol-rich oils. *Journal of the American Oil Chemists' Society* 87: 1–7.
Karawya, M.S., Awaad, K.E., Svab, J., and Fahmy, T. 1968. A histochemical study of *Matricaria chamomilla* L. *Planta Medica*, 2: 166–173.

Khanina, M. A., Serykh, E. A., and Berezovskaya, T. P. 1995. Development of a method for the quantitative determination of chamazulene in the essential oil of *Artemisia jacutica*. *Chemistry of Natural Compounds* 31(3): 420–421.

Kobayashi, Y., Takemoto, H., Asaka, Y., Zuqi, F., Takane, T., and Kitajo, H. 2013. Chemical analysis of the volatile and polyphenols components of the leaves and stems of German chamomile, and the comparison with flower heads components. *Aroma Research* 14(2): 155–159.

Kotnik, P., Škerget, M., and Knez, Ž. 2007. Supercritical fluid extraction of chamomile flowerheads: Comparison with conventional extraction, kinetics and scale-up. *Journal of Supercritical Fluids* 43(2): 192–198.

Krüger, H. 2010. Characterisation of chamomile volatiles by simultaneous distillation solid-phase extraction in comparison to hydrodistillation and simultaneous distillation extraction. *Planta Medica* 76(8): 843–846.

Kumar, S., Das, M., Singh, A., Ram, G., Mallavarapu, G. R., and Ramesh, S. 2001. Composition of the essential oils of the flowers, shoots and roots of two cultivars of *Chamomilla recutita*. *Journal of Medicinal and Aromatic Plants* 23(4): 617–623.

Kumar, S., Gopal Rao, M., Khanuja, S. P. S., Gupta, S. K., Das, M., and Shasany, A. K. 1999. The chamazulene-rich chamomile variety—Prashant (a new genotype). *Journal of Medicinal and Aromatic Plants* 21(4): 1096–1098.

Kunde, R. and Isaac, O. 1979a. On the flavones of chamomile (*Matricaria chamomilla* L.) and a new acetylated apigenin-7-glucoside. *Planta Medica* 37: 124–130.

Kunde, R. and Isaac, O. 1979b. Identification of racemic alpha-bisabolol in specialties made from chamomile extracts. *Planta Medica* 35: 71–75.

Lakszner, K., Szepesy, L., Torok, I., and Csapo Barthos, E. 1990. Quality control and classification of chamomile oils with oxygen sensitive flame-ionization detector. *Chromatographia* 30: 47–50.

Laskova, I. L. and Uteshev, B. S. 1992. [Immunomodulating action of heteropolysaccharides isolated from camomile flowers]. *Antibiot Khimioter* 37(6): 15–18.

Lawrence, B. M. 1987. Progress in essential oils. Chamomile oil. *Perfumer and Flavour* 12: 35–52.

Lawrence, B. M. 1996. Progress in essential oils. Chamomile oil. *Perfumer and Flavour* 47: 330–332.

Lin, L. and Harnly, J. M. 2012. LC-PDA-ESI/MS identification of the phenolic components of three Compositae spices: Chamomile, tarragon, and Mexican arnica. *Natural Product Communications* 7(6): 749–752.

Liu, J., Zhang, Y., Qu, J., Xu, L., Hou, K., Zhang, J., Qu, X., and Liu, Y. 2011. β-Elemene-induced autophagy protects human gastric cancer cells from undergoing apoptosis. *BMC Cancer* 11: 183.

Liu, X., Zhao, M., Luo, W., Yang, B., and Jiang, Y. 2009. Identification of volatile components in *Phyllanthus emblica* L. and their antimicrobial activity. *Journal of Medicinal Food* 12(2): 423–428. doi: 10.1089/jmf.2007.0679.

Magiatis, P., Michaelakis, A., Skaltsounis, A., and Haroutounian, S. A. 2001. Volatile secondary metabolite pattern of callus cultures of *Chamomilla recutita*. *Natural Product Letters* 15(2): 125–130.

Mann, C. and Staba, E. J. 1986. The chemistry, pharmacognosy and chemical formulations of chamomile. *Herbs Spices and Medicinal Plants* 1: 236–280.

Marczal, G. and Petri, G. 1979. Essential oil production and formation of its constituents during ontogeny in *Matricaria chamomilla* L. *Planta Medica* 36(3): 382–384.

Marczal, G. and Petri, G. 1988. Quantitative variations of some volatile compounds in various ontogenetical phases of *Matricaria chamomilla* L. (*Chamomilla recutita* L. Rauschert) grown in phytotron. *Acta Agronomica Hungarica* 37: 197–208.

Marczal, G. and Petri, G. 1989. Composition of Hungarian chamomile essential oils. *Acta Pharmaceutica Hungarica* 59(4): 145–146.

Martin, T. 2010. Cigarette additives. About.com Guide. http://quitsmoking.about.com/cs/nicotineinhaler/a/cigingredients.htm (Accessed March 15, 2013).

Martins, A., Hajdú, Z., Vasas, A., Csupor-Löffler, B., Molnár, J., and Hohmann, J. 2010. Spathulenol inhibits the human ABCB1 efflux pump. *Planta Medica* 76: 608.

Matos, F. J. A., Machado, M. I. L., Alenear, J. W., and Cravierom, A. A. 1993. Constituents of Brazilian chamomile oil. *Journal of Essential Oil Research* 5: 337–339.

Matricin. 2013. http://www.chemicalize.org/structure/#!mol=Matricin (Accessed March 17, 2013).

Mechler, E. 1979. The yield of essential oils according to the two different methods of European Pharmacopoeia and German Pharmacopoeia 7th edition. *Planta Medica* 36: 278–279.

Medić-Šarić, M., Stanić, G., Males, Z., and Sarić, S.1997. Application of numerical methods to thin-layer chromatographic investigation of the main components of chamomile (*Chamomilla recutita* (L.) Rauschert) essential oil. *Journal of Chromatography A* 776(2): 355–360.

Melo, M. J. 2009. History of natural dyes in the ancient Mediterranean world. In: T. Bechtold and R. Mussak (eds.) *Handbook of Natural Colorants*. Sussex: John Wiley & Sons, p. 15.

Meneses-Reyes, J. C., Soto-Hernández, R. M., Espinosa-Solares, T., and Ramírez-Guzmán, M. E. 2008. Optimization of the process of flavonoid extraction from chamomile (*Matricaria recutita* L.). *Agrociencia* 42(4): 425–433.

Menziani, E., Tosi, B., Bonora, A., Reschglian, P., and Gaetano, L. 1990. Automated multiple development high-performance thin-layer chromatographic analysis of natural phenolic compounds. *Journal of Chromatography A* 511: 396–401.

Merib, J., Nardini, G., Bianchin, J. N., Dias, A. N., Simão, V., and Carasek, E. 2013. Use of two different coating temperatures for a cold fiber headspace solid-phase microextraction system to determine the volatile profile of Brazilian medicinal herbs. *Journal of Separation Science* 36(8): 1410–1417.

Mincsovics, E., Tyihak, E., Nagy, J., and Kalász, H. 1979. Thin layer chromatographic investigation of components in essential oil of *Matricaria chamomilla* L. by means of classical and pressurized chamber systems. *Planta Medica* 36: 296.

Mirshekari, B. 2011. Effect of irrigation time and nitrogen fertilizer on growth period and chamazulene content of German chamomile (*Matricaria chamomilla* L.) in cold and semi-arid region. *Medicinal and Aromatic Plants* 27(1): Pe173–Pe176.

Mötl O, Felklová, M., Jasicova M, Lukes V. 1977. Zur GC Analyse and zu Chemischen Typen von Kamillenol (GLC analysis and chemical types of chamomile essential oil [author's transl]). *Arch Pharm* (Weinheim), 310(3): 210–5.

Motl, O. and Repcak, M. 1979. New components of chamomile essential oil. *Planta Medica* 36: 272–273.

Motl, O., Repcak, M., and Sedmera, P. 1978. Additional constituents of chamomile oil. II. The conversation of hydrocodone with 2,4-nitrobenzene in an alkaline medium. *Archiv der Pharmazie* 311: 75–76.

Mulinacci, N., Romani, A., Pinelli, P., Vinceri, F. F., and Prucher, D. 2000. Characterization of *Matricaria recutita* L. flower extracts by HPLC-MS and HPLC-DAD analysis. *Chromatographia* 51(5/6): 301–307.

Ness, A., Metzger, J. W., and Schmidt, P. C. 1996. Isolation, identification and stability of 8-desacetylmatricine, a new degradation product of matricine. *Pharmaceutic Acta Helvetiae* 71: 265–271.

Ness, A. and Schmidt, P. C. 1995. Kamillenpraparate. Biespiel fin stabilisierung von phytopharmaka. *Dtsch Apoth Ztg* 13(39): 30–40.

Nidagundi, R. and Hegde, L. 2007. Cultivation prospects of German chamomile in South India. *Natural Product Radiance* 6(2): 135–137.

Nováková, L., Vildová, A., Matius, J. P., Gonçalves, T., and Solich, P. 2010. Development and application of UHPLC–MS/MS method for the determination of phenolic compounds in chamomile flowers and chamomile tea extracts. *Talanta* 82(4): 1271–1280.
Nurzyńska-Weirdak, R. 2011. The essential oil of *Chamomilla recutita* (L.) Rausch. cultivated and wild growing in Poland. *Annales Universitatis Mariae Curie-Sklodowska Lublin-Polonia* 24(2): 197–206.
Orav, A., Kailas, T., and Ivask, K. 2001. Volatile constituents of *Matricaria recutita* L. from Estonia. *Proceedings of the Estonian Academy Sciences Chemistry* 50(1): 39–45.
Pachaly, P. 1982. TLC in pharmacy. *Deutsche Apotheker-Zeitung* 122: 2057–2061.
Padula, L. Z., Rondina, R. V. D., and Coussio, J. D. 1977. Quantitative determination of essential oil, total azulenes and chamazulene in German chamomile cultivated in Argentina. *Planta Medica* 30: 273–280.
Pakic, B., Lepojevic, Z., and Slavica, B. 1989. The determination of apigenin 7-0-beta-glucoside in the *Matricaria chamomilla* ligulate flowers. *Arhiv za farmaciju* 39(5): 163–168.
Park, S. N., Lim, Y. K., Freire, M. O., Cho, E., Jin, D., and Kook, J. K. 2012. Antimicrobial effect of linalool and α-terpineol against periodontopathic and cariogenic bacteria. *Anaerobe* 18(3): 369–372.
Pattiram, P. D., Lasekan, O., Tan, C. P., and Zaidul, I. S. M. 2011. Identification of the aroma-active constituents of the essential oils of Water Dropwort (*Oenanthe javanica*) and 'Kacip Fatimah' (*Labisia pumila*). *International Food Research Journal* 18(3): 1021–1026.
Paulsen, E., Otkjaer, A., and Andersen, K. E. 2010. The coumarin herniarin as a sensitizer in German chamomile [*Chamomilla recutita* (L.) Rauschert, Compositae]. *Contact Dermatitis* 62(6): 338–342.
Pekic, B., Zekivic, Z., and Lepojevic, Z. 1994. HPLC determination of apigenin and its glucosides in chamomile (*Matricaria chamomilla* L.). *Zbornik Matice Srpske za Prirodne Nauke* 86: 37–41.
Pereira, N. P., Cunico, M. M., Miguel, O. G., and Miguel, M. D. 2008. Promising new oil derived from the seeds of *Chamomilla recutita* (L.) Rauschert produced in Southern Brazil. *Journal of the American Oil Chemists Society* 85: 493–494.
Petronilho, S., Maraschin, M., Delgadillo, I., Coimbra, M. A., and Rocha, S. M. 2011. Sesquiterpenic composition of the inflorescences of Brazilian chamomile (*Matricaria recutita* L.): Impact of the agricultural practices. *Industrial Crops and Products* 34(3): 1482–1490.
Petruľová-Poracká, V., Repčák, M., Vilková, M., and Imrich, J. 2013. Coumarins of *Matricaria chamomilla* L.: Aglycones and glycosides. *Food Chemistry* 141(1): 54–59.
Piccaglia, R., Marotti, M., Giovanelli, E., Deans, S. G., and Eaglesham, E. 1993. Antibacterial and antioxidant properties of Mediterranean aromatic plants. *Industrial Crops Products* 2: 47–50.
Pickett, J. A. and Griffiths, D. C. 1980. Composition of aphid alarm pheromones. *Journal of Chemical Ecology* 6(2): 349–360.
Pículo, F., Guiraldeli Macedo, C., de Andrade, S. F., and LuisMaistro, E. 2011. In vivo genotoxicity assessment of nerolidol. *Journal of Applied Toxicology* 31(7): 633–639.
Pöthke, W. and Bulin, P. 1969. Phytochemical investigation of a new hybrid chamomile variety. II. Essential oil. *Pharm Zentralle*, 108: 813–823.
Raal, A., Arak, E., Orav, A., and Ivask, K. 2003. Comparison of essential oil content of *Matricaria recutita* L. from different origins. *Ars Pharmaceutica* 44(2): 159–165.
Raal, A., Kaur, H., Orav, A., Arak, E., Kailas, T., and Müürisepp, M. 2011. Content and composition of essential oils in some Asteraceae species. *Proceedings of the Estonian Academy of Sciences* 60(1): 55–63. doi: 10.3176/proc.2011.1.06.
Rafieiolhossaini, M., Adams, A., Sodaeizadeh, H., Van Damme, P., and De Kimpe N. 2012. Fast quality assessment of German chamomile (*Matricaria chamomilla* L.) by headspace solid-phase microextraction: Influence of flower development stage. *Natural Product Communications* 7(1): 97–100.

Ramaswami, S. K., Briseese, P., Garguillo, R. J., and Gelderen, T. 1988. Sesquiterpene hydrocarbons: From mass confusion to orderly line up. In: B.M. Lawrence, B.D. Mookherjee, B.J. Willis (eds.) *Flavors and Fragrances: A World Perspective*. Amsterdam, the Netherlands: Elsevier Science Publishers BV, pp. 951–980.

Redaelli, C., Formentini, L., and Santaniello, E. 1980. Apigenin-7-glucoside and its 2″ and 6″ acetates from ligulate flowers of *Matricaria chamomilla*. *Phytochemistry* 19: 985–986.

Redaelli, C., Formentini, L., and Santaniello, E. 1981a. Reversed phase high performance liquid chromatography analysis of apigenin and its glucosides in flowers of *Matricaria chamomilla* and chamomile extracts. *Planta Medica* 42: 288–292.

Redaelli, C., Formentini, L., and Santaniello, E. 1981b. HPLC determination of the coumarins of *Matricaria chamomilla*. *Planta Medica* 43(4): 412–413.

Redaelli, C., Formentini, L., and Santaniello, E. 1982. Apigenin 7-glucoside diacetates in ligulate flowers of *Matricaria chamomilla*. *Phytochemistry* 21(7): 1828–1830.

Redaelli, C., Santaniello, E., and Formentini, L. 1979. Phytochemical screening of the components of petals of *Marticaria chamomilla* L. *Planta Medica* 36: 284. (Abstracts Section.)

Reich, E. 2000. Chromatography: Thin-layer (planar) historical development. In: I.D. Wilson (ed.) *Encyclopedia of Separation Science*. New York: Academic Press, pp. 834–839.

Reichling, J. and Beiderbeck, R. 1991. Chamomilla recutita (L.) Rauschert(Camomile): In vitro culture and production of secondary metabolites. In: Y.P.S. Bajaj (ed.) *Biotechnology in Agriculture and Forestry: Medicinal and Aromatic Plants-III*. Heidelberg, Berlin: Springer Verlag, pp. 157–175.

Reichling, J., Beiderbeck, R., and Becker, H. 1979. Vergleichende Untersuchungen über sekundäre Inhaltsstoffe bei Pflanzentumoren, Blüte, Kraut und Wurzel von *Matricaria chamomilla* L. [Comparative Studies on Secondary Products from Tumors, Flowers, Herb and Roots of *Matricaria chamomilla* L.]. *Planta Medica* 36(8): 322–332.

Reichling, J., Bission, W., Becker, H., and Schilling, G. 1983. Composition and accumulation of essential oil in Matricariae radix. (2. Communication). *Zeitschrift für Naturforschung* 38C: 159.

Repcak, M., Cernaj, P., and Martonfi, P. 1993. The essential content and composition of diploid and tetraploid *Chamomilla recutita* during the ontogenesis of anthodia. *Journal of Essential Oil Research* 5(3): 297–300.

Retamar, J., Malinskar, G., and Santi, M. 1989. Essential oil of *Matricaria recutita*, 2nd Communication. *Essenze e Derivati Agrumari* 59: 40–43.

Reverchon, E. and Senatore, F. 1994. Supercritical carbon dioxide extraction of chamomile essential oil and its analysis by gas chromatography–mass spectrometry. *Journal Agriculture and Food Chemistry* 42(1): 154–158.

Ristić, M. S., Dordevic, S. M., Dokovic, D. D., and Tasic, S. R. 2007. Setting a standard for the essential oil of chamomile originating from Banat. *ISHS Acta Horticulturae* 749: 127–140.

Rivero-Cruz, B., Rivero-Cruz, I., Rodriguez, J. M., Cerda-Garcia-Rojas, C. M., and Mata, R. 2006. Qualitative and quantitative analysis of the active components of the essential oil from *Brickellia veronicaefolia* by nuclear magnetic resonance spectroscopy. *Journal of Natural Products* 69(8): 1172–1176.

Rogerio, A. P., Andrade, E. L., Leite, D. F., Figueiredo, C. P., and Calixto, J. B. 2009. Preventive and therapeutic anti-inflammatory properties of the sesquiterpene alpha-humulene in experimental airways allergic inflammation. *British Journal of Pharmacology* 158(4): 1074–1087.

Rubiolo, P., Belliardo, F., Cordero, C., Liberto, E., Sgorbini, B., and Bicchi, C. 2006. Headspace-solid-phase microextraction fast GC in combination with principal component analysis as a tool to classify different chemotypes of chamomile flower-heads (*Matricaria recutita* L.). *Phytochemical Analysis* 17(4): 217–225.

Salamon, I. 1994. Growing conditions and the essential oil of *Chamomilla recutita* (L.) Rausch. *Journal of Herbs, Spices and Medicinal Plants* 2: 31–37.

Salamon, I., Ghanavati, M., and Khazaei, H. 2010. Chamomile biodiversity and essential oil qualitative-quantitative characteristics in Egyptian production and Iranian landraces. *Emirates Journal of Food and Agriculture* 22(1): 59–64.

Scalia, S., Giuffreda, L., and Pallado, P. 1999. Analytical and preparative supercritical fluid extraction of chamomile flowers and its comparison with conventional methods. *Journal of Pharmaceutical and Biomedical Analysis* 21(3): 549–558.

Schilcher, H. 1977. [Biosynthesis of (−)-α-bisabolol and α-bisabololoxide. Ist Report; in vivo tracer studies with 14C-precusors]. *Planta Medica* 31(4): 315–321.

Schilcher, H., Imming, P., and Goeters, S. 2005. Active chemical constituents of *Matricaria chamomilla* L. syn. *Chamomilla recutita* (L.) Rauschert. In: R. Franke and H. Schilcher (eds.) *Chamomile: Industrial Profiles*. Boca Raton, FL: CRC Press, pp. 55–76.

Schmidt, P. C. and Ness, A. 1993. Isolierung und charakterisierung eines Matricin-Standards. *Pharmazie* 48: 146–147.

Schmidt, P. C., Weibler, K., and Soyke, B. 1991. Kamillenblüten und-extrakte. Matricin—und chamazulenbestimmung-Vergleich von GC, HPLC un photometrischen Methoden. *Deutsche Apotheker Zeitung* 131(5): 175–181.

Schreiber, A., Carle, A., and Reinhard, E.1990. On accumulation of apigenin in chamomile flowers. *Planta Medica* 56: 179–181.

Senthilkumar, A., Jayaraman, M., and Venkatesalu, V. 2013. Chemical constituents and larvicidal potential of *Feronia limonia* leaf essential oil against *Anopheles stephensi*, *Aedes aegypti* and *Culex quinquefasciatus*. *Parasitology Research* 112(3): 1337–1342.

Setzer, W. N. 2008. Germacrene D cyclization: An *ab initio* investigation. *International Journal of Molecular Sciences* 9(1): 89–97.

Shaath, N. A. and Azzo, N. R. 1993. Essential oils of Egypt. In: G. Charalmbous (ed.) *Foods, Flavours, Ingredients and Composition*. Amsterdam, the Netherlands: Elsevier Science Publishers BV, pp. 591–609.

Snuparek, V., Varga, I., Frimm, R., Gattnar, O., and Minczinger, S. 1988. Ethanolic chamomile extract for use in pharmaceuticals and cosmetics. Czech CS 252,992 (CI A 61 k35/78), October 15, 4988.Appl 86/233. January 10, 1986, p. 4.

Šorm, F., Novák, J., and Herout, V. 1953. On terpenes. 51. The composition of chamazulene. *Collection of Czechoslovak Chemical Communications* 18: 527–529.

Sun, J. 2007. D-Limonene: Safety and clinical applications. *Alternative Medicine Review* 12(3): 259–264.

Surburg, H., Guentorte, M., and Harder, H. 1996. Volatile compounds from the flowers: Analytical and olfactory aspects. In: *Bioactive Volatile Compounds from Plants*; Teranish, R., Buttery, R. G., and Sugisawa, H. (eds.); ACS Symposium Series 525; American Chemical Society: Washington, DC, pp. 168–186.

Švehlíková, V., Bennett, R. N., Mellon, F. A., Needs, P. W., Piacente, S., Kroon, P. A., and Bao, Y. 2004. Isolation, identification and stability of acylated derivatives of apigenin7-O-glucoside from chamomile (*Chamomilla recutita* [L.] Rauschert). *Phytochemistry* 65(16): 2323–2332.

Szöke, E., Máday, E., Gershenzon, J., Allen, J. L., and Lemberkovics, E. 2004a. Beta-eudesmol, a new sesquiterpene component in intact and organized root of chamomile (*Chamomilla recutita*). *Journal of Chromatographic Science* 42(5): 229–233.

Szöke, E., Máday, E., Kiss, S. A., Sonnewend, L., and Lemberkovics, E. 2004b. Effect of magnesium on essential oil formation of genetically transformed and non-transformed chamomile cultures. *Journal of the American College of Nutrition* 23(6): 763S–767S.

Szöke, E., Máday, E., Tyihák, E., Kuzovkina, I. N., and Lemberkovics, E. 2004c. New terpenoids in cultivated and wild chamomile (*in vivo* and *in vitro*). *Journal of Chromatography B Analytical Technologies Biomedical and Life Sciences* 800(1–2): 231–238.

Takahashi, K., Muraki, S., and Yoshida, T. 1981. Synthesis and distribution of (−)-mintsulphide, a novel sulfur-containing sesquiterpene. *Agricultural and Biological Chemistry* 45: 129–131.

Takei, M., Umeyama, A., and Arihara, S. 2006. T-cadinol and calamenene induce dendritic cells from human monocytes and drive Th1 polarization. *European Journal of Clinical Pharmacology* 537(1–3): 190–199.

Tanaka, T., Okemoto, H., and Kuwahara, N. 1993. Quercetin containing coloring. US 5445842. http://www.patentlens.net/patentlens/patents.html?patnums=US_5445842&language=&#tab_1 (Accessed March 22, 2013).

Tandon, S., Ahmad, J., and Ahmad, A. 2013. GC-MS analysis of the steam and hydrodistilled essential oil of *Matricaria recutita* L. flowers of north east region on India. *Asian Journal of Chemistry* 25(11): 6048–6050.

Thune, P. O. and Solberg, Y. J. 1980. Photosensitivity and allergy to aromatic lichen acids. Compositae oleoresins and other plant substances. *Contact Dermatitis* 6: 81–87.

Trinh, H. T., Lee, I. A., Hyun, Y. J., and Kim, D. H. 2011. Artemisia princeps Pamp. Essential oil and its constituents eucalyptol and α-terpineol ameliorate bacterial vaginosis and vulvovaginal candidiasis in mice by inhibiting bacterial growth and NF-κB activation. *Planta Medica* 77(18): 1996–2002.

Tschiggerl, C. and Bucar, F. 2012. Guaianolides and volatile compounds in chamomile tea. *Plant Foods for Human Nutrition* 67(2): 129–135.

Tsutsulova, A. L. and Antonova, R. A. 1984. Analysis of Bulgarian chamomile oil. *Maslo-žirovaâ promyšlennost'* 11: 23–24.

Tyihák, E. 1987. Overpressured layer chromatography and its applicability in pharmaceutical and biomedical analysis. *Journal of Pharmaceutical and Biomedical Analysis* 5(3): 191–203.

Tyihák, E., Mincsovics, E., and Móricz, A. M. 2012. Overpressured layer chromatography: From the pressurized ultramicro chamber to BioArena system. *Journal of Chromatography A* 6(1232): 3–18.

Uteshev, B. S., Laskova, I. L., and Afanas'ev, V. A. 1999. [The immunomodulating activity of the heteropolysaccharides from German chamomile (*Matricaria chamomilla*) during air and immersion cooling]. *Eksperimental'naia i Klinicheskaia Farmakologiia* 62(6): 52–55.

Vaverkova, S. and Herichova, A. 1980. Histochemical proof of prochamazulene at different stages of development of flower crown of *Matricaria chamomilla* L. *Biologia* 35: 753–758.

Vogel, K. and Schmidt, P. C. 1993. An ethyl alcohol free liquid extract for *Chamomilla recutita* (L.) Rauschert for oral and topical application. *Pharmaceutical and Pharmacological Letters* 3: 153–155.

Vuorela, H., Holm, Y., and Hiltunen, R. 1989. Application of headspace gas chromatography in essential oil analysis. Part VIII. Assay of matricine and chamazulene. *Flavour and Fragrance Journal* 4(3): 113–116.

Vuorela, H., Holm, Y., Hiltunen, R., Harvala, T., and Laitinen, A. 1990. Extraction of volatile oil in chamomile flowerheads using supercritical carbon dioxide. *Flavour and Fragrance Journal* 5(2): 81–84.

Wagner, H. and Bladt, S. 1996. *Plant Drug Analysis: A Thin Layer Chromatography Atlas*. Munich, Germany: Springer, pp. 186, 212.

The Wealth of India. 1962. *Raw Materials (L-M): Vol. 6*. Botany of Matricaria Chamomilla. New Delhi: CSIR, pp. 308–309.

Yamazaki, H., Miyakado, M., and Mabry, T. J. 1982. Isolation of a linear sesquiterpene lactone from *Matricaria chamomilla*. *Journal of Natural Products* 45(4): 508.

Zeković, Z., Pekić, B., Lepojević, Ž., and Petrović, L. 1994. Chromatography in our investigations of camomile (*Matricaria chamomilla* L.). *Chromatographia* 39(9–10): 587–590.

Zenkevich, I. G., Makarov, V. G., Pimenov, A. I., Kosman, V. M., Pozharitskaya, O. N., and Shikov, A. N. 2002. Identification and quantification of main components of *Chamomilla recutita* (L.) Rauschert oily extracts. *Revista de Fitoterapia* 2(S1): B027.

Zou, B., Quentin Li, Q., Zhao, J., Li, M. J., Cuff, C. F., and Reed, E. 2013. β-Elemene and taxanes synergistically induce cytotoxicity and inhibit proliferation in ovarian cancer and other tumor cells. *Anticancer Research* 33(3): 929–940.

5 Genetics and Breeding of Chamomile

5.1 INTRODUCTION

The chamomile plant is morphologically heterogeneous, both in the wild and when cultivated. Chamomile plants originating from different areas show a great variation. There are differences in the sizes of the habit, flowers, and the content of essential oil and its compounds (Franz et al. 1985). The variability observed in chamomile could largely be attributed to the ecological and environmental conditions, as studies indicate (Franz et al. 1985; Massoud and Franz 1990a).

In chamomile, morphological markers, or morphotypes, have not yet been described. However, tetraploid chamomile plants can be distinguished from the diploid plants by their relatively larger dimensions of the morphological characters and higher essential oil yield.

Genetics of the morphological characteristics of chamomile, such as height of the plant, branching, and flower size, is restricted at present to the comparative study of the diploid and tetraploid varieties of chamomile.

Chemical markers exist as chemotypes, which have been deployed in the chamomile breeding experiments. Each chemotype is described on the basis of the most abundant chemical compound in its essential oil. Thus, there are chamazulene, bisabolol, bisabolol oxide A, bisabolol oxide B, and bisabolone oxide A types of populations (Hörn et al. 1988). Chamazulene-free chemotype has also been reported with the color of the oil being yellow instead of the usual blue (Franz et al. 1985; Meriçli 1985; Salamon et al. 2010).

5.2 VARIABILITY IN CHAMOMILE POPULATION

Chamomile is believed to have originated as a weed in the Middle East Asian region of Iran (Salamon et al. 2010). Thereafter, it spread to different parts of Europe and Asia.

The chamomile plant is extremely heterogeneous in its morphology. Franz et al. (1978a) carried out a study of the morphological variation of some populations and varieties of chamomile in Germany. They found that the spectrum of variation ranged from small flower heads in wild plants to big flowers in tetraploid varieties. Later, Franz et al. (1985) reported the morphological characteristics of a diploid variety and compared it with a tetraploid variety developed by them, as presented in Table 5.1. The other morphological characters studied were the weight of the flowers, ramification of the branches, and large flowers at uniform height.

TABLE 5.1
Morphological Characteristics of Diploid and Tetraploid Chamomile Plants

S. No.	Morphological Characteristics	Degumille (Diploid Variety)	Manzana (Tetraploid Variety)
1.	Growth	Upright, ramified	Upright, ramified
2.	Height at flowering	50–60 cm	65–75 cm
3.	Herb to flower ratio	5.4	4

Popova and Peneva (1987) studied 14 morphological characters of chamomile and found that these characters were influenced by ploidy and growing conditions. They proposed that based on the characters of the flowers the chamomile plants could be classified. In other words, they attempted to describe chamomile morphotypes based on the flower characteristics.

The heterogeneity in the branch numbers and flower number and size also reflects the heterogeneity in the yield of the plant. Dragland et al. (1996) grew one chamomile cultivar at 42 locations in the southern, central, and northern regions of Norway. They found that chamomile grown in the southern and central regions of Norway gave a reasonably high flower yield, but in northern Norway the yield was lower.

The morphological characters of chamomile have been documented in detail by Das (1999) and have been described in Section 2.10. These include 40 morphological features, such as habit, canopy, height, leaf characteristics, and flower characteristics. As can be observed from Table 2.6, the diversity within each plant part is huge. For example, the habit of the plants was found to range from erect, semierect to prostrate. The canopy ranged from less than 315 cm^2 to ~916 cm^2. The height of the plants was also diverse ranging from 50 cm to almost double, that is, 110 cm. The leaf characteristics, such as shape, number, and coating of surface also varied to a large extent. The flower characteristics, such as the shape and number of ligulate florets and size of disc florets, varied immensely.

Aiello et al. (2004) analyzed the morphological characteristics of a few varieties of chamomile, such as Ippo, Bona, Degumille, Benary, Ingegnoli, and Bornträger Wild growing in Italy. They found that the Ippo variety had smaller size and weight of flowers compared with Bona. The morphological variation among the varieties was quite high.

Gosztola et al. (2008b) studied in detail the morphological characteristics of chamomile populations and their progenies. The characters studied were plant height, diameter of flowers and discus (receptacle), and the size of the ray florets. They found that there was a 52%–69% morphological variation in the progenies from their parents. About 50% progenies had lower plant height than their parents. About 75% of the progenies had smaller size of flowers. However, more morphological homogeneity was observed in the progenies than in the parent populations.

Solouki et al. (2008) studied the genetic diversity of chamomile population based on morphological traits. They found that there was minimum variation in the flower diameter and plant height and maximum variation in the number of flowers per plant.

Genetics and Breeding of Chamomile

However, these morphological characteristics have been found to be mainly controlled by the environmental factors. As a result, they are unstable and are not inherited. The characteristics such as plant height, number of flowers and weight of flowers, root, stem, and leaves have been found to be controlled by physio-ecological conditions. Therefore, it is not possible to determine the characteristics of a diploid plant beforehand (Salamon 2009). The method of improving such a population through breeding, therefore, chiefly depends on artificial selection and recurrent selection from the variable population. The chemical constituents of the oil remain the chief consideration for selection of chamomile plants for the breeders. Some researchers have also worked on the levels of compounds in the content of the extracts and attempted to assign the plants a distinct chemotype (Repcak et al. 2001).

5.2.1 Variation in Oil Content

The earliest report on the variation of oil content may be traced back to Debska in 1958 in Poland, who reported that dried chamomile flowers contained 0.46%–0.67% essential oil (Lawrence 1987). Subsequently, many reports emerged that substantiated the observation on the variation of the essential oil content. In Yugoslavia, Kusrack and Benzinger, in 1977, found that the essential oil content in dried flowers ranged from 0.24% to 1.9% (Lawrence 1987). The variation in the essential oil content has been described in detail in Section 4.2.

5.2.2 Variation in Oil Composition

In 1973, Schilcher from Germany reported the composition of the oil in some commercial samples of chamomile that included the levels (in percentage) of chamazulene, α-bisabolol, bisabolol oxide A, bisabolol oxide B, and spiroether (Schilcher 1973; Lawrence 1987). Subsequently, Motl et al. (1977), also from Germany, provided the composition of the essential oil in some commercial samples of chamomile (Lawrence 1987). The findings of Schilcher (1973) and Motl et al. (1977) are reproduced in Table 5.2.

As can be observed from the table, there is a large variation in the percentages of individual compounds reported by Schilcher. Similarly, the range of percentage of

TABLE 5.2
Levels of Pharmacologically Active Compounds in Chamomile as Described by Schilcher (1973) and Motl et al. (1977)

S. No.	Compound	Schilcher	Motl et al.
1.	Chamazulene	1.45–17.69	2.16–35.59
2.	α-Bisabolol	4.37–77.21	1.72–67.25
3.	α-Bisabolol oxide A	2.13–52.25	0–55
4.	α-Bisabolol oxide B	3.17–58.85	4.35–18.93
5.	Spiroether	1.92–12.00	—
6.	Bisabolone oxide	—	0–63.85

individual compounds in the oil reported by Motl et al. is also huge. Further, when the results of Schilcher are compared with the results of Motl et al., it is found that there is a wide variation in the levels of the compounds between the two studies.

Franz (1980) reported the content of compounds in the essential oils of four cultivars H-29, CH-29, BK-2, and Menemen in different stages of flower development. He found that the flower buds contained more farnesene and α-bisabolol, whereas in fully developed flowers, chamazulene and α-bisabolol oxides are found. The results are summarized in Table 5.3. The maximum values of the compounds as estimated from the figures present in the article are provided.

The results of this study showed that the three cultivars CH-29, BK-2, and Menemen did not contain α-bisabolol at all. The variety Menemen did not contain chamazulene as well as α-bisabolol. However, it contained the highest levels of en-yn-dicycloethers.

In 1991, Galambosi et al. reported a large variation in the essential oils among four chamomile cultivars, namely Budakalaszi-2, Degumille, Csomori, and Bona grown in Finland (Galambosi et al. 1991). The results are presented in Table 5.4.

Evidently, the varieties Degumille, Csomori, and Bona are rich in chamazulene and α-bisabolol, whereas the variety Budakalaszi-2 has high levels of chamazulene and α-bisabolol oxide A.

A similar variation in the composition of the oil was reported in 1993. A study on seven cultivars of chamomile, which were cultivated at Ozzano, Italy, compared the variability in the oil composition (Piccaglia et al. 1993). The cultivars were sourced from Egypt, Argentina, Hungary, Germany, Holland, Czechoslovakia, and Yugoslavia. The oils were found to vary in their composition (Table 5.5). Of the seven cultivars, the cultivar from Czechoslovakia was rich in α-bisabolol (58.5%) and chamazulene (17.5%).

Das et al. (1999) reported a comparative account of the composition of the oil in two varieties (Vallary and CR-3A) and a selection (CR-SPL) grown in India. The findings are reproduced in Table 5.6.

The findings revealed variability in the composition of the essential oil, as can be observed in Table 5.5. Whereas Vallary was rich in α-bisabolol oxide A, CR-3A is rich in chamazulene.

TABLE 5.3
Comparison of the Levels (in Percentage) of Pharmacologically Active Compounds in Different Cultivars

S. No.	Compound	H 29/1 (%)	CH-29 (%)	BK-2 (%)	Menemen (%)
1.	Chamazulene	30	22	22	0
2.	α-Bisabolol	70	0	0	0
3.	α-Bisabolol oxide A	5	4	60	6
4.	α-Bisabolol oxide B	5	70	7	6
5.	Spiroether	40	32	10	70
6.	Bisabolone oxide	0	0	33	15

TABLE 5.4
Concentration of the Main Components in the Capitula Oils of Four Cultivars of Chamomile over Three Seasons 1985–1988 as Reported by Galambosi et al. (1991)

S. No.	Compound	Range of Concentration (%) in the Oil of the Variety			
		Budakalaszi-2	Degumille	Csomori	Bona
1.	Chamazulene	12.0–17.6	17.8–20.8	10.8–18.3	11.6–21.8
2.	α-Bisabolol	1.3–3.3	34.6–40.9	32.4–57.6	33.4–60.1
3.	α-Bisabolol oxide A	33.4–39	2.2–6.2	0.0–1.7	0.4–1.4
4.	α-Bisabolol oxide B	2.4–5.1	2.1–7.5	0.2–2.4	0.0–2.2
5.	Bisabolone oxide	0.4–3.7	Traces	Traces	Traces

TABLE 5.5
Oil Content and Composition of Seven Chamomile Cultivars from Different Countries Grown at Ozzano (Bologna)

S. No.	Compound	Concentration (%) in the Oil of the Chamomile Accession						
		Egypt	Argentina	Hungary	Germany	Holland	Czecho-slovakia	Yugoslavia
1.	Chamazulene	2.7	10.3	19.7	7.6	3.5	17.5	15.8
2.	α-Bisabolol	2.9	6.8	5.9	1.6	2.8	58.5	1.7
3.	α-Bisabolol oxide A	53.6	11.5	35.0	44.7	51.6	2.5	28.6
4.	α-Bisabolol oxide B	9.5	50.5	8.4	13.5	10.7	3.9	21.3
5.	α-Bisabolone oxide A	8.5	1.7	5.8	12.0	8.5	0.4	12.8
6.	(Z)-and (E)-Spiroether	5.9	4.4	6.4	5.9	7.0	5.8	6.1

TABLE 5.6
Comparative Account of the Oil Composition in Two Varieties (Vallary, CR-3A) and Selection (CR-SPL) Grown in India

S. No.	Compound	Vallary	CR-3A	CR-SPL
1.	Chamazulene	2.4	18.2	3.5
2.	α-Bisabolol	14.8	2.2	2.2
3.	α-Bisabolol oxide A	38.0	13.7	15.5
4.	α-Bisabolol oxide B	6.4	16.1	6.3
5.	Spiroether	4.4	7.7	8.8

Raal et al. (2003) in Estonia made a comparative study of chamomile populations obtained from Estonia, France, Hungary, Belgium, and Britain. They analyzed the essential oils of these populations and found that there was a variation in the content levels of compounds in the oil. The sample from Hungarian origin contained the highest content of α-bisabolol (24%). The highest content of chamazulene (14%) and bisabolol oxide B (25%) was found in the oil of the sample from Britain. The highest content of α-bisabolol oxide A (43%–55%) was found in the samples of French, Estonian, and Belgian origin.

Šalamon (2004) compared the essential oil composition of 42 different populations collected from natural sites in East Slovakia with the diploid cultivars (Bona and Novbona) and tetraploid cultivars (Lutea and Goral). The findings are summarized in Table 5.7.

Sashidhara et al. (2006) carried out a study on the essential oil composition of chamomile grown in the lower reaches of the Himalayas with those grown in the plains of India. They found that the main constituents, except for α-bisabolol oxide B, were found in higher concentration in chamomile oil from the foothills of the Himalayas than in that from the northern Indian plain. The main constituents reported were chamazulene (5.6%), α-bisabolol (16%), α-bisabolol oxide A (36.5%), α-bisabolol oxide B (8.6%).

Jalalil et al. (2008) reported that in Iran, the main constituents found in chamomile oil were chamazulene (2.6%–10.6%), α-bisabolol oxide A (3.3%–15.4%), and bisabolene oxide A (63.4%–92.4%). This is perhaps the highest levels of bisabolene oxide reported in chamazulene so far.

Another study in Iran by Salamon et al. (2010) reported wide variability in the wild chamomile populations growing in different parts of the country. They studied 20 landraces and reported chamomile populations with very high α-bisabolol (55%–61% from Baba Mydan) and α-bisabolol oxide A (56%–62% from Esfahan/20) contents in the essential oils. On the basis of this study, they proposed the chemotypes of chamomile that were found in Iran.

TABLE 5.7
Composition of Essential Oil of Bona, Novbona, Lutea, and Goral as Reported by Salamon in 2004

S. No.	Compound	Bona	Novbona	Lutea	Goral
1.	Chamazulene	20	12	16	19
2.	α-Bisabolol	42	39	48	30
3.	α-Bisabolol oxide A	2	1	0.5	16.6
4.	α-Bisabolol oxide B	3	2	2	9
5.	α-Bisabolone oxide A	0.5	0.7	0.6	0.5
6.	cis-Spiroether	13	18	12	7
7.	trans-Spiroether	0.1	0.1	0.8	1.3

TABLE 5.8
Results of the Comparative Analysis of the Essential Oils of Wild Type and Zloty Lan Variety of Chamomile

S. No.	Compound	Wild Type	Zloty Lan
1.	Chamazulene	15.6	24.9
2.	α-Bisabolol	—	—
3.	α-Bisabolol oxide A	31.7	17.2
4.	α-Bisabolol oxide B	17.1	24
5.	Bisabolene oxide A	15.7	17

Szabó et al. (2010) studied the composition of the essential oil in Hungary. They collected five populations of chamomile growing in the wild and extracted the essential oil. They found that the chamazulene content varied from 4.6% to 12.2%. The α-bisabolol content varied from 7.5% to 50%.

Orav et al. (2010) studied the essential oil composition of chamomile flowers from 13 populations collected from different European countries. These countries were Estonia (three samples), Greece, Scotland, Armenia, Moldova, Russia, Germany, Czech Republic, England, Latvia, and Ukraine. They found that eight chamomile samples contained high levels of α-bisabolol oxide A (27.5%–56%). These samples were from Estonia (three samples), Greece, Scotland, Germany, England, and Latvia. The oil samples of Moldova, Russia, and Czech Republic had high content of α-bisabolol. The sample from Armenia contained high content of α-bisabolol oxide B (27.2%) and chamazulene (15.3%).

Nurzyńska-Weirdak (2011) carried out a comparative analysis of the essential oils of a wild-type chamomile plant population and a cultivar Zloty Lan. It was reported that the wild-type sample has high content of α-bisabolol oxide A in the essential oil, whereas Zloty Lan had high content of chamazulene. The results did not indicate the presence of α-bisabolol in both the samples (Table 5.8).

Petronilho et al. (2011) in Brazil studied the composition of essential oil of 13 samples. Their results revealed that 11 samples were rich in bisabolol oxide B and 2 were rich in bisabolol oxide A.

5.2.3 CHEMICAL TYPES OF CHAMOMILE

It was Schilcher in 1973, who found that the essential oils could be divided into four types based on the predominant compound found in the oil. The plants yielding such oils were designated to be the chemical types of chamomile. He recognized four chemical types of chamomile as follows (Lawrence 1987):

1. Type A: α-Bisabolol oxide B > α-bisabolol oxide A > α-bisabolol
2. Type B: α-Bisabolol oxide A > α-bisabolol oxide B > α-bisabolol
3. Type C: α-Bisabolol > α-bisabolol oxide B > α-bisabolol oxide A
4. Type D: α-Bisabolol oxide B ≡ α-bisabolol oxide A ≡ α-bisabolol

Type A was characterized by high levels of α-bisabolol oxide B ranging from 22.43% to 58.85%. In this study, the levels of α-bisabolol oxide A are from 4.74% to 15.68% and α-bisabolol from 4.37% to 15.41%. The chamazulene content was from 2.7% to 17.69%.

Type B was characterized by a predominance of α-bisabolol oxide A ranging from 31.07% to 52.25%. The levels of α-bisabolol oxide B ranged from 5.27% to 8.79% and those of α-bisabolol from 8.81% to 12.92%. The level of chamazulene was from 5.4% to 7.95%.

Type C was characterized by high levels of α-bisabolol, which ranged from 24.18% to 77.21%. The levels of α-bisabolol oxide A ranged from 2.13% to 18.5% and those of α-bisabolol oxide B from 4.37% to 15.41%. The chamazulene content ranged from 1.45% to 14.9%.

Type D had α-bisabolol, α-bisabolol oxide A, and α-bisabolol oxide B in equivalent levels. The levels of α-bisabolol oxide B ranged from 10.23% to 24.2%, of α-bisabolol oxide A from 9.62% to 25.83%, and of α-bisabolol from 8.49% to 19.58%. The levels of chamazulene ranged from 1.91% to 7.89%.

All these chemical types contained chamazulene as well as spiroether in the essential oil.

Honcariv and Repcak, in 1979, also found the occurrence of similar chemotypes based on their study of essential oils of chamomile plants of diverse origin, such as from Hungary, Czechoslovakia, Germany, and Spain (Lawrence 1987).

Franz et al. (1985) classified the chemical types (chemotypes) into three categories:

1. Free of chamazulene, rich in bisabolone oxide: These plants are endemic to Southeast Europe and Asia Minor.
2. Chamazulene and bisabolol oxides: These chemotypes are found in Middle Europe.
3. Chamazulene and α-bisabolol: These chemotypes are found in Northeast Spain.

Franz et al. (1985) thus described a new chemotype in addition to what Schilcher had described. This chemotype was a plant that was devoid of chamazulene but containing high levels of bisabolone oxide. The third type, that is, chamazulene and α-bisabolol, is considered to be the most desirable chemotype.

Gosztola et al. (2007) compared eight populations of the Hungarian chamomile, both cultivated and wild, and found that the populations could be classified into α-bisabolol oxide A and α-bisabolol.

Salamon et al. (2010) also described the chemotypes recognized by Schilcher in Iranian chamomile plants and also in Egyptian chamomile flowers. They found that the Iranian plants belonged to Types A, B, C, and D. The Egyptian plants belonged to Type B with predominance of α-bisabolol oxide A in the essential oil. They classified Type A to be South American Collection; Type B to be Egypt and Central European; Type C to be Spain/Catalonia, Malta, Crimea; and Type D to be Southeast Europe and Turkey.

Gosztola et al. (2010) studied the variability of the essential oil components in 4 cultivars and 28 wild populations in Hungary. They classified the plants having high content of α-bisabolol oxide A as chemotype A. In a previous study by Gosztola et al. (2006), the classification was done in the same way, where plants having high content of α-bisabolol oxide A were classified as chemotype A. Petronilho et al. (2011) analyzed 13 populations of chamomile growing in Brazil. They found that there was a predominance of chamomile plants that has high levels of bisabolol oxide B. They classified 11 samples as chemotype B (rich in α-bisabolol oxide B) and 2 as chemotype A (rich in α-bisabolol oxide A). This classification does not appear to be in accordance with what Schilcher proposed (Lawrence 1987; Salamon et al. 2010), and perhaps needs to be reviewed.

5.2.4 CHEMOTYPES BASED ON FLOWER EXTRACT

Several chemotypes of chamomile have been reported based on the predominance of the compounds in the extracts of chamomile flowers. Chemotypes such as the apigenin chemotype, jaceidin chemotype, and chrysosplenetin chemotype have been reported.

Letchamo and Vömel (1990) reported that the tetraploids had twice the apigenin content (103 mg/100 g ligulate florets) as compared to diploids (55 g apigenin/100 g ligulate florets).

Repčak and Martonfi (1995) reported that the tetraploids had 15% more apigenin content than the diploid cultivars. In 2000, Švehlíková and Repčák suggested that chemotypes could be identified based on the content of apigenin in the ligulate florets. They reported that apigenin in high levels were found in a diploid cultivar in comparison with the tetraploid cultivars. The apigenin content in this diploid cultivar was stable and was not influenced by varying environmental conditions (Švehlíková and Repčák 2000).

In 1999, Repčak et al. reported two new chemotypes of chamomile based on the compounds jaceidin (quercetagetin-3,6,3'-trimethyl ether) and chrysosplenetin (quercetagetin-3,6,7,3'-tetramethyl ether). These compounds are found in the tubular florets but absent in the ligulate florets and the involucral bracts. The jaceidin and chrysosplenetin chemotypes were identified in both diploid and tetraploid varieties of chamomile, although the content of these compounds were twice as high in the tetraploids. The researchers reported that the chrysosplenetin chemotype predominate in a population, whereas the jaceidin chemotype is found only in 20% of the population (Repčak et al. 1999).

Repčak et al. (2001) also indicated that levels of the precursors of herniarin, namely (Z)- and (E)-2-β-D-glucopyranosyloxy-4-methoxycinnamic acids (GMCA), might be helpful in determining chemotypes, which could be useful for breeding purposes.

Gosztola et al. (2006) reported high levels of apigenin (10–13 mg/g) in the flowers of chamomile grown in Hungary.

The chemotypes are important in breeding of plants containing phytomedicines, as a directed breeding program can be done if the chemotypes are known (Salamon et al. 2010).

5.2.5 Variability Due to Ontogeny

The content and composition of the essential oil show a variation depending on the developmental stages of the flower (Franz et al. 1978b). Franz (1980) reported that 1 week after the beginning of flowering, the oil content was found to reach the highest levels. The level of oil content fell thereafter. The levels of the compounds in the essential oil varied depending on the developmental stage of the flower. For example, the flower buds had a high content of α-bisabolol and farnesene as compared to the fully developed flowers. Similarly, chamazulene and α-bisabolol oxide levels increased with the age of the flowers.

Marczal and Verzár-Petri (1980) also reported the ontogenetic variation of the oil content and composition in cultivars rich in chamazulene and α-bisabolol.

Repcak et al. (1993) reported that the oil content was highest in the young flowers in diploid and tetraploid cultivars. The levels of chamazulene increased in the terminal flowering stages in tetraploid cultivars.

Ohe et al. (1995) estimated the accumulation of glycosides (*cis*- and *trans*-GMCA and apigenin 7-*O*-β-D-glucopyranoside) in the flowers of chamomile during their ontogeny. They found that during the early stages of flowering, the *cis*- and *trans*-GMCA levels were higher in the ligulate flowers. However, the *cis*-GMCA levels started declining rapidly, whereas the *trans*-GMCA levels reduced slowly. Apigenin 7-*O*-β-D-glucopyranoside levels were high in the initial stages of flowering.

5.2.6 Variability Due to Environmental Conditions

The environmental conditions depend on a variety of conditions, such as geographical location, cultivation practices, biotic and abiotic stresses, and interspecific competition. The variation can be location specific, such as the type of soil, temperature, light intensity, duration of light, and rainfall. It could also be due to cultivation practices such as irrigation and fertilization regime, harvest practices, and postharvest treatment.

Several physiological experiments were carried out to determine the effect of temperature and light on the yield and essential oil content and quality of flowers. Sváb et al. in 1967 carried out a study that showed that with increasing temperature, the weight of the flowers reduced, but there was no effect on the essential oil composition (Franz et al. 1986). Subsequently, in 1971 Saleh studied the effect of air temperature and thermoperiod on the quality and quantity of chamomile essential oil in chamomile plants grown in a phytotron. It was found that when the plants were exposed to 16 hours light of 9.5×10^4 erg/cm^2/sec and 25°C day temperature and 15°C night temperature, the maximum chamazulene synthesis took place (Saleh 1971). In another experiment by Saleh (1973), the chamomile plants were grown in a phytotron under a constant temperature of 20°C and relative humidity at 72%. The light intensities were varied. It was found that when the light intensities were decreased from 90,000 to 32,000 erg/cm^2/sec, the flower number, flower size, dry weight of flowers, oil content, and chamazulene content decreased. Two chamomile cultivars were grown under varying photoperiods (light periods) of 8, 13, and 18 hours per 24-hour cycle. Best performance in terms of flower and oil yield in both the cultivars was observed when the photoperiod was of 18 hours.

Sváb et al. (1980) investigated the effect of early spring frost on the flower yield and compounds in the essential oil in phytotron conditions. They reported that cold treatment did not reduce the flower production of chamomile. The effect of freezing resulted in a shorter flowering period. The chamazulene and α-bisabolol content did not show any significant difference between treated and untreated plants, whereas frost treatment increased bisabolol oxide A content significantly.

Franz et al. (1986) reported that when the chamomile plants were grown under 18 hours day length, the chamazulene content increased in the essential oil. They also observed that bisaboloids reacted on day length in phytotron conditions.

Repcak et al. (1980) found that the chamomile plant accumulated the sesquiterpenes (α-bisabolol, α-bisabolol oxide A, α-bisabolol oxide B, chamazulene, and spathulenol) by following a circadian rhythm. α-Bisabolol accumulated twice a day following a 12-hour period. The maximum accumulation of α-bisabolol occurred during the night. The other sesquiterpenes (i.e., α-bisabolol oxide A, α-bisabolol oxide B, chamazulene, and spathulenol) followed a 24-hour circadian rhythm.

Bettray and Vömel (1992) reported that under phytotron conditions, an increasing temperature has positive effects on flower drug yield, flowering speed, and essential oil content. (–)-α-Bisabolol, chamazulene, and apigenin content were determined with increasing temperature. However, the plant weight and flower weight decreased with the increase in temperature.

Salamon (1994) reported that the environment has an impact on the quantity and constituents of the essential oil.

Letchamo and Vömel (1990) grew diploid and tetraploid chamomile plants of German origin in two locations of extremely varying ecological conditions, namely Addis Ababa in Ethiopia and Germany, to study the levels of apigenin in the ligulate florets. They found that the plants growing in Ethiopia had high concentrations of apigenin compared with those growing in Germany.

Gašić and Lukić (1990) from Yugoslavia reported that the oil content was higher in spring-sown plants as compared to autumn-sown plants. The content of chamazulene and α-bisabolol was found to be variable. Apigenin has been reported to be not influenced by environmental conditions (Švehlíková and Repčák 2000).

Gosztola et al. (2008a) studied the effect of ecological conditions on the morphological characteristics of chamomile populations in Hungary. They found the height was almost doubled in the progeny than in the parents. Further the size of the flowers increased by 30% in the progeny population. This change in morphological characters was attributed to higher precipitation in the year of growing of the progeny population.

Repcak et al. (2009) reported the circadian rhythm of herniarin and its precursor GMCA. The highest levels of herniarin were found to accumulate maximum in the afternoon. The highest levels of GMCA were found at daybreak.

Houshmand et al. (2011) observed chamomile plants growing under drought conditions and found that due to drought the flowering stage was reduced by 7 days, resulting in an early onset of flowering. Further, there was a significant reduction in plant weight, and flower number and weight.

5.3 GENETICS

The genetic component of the morphological and chemical variability has been studied well.

The various selfing and crossing experiments using the available chemotypes have helped to elucidate a part of the biosynthetic pathway of the major secondary metabolites (Holz 1979). The genes controlling the conversion of precursor components into their products have also been defined.

5.3.1 Genetics of Chamazulene and Bisaboloids

Biochemical studies have revealed that farnesene is the precursor of both chamazulene and bisabolol. Subsequent oxidation of bisabolol forms bisabolol epoxide, which further gets converted into α-bisabolol oxide A or α-bisabolol oxide B. α-Bisabolol oxide A gets further converted into α-bisabolone oxide (Figure 5.1).

It has been found that individual Mendelian genes control the presence or absence of chamazulene and bisabolol. The most desirable genotype with high chamazulene and bisabolol content is a double recessive (Horn et al. 1988). It was also elucidated that although both matricin and bisaboloids are derived from farnesene, they are formed independently (Franz 1990).

In 1961, Tétényi, on the basis of breeding experiments, hypothesized that matricin (pro-chamazulene) was produced in plants that contained recessive genes for matricin (Horn et al. 1988). It meant that the genes, when present in dominant form did not allow the production of matricin. This viewpoint was supported by Svab and Sarkany in 1975 and Franz and Wickel in 1985 (Horn et al. 1988). In 1988, Horn et al. reported an elaborate breeding experiment in which they had analyzed 40 progenies obtained from selfing, crossing, and backcrossing chamomile chemotypes. These chemotypes were recognized as follows (Horn et al. 1988):

1. Type I: α-Bisabolol and pro-chamazulene as the main components; free from bisabolone oxide
2. Type IIa: α-Bisabolol oxide A and pro-chamazulene as the main components; bisabolol, bisabolol oxide B, and bisabolone oxide in traces
3. Type IIb: α-Bisabolol oxide B and pro-chamazulene as the main components; bisabolol, bisabolol oxide A, and bisabolone oxide in traces
4. Type III: α-Bisabolone oxide as the major component; free from bisabolol and pro-chamzulene

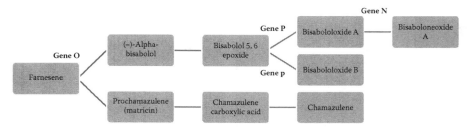

FIGURE 5.1 Gene action on the formation of bisaboloids from farnesene

The results revealed that the bisabolol chemotype was controlled by a recessive gene. Therefore, the desirable chamomile genotype with high chamazulene and high α-bisabolol contents is a double recessive.

This breeding experiment also revealed that bisabolol is converted to bisabolol epoxide due to the presence of a dominant gene O. A dominant epistatic gene P controls the formation of bisabolol oxide A from bisabolol-5,6-epoxide. A recessive epistatic gene p controls the formation of bisabolol oxide B from bisabolol-5,6-epoxide. A dominant gene N controls the conversion of α-bisabolol oxide A into bisabolone oxide.

Mader et al. (2007) reported that the bisabolol oxide content is transmitted dominantly in the progenies of a cross between α-bisabolol cultivar (Manzana) and bisabolol oxide (A and B) cultivar (BK-2).

The genotypes assigned to different chemotypes are bisabolol (OO), bisabolol epoxide (OOPP), bisabolol oxide B (O-pp), bisabolol oxide A (O-P-nn), and bisabolone oxide (O-P-N) (Franz 1990) types.

5.3.2 MOLECULAR TECHNIQUES TO DETECT GENETIC VARIATION IN CHAMOMILE

The techniques of molecular biology, such as random amplified polymorphic DNA (RAPD) and amplified fragment length polymorphisms (AFLPs), are being used to detect the polymorphisms in the chamomile plants at the DNA level. In addition to facilitating the detection of polymorphisms, these techniques help in the development of molecular markers for several characteristics of the plant.

Kumar et al. (1999) used RAPD analysis to compare the 2 varieties Vallary and Prashant using 18 random primers. The similarity index calculated on the basis of the banding patterns showed a difference of 33% at the DNA level of the two varieties.

Wagner et al. (2004) used RAPD and AFLP to identify genetic similarity and polymorphism in chamomile populations, including self-pollinating lines that had variable α-bisabolol content. An F_2 population was obtained by crossing between these lines for obtaining molecular markers for α-bisabolol. They discovered two AFLP markers linked to α-bisabolol. In a further work, they (Wagner et al. 2005a) used the bulk segregant analysis. They used several chemotypes such as (1) rich in α-bisabolol (traces of α-bisabolol oxide B and free of α-bisabolol oxide A), (2) rich in α-bisabolol oxide B (traces of α-bisabolol and free of α-bisabolol oxide A), (3) rich in α-bisabolol oxide A and B (traces of α-bisabolol and free of α-bisabolol oxide A), (4) rich in α-bisabolol oxide A (traces of α-bisabolol and α-bisabolol oxide B), (5) rich in chamazulene, and (6) free of chamazulene. The polymorphisms in these plants were detected using RAPD, and potential markers were tested using the AFLP method. The markers were then mapped to 92 F_2 individuals. They could detect the AFLP markers at a distance of 11.9–15.2 cM from the α-bisabolol locus. They could also detect eight AFLP markers at a distance of 2.0–24.5 cM from the chamazulene locus. Wagner et al. (2005b) reported the detection of 3 AFLP markers linked to the bisabolol locus and 17 AFLP markers linked to the chamazulene locus in tetraploid chamomile plants.

Solouki et al. (2008) screened 20 landraces in Iran by RAPD with 29 primers. Of the 369 bands detected, 314 were polymorphic. They observed that the genetic diversity was not according to the geographical diversity.

Pirkhejri et al. (2010) analyzed 25 chamomile populations using RAPD. The 18 primers generated 220 bands, of which 205 were polymorphic. The similarity matrix ranged from 0.15 to 0.78. The unweighted pair group method with arithmetic mean (UPGMA) of cluster analysis for genetic similarity generated two dendrograms. The genetic diversity was found to be large.

The RAPD method was used by Okoń et al. (2011) to detect genetic similarity in 7 cultivars and 13 wild plants of chamomile. They used 12 primers that generated 157 fragments. Of the 157 fragments, 149 were found to be polymorphic. The analysis using the Dice's coefficient showed that the genetic similarity was between 0.46 and 0.832. The UPGMA method generated two dendrograms. They observed that the RAPD markers provide an effective method to evaluate the genetic similarity and relationships, and could also help in identifying genotypes.

Irmisch et al. (2012) isolated five partial terpene synthase genes *MrTPS1, MrTPS2, MrTPS3, MrTPS4*, and *MrTPS5* that coded for the terpene synthase enzymes MrTPS1, MrTPS2, MrTPS3, MrTPS4, and MrTPS5, respectively, in chamomile. The open-reading frames were obtained and were cloned to a vector pASK-IBA7 and expressed in *Escherichia coli*. The MrTPS1, MrTPS2, MrTPS3, and MrTPS5 showed highest activity with farnesyl diphosphate and so these were characterized as sesquiterpene synthases. MrTPS1 produced (E)-β-caryophyllene as a major product and traces of α-humulene. MrTPS2 produced α-isocomene, β-isocomene, silphinene, and modeph-2-ene in addition to (E)-β-caryophyllene and α-humulene, thus exhibiting a broad spectrum. MrTPS3 produced germacrene A and traces of β-elemene, which is the degradation product of germacrene A. MrTPS5 produced germacrene D as a major compound. MrTPS4 acted on geranyl diphosphate and produced (E)- and (Z)-β-ocimene, the acyclic monoterpenes. The gene expression experiments revealed that *MrTPS1* and *MrTPS3* showed high accumulation of the transcripts in the disk florets, moderate transcripts in the ray florets, and trace accumulation in the shoot. Both these genes were not expressed in the roots. *MrTPS4*, and *MrTPS5* showed highest transcript accumulation in the shoots of the plant. *MrTPS2* was found to be highly expressed in the roots, with low levels of expression in the stem, leaves, and disk florets.

With the enhancements in DNA extraction and amplification methods (Schmiderer et al. 2013), more DNA exploration studies are expected in chamomile.

Inter simple sequence repeat (ISSR) polymorphism technique has been used in chamomile to detect the genetic variation. Okoń et al. (2013) used ISSR analysis to screen 20 primers, of which 5 yielded polymorphic and repeatable fragments. The fragments produced by these 5 primers were 48 in number, of which 41 (85.4%) were polymorphic. The genetic similarity value based on the ISSR markers was found to be 0.653.

5.3.3 Patents on Cloned Genes

The α-tubulin gene and the farnesyl diphosphate synthase genes have been cloned and patented. Dai and Huang (2012) cloned the α-tubulin gene (*C1TUA* gene) in chamomile by extracting the total RNA, followed by its reverse transcription, synthesizing cDNA, obtaining the homologous sequence of the α-tubulin gene and amplifying

Genetics and Breeding of Chamomile

the sequence. The inventors opine that the gene could be used for studying stress tolerance of chamomile.

Yuan et al. (2012) cloned the farnesyl diphosphate synthase gene of chamomile along with constructing a recombinant expression vector. The inventors suggest that the clone could be used for improving yields of chamazulene.

5.4 BREEDING

Paleobotanical evidence from archaeological excavations in Europe shows that wild chamomile plants were brought into cultivation as early as the Neolithic period (9000–7000 BC) and chamomile was used as a medicinal plant in the advanced ancient civilizations (Salamon 1993).

Chamomile is native to the Near East and South-Southeast Europe (Galambosi and Galambosi-Szebeni 1992; Franke and Schilcher 2007). It is found in the wild in these areas. The plant has gradually spread over to many European and Asian countries. Today, it is found in all the continents (Holm 1997). A study by Galambosi and Galambosi-Szebeni (1992) indicated that in Finland, chamomile cultivation perhaps began in 1915, as evident from the documentation of the collection and cultivation by Grotenfelt in 1915.

In Europe, till about 60 years ago, the chamomile flowers were collected from the wild and sold.

By 1950s, many landraces of chamomile had been identified whose seeds had been in use for the crop's cultivation in Europe. In 1960s, these landraces were compared with wild plants assembled through mass collection from different areas to reveal the morphological and essential oil variability in the plants. Careful selection from the variable populations was practiced and the process of chamomile cultivation began, followed by breeding practices and isolation of several breeding lines (Salamon 1994; Franke and Schilcher 2007).

The breeding experiments mainly involved selfing and/or crossing within/ between the variants followed by screening. The breeding programs also incorporated variability of the mutants by the use of chemical or physical (γ-rays) mutagens and tetraploids induced by the use of colchicine in the different varieties.

The aim of breeding chamomile is to develop high-yielding varieties. The important goals of breeding are to develop varieties that have the following characteristics (Franz et al. 1985; Franke et al. 2005):

1. Large flowers
2. Low proportion of herb to flowers, that is, more flowers
3. Flowers at the same level in the plant
4. Firm flowers with low disintegration
5. Uniform plant height for easy mechanical harvest
6. Uniform flowering time
7. High essential oil content
8. High levels of chamazulene
9. High levels of α-bisabolol
10. Low levels of bisabolol oxides

11. High levels of flavonoids
12. High levels of seed germination
13. Resistance to diseases

The selected progenies showing best performance were released as varieties, which included the diploid and tetraploid varieties (Höelzl et al. 1989). The tetraploid plants differed from the diploids in several biochemical and morphological traits (Peneva et al. 1989a; Repcak and Martonfi 1995). Morphologically, the tetraploid varieties had twice the height, larger capitulum size and weight, higher dry seed weight, and higher drug and essential oil yields than the diploid varieties (Salamon 1994).

The tetraploids produced more quantity of oil (Schilcher 1974), more bisaboloids (Franz and Isaac 1986), more chamazulene (Repcak et al. 1993), more apigenin (Repcak et al. 1994), and more flavonoids (Peneva et al. 1989b). Cultivation of tetraploid varieties was recommended in Europe so as to avoid natural crossing between the cultivated and wild plants.

In Germany, by the 1950s, chamomile cultivation started and the varieties were Holsteiner Marschenkamille, Quedlinburger Großblütige Kamille, and Ehfurter Kleinblütige Kamille. In 1962, the tetraploid variety Bodegold was developed and released as a variety (Franke and Hannig 2005). In the subsequent years, other varieties, such as Kor, Offstein, Camoflora, Germania, Mabamille, Euromille, Robumille, and Degumille were developed (Bundessortenamt 2002; Gosztola 2012).

In the Czech Republic, chamomile cultivation began in the 1960s (Holubář 2005). The variety Bohemia was developed followed by the variety Pohorelicky Volkosvety. In addition, the foreign varieties, namely Bona, Goral, Lutea, and Novbona are also registered (Holubář 2005).

In Slovakia, breeding experiments began in the 1970s. The original indigenous chamomile was Bohemia, and it was crossed with the Spanish chamomile plants that were rich in α–bisabolol. These breeding experiments resulted in the development of Kosice 1, Kosice 2, and Kosice 3. Kosice 3 is a diploid variety, whereas Kosice 1 and Kosice 2 are tetraploid varieties. Further breeding experiments in the 1980s resulted in the development of the diploid variety Bona, which was released in 1984. Kosice 2 was later named as Goral and was released as a variety in 1995. Another cultivar Novbona was released in 1995 in Slovakia. However, it did not become popular as Bona. Another tetraploid Lutea was released in 1995 (Oravec et al. 2007).

In Poland, before the 1970s, plant material was collected from the wild. However, in 1972, breeding experiments led to the development of a cultivar named Zloty Lan. It is a tetraploid variety. A diploid variety Promyk was released in 1992 and registered as a variety. In 2006, two varieties rich in α–bisabolol Mastar (a diploid) and Dukat (a tetraploid) were registered in Poland (Seidler-Lozykowska 2007).

In Romania, the varieties were developed taking Zloty Lan and the parent material. Two tetraploid varieties were developed in Romania, Flora, and Margaritar. Flora gives a higher flower yield and essential oil content as compared to Margaritar (Bernáth and Németh 2005).

In Hungary, breeding of chamomile started in the early1980s, following the preliminary work of screening and cultivation that happened in the 1950s. The breeding experiments in the late 1970s led to the development of several high-yielding varieties in the 1980s (Sváb 1983), such as Budakalaszi-2. Soroksári is another variety developed from local cultivars (Bernáth and Németh 2005).

In Bulgaria, the first tetraploid variety Lazur was selected from the locally growing plants by Stanev and Mihailova (Nedkov 2006). Lazur was released as a variety in 1980 (Bundessortenamt 2002). It is a tetraploid. A study in Bulgaria in the year 1994–95 compared the essential oil composition of 29 local populations with the standard cultivars Lazur and Bodegold. The populations containing 13.7%–73.7% bisabolol were identified as α-bisabolol types (Stanev et al. 1996).

In erstwhile Yugoslavia, selection breeding programs were carried out as early as 1940. These programs included the survey of the plants growing in Yugoslavia and also the study of the plants outside the country. The propagation of the plants for flower production was extensively carried out. Several selections of high-yielding chamomile plants were made (Adamović 2000). In the 1980s, the breeding program was revitalized and selections were made based on the yield and essential oil content. Later on, the emphasis shifted to selecting plants with high levels of compounds (chamazulene and α-bisabolol in the oil). Various cultivars were developed during this time in Yugoslavia such as Banatska, Tip-29, and Tetraploidina. In Serbia, the Banatska variety is still predominantly cultivated (Ristic et al. 2007). However, a new cultivar Mina was developed and registered in 2010 (National List 2010). It had an oil content of 0.8%–1.1% and chamazulene content of 16%, which was twice as much as the Banatska variety (Drazic and Drazic 2011). In Slovenia, chamomile research began in 1964. The production started in 1980 (Wagner 1993). A variety Tetra was developed.

In Spain, chamomile is known as Manzanilla and is mostly collected from the wild. The variety Adzet has been reported (Franke et al. 2005).

In Austria, Franz and Isaac (1985) developed a new tetraploid variety Manzana, which was rich in chamazulene and bisabolol. The variety was obtained by tetraploidization of the diploid variety Degumille followed by selection and multiplication. Franz and Isaac (1986) also developed and patented the process for developing tetraploid chamomile varieties.

In Argentina, the variety Bodegold is mostly cultivated (Fogola 2005).

In Brazil, in 1995, Corrêa Júnior reported the development of a new cultivar Mandirituba (Corrêa Júnior 1995).

In Italy, several varieties are grown, such as Ingegnoli, Bornträger, and Benary. In 2003, a high oil yielding selection, Ippo, was identified as more suited for industrial use (Aiello et al. 2004).

In India, in 1993, a mutation breeding experiment using γ-rays on the chamomile seeds of German origin generated an agronomically superior plant, which was released as the variety Vallary (Lal et al. 1993, 1996). Lal and Khanuja (2007) further carried experiments on Vallary by irradiating with γ-rays to induce mutagenesis with an aim to develop a better variety. They isolated two promising mutants and released an improved version of Vallary. Another variety Prashant was reported by Kumar et al. (1999).

The names of the different varieties mentioned in the literature are compiled in Table 5.9.

TABLE 5.9
Chamomile Varieties Developed in Different Countries

S. No.	Variety	Country	References
1.	Bezzi	Argentina	Dellacecca (1994)
2.	Argentina	Argentina	Franke et al. (2005)
3.	Manzana (4×)	Austria	Franz and Isaac (1985)
4.	Mandirituba	Brazil	Franke et al. (2005)
5.	Lazur (4×)	Bulgaria	Nedkov (2006)
6.	Sregez	Bulgaria	Franke et al. (2005)
7.	Bisabolol	Bulgaria	Franke et al. (2005)
8.	Manzanilla Primavera Puelche	Chile	Franke et al. (2005)
9.	Bohemia	Czech Republic	Franke et al. (2005)
10.	Pohorelicky Velkosvety	Czech Republic	Franke et al. (2005)
11.	Minardi	Egypt	Dellacecca (1994)
12.	MA.VS.1	France	Franke et al. (2005)
13.	Offstein	Germany	Dellacecca (1994)
14.	Kor	Germany	Dellacecca (1994)
15.	Leipzig	Germany	Franke et al. (2005)
16.	Bodegold	Germany	Franke et al. (2005)
17.	Camoflora	Germany	Franke et al. (2005)
18.	Chamextrackt	Germany	Franke et al. (2005)
19.	Degumill	Germany	Franke et al. (2005)
20.	Euromille	Germany	Franke et al. (2005)
21.	Mabamille	Germany	Franke et al. (2005)
22.	Robumille	Germany	Franke et al. (2005)
23.	Dotto	Holland	Dellacecca (1994)
24.	Budakalaszi-2	Hungary	Franke et al. (2005)
25.	Soroksari	Hungary	Bernáth and Németh (2005)
26.	Vallary	India	Lal et al. (1996)
27.	Prashant	India	Kumar et al. (1999)
28.	Ingegnoli	Italy	Dellacecca (1994)
29.	Minardi	Italy	Franke et al. (2005)
30.	Olanda	Italy	Franke et al. (2005)
31.	Ciclo1	Italy	Gosztola (2012)
32.	Ippo	Italy	Aiello et al. (2004)
33.	Borntrager Wild	Italy	Aiello et al. (2004)
34.	Zloty Lan	Poland	Franke et al. (2005)
35.	Tonia	Poland	Franke et al. (2005)
36.	Promyk	Poland	Franke et al. (2005)
37.	Mastar	Poland	Seidler-Lozykowska (2007)
38.	Ducat	Poland	Seidler-Lozykowska (2007)
39.	Margaritar	Romania	Franke et al. (2005)
40.	Flora	Romania	Franke et al. (2005)

Genetics and Breeding of Chamomile

TABLE 5.9 (*Continued*)
Chamomile Varieties Developed in Different Countries

S. No.	Variety	Country	References
41.	Banatska	Serbia	Ristic et al. (2007)
42.	Mina	Serbia	National List (2010)
43.	Bona	Slovakia	Franke et al. (2005)
44.	Lutea	Slovakia	Franke et al. (2005)
45.	Goral	Slovakia	Franke et al. (2005)
46.	Novbona	Slovakia	Franke et al. (2005)
47.	Tetra	Slovenia	Dellacecca (1994)

5.5 COMPARATIVE ACCOUNT OF TETRAPLOID AND DIPLOID VARIETIES OF CHAMOMILE

There are more than 45 varieties of chamomile cultivated in the world, and of these, ~11 varieties are tetraploids. These are Manzana, Lazur, Goral (Kosice II), Lutea, Bodegold, Mabamille, Robumille, Budakalaszi-2, Zloty Lan, Margaritar, and Flora. The tetraploids have been created by the breeders by mutagenic treatment using colchicine.

5.5.1 MORPHOLOGICAL CHARACTERS

Czabajska (1960) observed that compared with the diploids, the tetraploids had a greater plant size, larger leaves, larger stomata, a broader flower diameter, larger pollen grains, a higher number of glandular hairs on the ovaries of ray and disc florets, and a higher 1000-seed weight.

Franz et al. (1985) developed a tetraploid Manzana and compared the characteristics of the tetraploid variety with a diploid (Degumille) variety. The characteristics studied were plant growth pattern, herb development, plant height at flowering, flower size at start of flowering, dry weight of flowers, flower yield, flower to herb ratio, essential oil quantity and quality. The plant growth pattern was found to be upright and ramified, and the herb development was medium in both the tetraploid and diploid varieties. The height of the tetraploid variety was 70 ± 5 cm, whereas that of the diploid variety ranged from 50 to 60 cm. The diploid flowered one fortnight earlier than the tetraploid. The flower diameter of the diploid plant was ~1 cm, whereas the tetraploid flowers measured 1.5 cm. The herb to flower ratio was reported to be 5.4 in the diploid and 4 in the tetraploid plants. In both the diploid and tetraploid varieties, the oil content was the same, that is, 1%. The tetraploid variety had a higher chamazulene (150–240 mg%) content than the diploid variety (150 mg%). The tetraploid variety had also a higher α-bisabolol content (300–500 mg%) as compared to the diploid (200 mg%).

Salamon (2004) reported that the tetraploid plants are characterized by height, a larger flower diameter, and a higher flower yield as compared to the diploids. This is presented in Table 5.10.

TABLE 5.10
Comparative Account of Some Morphological Characters and Oil Content between Bona and Goral

S. No.	Characteristics	Diploid Cultivar (Bona)	Tetraploid Cultivar (Goral)
1.	Plant height (cm)	15–30	40–60
2.	Flower diameter (cm)	1.5	3.5
3.	Flower weight (g)	0.0020	0.0045
4.	Oil content (%)	0.6	1.1

Azizi et al. (2007) studied and compared the diploid varieties (Germana and Bona) and the tetraploid varieties (Goral and Bodegold) growing in Iran. They found that the height of the tetraploids (Goral = 85.5 cm and Bodegold = 77.2 cm) were slightly more than the heights of the diploids (Germania = 74.2 cm and Bona = 70.4 cm). The tetraploid Bodegold had a flower diameter of 22 mm and Goral had the smallest diameter of 19 mm. The essential oil content of the tetraploids was not different with tetraploids (Goral = 0.63%–0.95% and Bodegold = 0.68%–0.82%) having the same levels as the diploids (Germania = 0.52%–1.08% and Bona = 0.53%–1.02%).

5.5.2 Determination of Ploidy Levels

Letchamo et al. (1995) elucidated the alternative methods that could rapidly determine the ploidy levels in chamomile. The method involved studying the germ pore number and diameter of pollen grains; chloroplast numbers; stomata numbers and length of stomatal guard cells; width, number, and length of ligulate florets; weight of 100 flowers; and height of plants. Their results indicated that the number of pollen germ pores, number of stomata, and number of chloroplasts provided a distinct, accurate, rapid, and convenient identification method for tetraploid and diploid plants. Table 5.11 presents these characteristics.

The diameter of pollen; length of stomatal guard cells; width, number and length of ligulate florets; weight of 100 flowers; and height of plants were found to be overlapping and so were not reliable in determining the ploidy levels.

However, Seidler-Łożykowska (2003) studied several morphological characters of five diploid strains and five tetraploid strains and compared them with the diploid (Bona and Promyk) and tetraploid (Goral and Złoty Łan) cultivars. She reported that

TABLE 5.11
Comparison of Some Characteristics of Diploid and Tetraploid Chamomile Plants as Reported by Letchamo et al. (1995)

S. No.	Characteristics	Diploid Plant	Tetraploid Plant
1.	Number of pollen germ pores	3	4
2.	Number of chloroplasts in the pair of guard cells	<22	>23
3.	Number of stomata per mm^2	>31.2	<11.7

TABLE 5.12
Comparison of Some Characteristics of Diploid and Tetraploid Chamomile Plants as Reported by Seidler-Łożykowska in 2003

S. No.	Characteristics	Diploid Plants	Tetraploid Plants
1.	Weight of 100 fresh flowers (g)	12.26–15.69	20.75–24.80
2.	Weight of 1000 seeds (g)	0.058–0.067	0.096–0.105
3.	Length of stomata (μm)	39.04–44.28	53.88–59.95
4.	Number of chloroplasts per guard cell	5.81–8.42	11.07–15.36

the morphological characters of stomata length, number of chloroplasts per guard cells, 100-flower weight, and 1000-seed weight were significantly higher in the tetraploids than in the diploids. Therefore, these could be reliably used to distinguish the tetraploids from the diploids. The findings are presented in Table 5.12.

Seidler-Łożykowska reported that the flower diameter and pollen diameter were not significantly different in diploids and tetraploids. The flower diameters were 1.75–1.93 cm in diploids and 1.94–2.24 cm in tetraploids. The pollen diameters were 19.38–22.13 μm in diploids and 24.73–26.75 μm in tetraploids. These characteristics are therefore not reliable.

5.5.3 COMPOUNDS IN THE ESSENTIAL OIL

Salamon (1994, 2004) reported the chemical content of the essential oil compounds in the diploid (Bona) and tetraploid (Goral) varieties grown in Slovakia (Table 5.13). The diploid Bona and the tetraploid Goral both had high oil content in flowers with differentiated involucral bracts. Their oil contents varied from 0.55% to 1.42% (dry weight of flowers). The oil contents, however, decreased with the development and maturity of the capitula. Oil content of Goral was found to be higher as compared to

TABLE 5.13
Comparative Account of the Composition of the Essential Oil of Diploid and Tetraploid Chamomile Varieties in the Years 1994 and 2004

S. No.	Characteristics	Diploid Cultivar (Bona) in 1994	Tetraploid Cultivar (Goral) in 1994	Diploid Cultivar (Bona) in 2004	Tetraploid Cultivar (Goral) in 2004
1.	Farnesene (%)	17.3	21.5	4	9
2.	α-Bisabolol oxide B (%)	1.2	9.5	3	9
3.	α-Bisabolone oxide A (%)	0.06	0.4	0.5	0.5
4.	α-Bisabolol (%)	57.3	31.7	42	30
5.	Chamazulene (%)	1.7	13.3	20	19
6.	α-Bisabolol oxide A (%)	16.4	14.8	2	16.6
7.	cis-Dicycloether (%)	4.7	1.5	13	7
8.	trans-Dicycloether (%)	—	—	0.1	1.3

that of Bona, although the oil of Bona had higher contents of α-bisabolol and chamazulene than that of Goral (Salamon 1994).

5.5.4 COMPOUNDS IN THE EXTRACT

Letchamo and Vömel (1990) reported that the levels of apigenin were high in the tetraploids than in the diploids. The tetraploids had 103 mg apigenin per 100 g ligulate florets as compared to diploids, which had 55 g apigenin per 100 g ligulate florets. These plants were grown in Addis Ababa, Egypt.

Repcak and Martonfi (1995), in Slovakia, studied the levels of the compounds in the flower extracts of diploid (Bona) and tetraploid (Lutea) chamomile. They found that the total apigenin aglycone content was higher in the ligulate florets of the tetraploid varieties.

5.6 BIOMETRICS

The statistical treatment of the data (Chetvernya et al. 1987) from the breeding experiments has revealed correlations between different biochemical components, the nature of the heritabilities of different characters, and has uncovered positive response to selection.

Adamovic et al. (1982) in Yugoslavia reported genetic variability in the morphological characters, contents of essential oil, bisabolol and chamazulene, and the yield of oil and flowers.

Massoud and Franz (1990a) in Germany evaluated the genetic component of the total phenotype of chamomile in four locations (three in Germany and one in Egypt) using 10 diploid clones as parents. They measured the genotypic variation in the flower weight, oil content, and chamazulene and α-bisabolol levels. They found that the flower weight, the oil content, and the level of chamazulene and α-bisabolol showed high heritability coefficients and, therefore, genetically controlled. They concluded that the high genetic variation of these characters would aid the breeders to carry out selection for developing high-yielding cultivars. They concluded that cultivars possessing these two characters could be bred through selection because of their high heritability.

Massoud and Franz (1990b) further reported that the highest yield of essential oil and α-bisabolol was obtained in Egypt, whereas the highest yields of chamazulene were obtained in the three locations of Germany. They also reported that the general combining ability was predominant in all the characteristics, leading to assumption that there was additive genetic variance occurring. The specific combining ability was found to be negative, indicating absence of dominance variance and insignificant heterosis. They suggested that the predominance of general combining ability and lack of specific combining ability confirmed heterogeneity in the parent populations and in such a case selection may produce better results than hybridization.

Das (1999) found that plant height was correlated significantly with flower number, fresh weight of plant, and fresh weight of flowers. It was also found that there was a significant correlation between the oil yield and fresh weight of plant and fresh weight of flowers. It appears that the plant height and fresh weight of the plant, which signify the growth of the plant, as well as the flower number and flower weight, which signify the yield of the flowers, are concerned, to various degrees, with the determination of oil content in the chamomile plant.

Lal et al. (2000) studied the nature and extent of variability, association, co-heritability, and path coefficients for eight economic traits, namely plant height, spread area, branches per plant, fresh flower yield, dry flower yield, oil content, and oil yield. They found higher genotypic correlations than the phenotypic correlations. They reported that the fresh flower yield was significantly and positively correlated with branches per plant, dry flower yield, spread area, and oil yield at both genotypic and phenotypic levels. Fresh flower yield was found to be a direct contributor of the oil yield. The correlations between plant height and oil content had the maximum co-heritability value followed by days to 50% flowering and dry flower yield.

Dražić (2000), in Yugoslavia, studied the characters of height, number of flower heads, yield of leaf and essential oil, and harvest on four varieties of chamomile over 2 years. He reported a high genetic control of the plant height and yield of leaf and essential oil as reflected by a high heritability of these characters. A high genetic and phenotypic correlation was observed.

Pirkhezri et al. (2008) in Iran measured 16 quantitative characteristics, such as yield, flower number per plant, anthodium diameter, receptacle diameter, 100-flower weight, essential oil percent, pollen diameter, stomata length, leaf length, leaf width, seed weight, vegetative period, height, generative period, dry flower percent, and ligulae flower number in 26 populations. A positive correlation was reported between the essential oil content and the characters such as flower number, flower yield, 100-flower weight, days to flowering, and plant height.

Gosztola et al. (2008b) did not observe any significant correlation between plant height and diameter of flowers, but could find a significant correlation between the size of the disc and the ray florets. They reported that the morphological homogeneity increased in the I_2 generation.

Golparvar et al. (2011, 2012), in Iran, found a positive correlation between the dry flower yield and the characters such as flower number, fresh flower yield, days to budding, days to 50% flowering, days to 100% flowering, and number of flowering branches. On the basis of the results given by path analysis, they suggested that the days to 50% flowering and flower number could be efficient indirect selection criteria.

Mohammadi et al. (2012), in Iran, found that the flower yield was positively correlated with flower diameter, plant height, flower number, branch number, number of sepals per flower, biological yield per plant, flower yield per plant, biological yield, flower yield, and harvest index. The flower yield was negatively correlated with flower height, days to start flowering, and days to end flowering. Using a sequential path analysis, they found a positive and significant direct effect of flower number and plant height on harvest index. They also found a positive and significant direct effect of harvest index on flower yield per plant.

There appeared to be no correlation between chamazulene and α-bisabolol (Schilcher 1974). A positive genetic correlation was noted between the oil and flavonoid contents (Peneva et al. 1989b). The main components of chamomile oil have been shown to be highly heritable (Franz 1990). Variance analyses in the populations for the essential oil yield and chamazulene and bisabolol content showed that the additive gene effects were high whereas dominant gene effects were low. In an experiment, the general combining ability for the essential oil yield and chamazulene and bisabolol contents proved to be high, although the specific combining ability was low (Massoud and Franz 1990a). These biometrical evidences (Massoud and Franz 1990a,b) appear to show that for breeding high-value genotypes, selection method in the natural variable populations and those derived by hybridization could both be advantageous.

The high degree of variability (Popova and Peneva 1987) among individual chamomile plants in a population gives high adaptability to the population and is the reason for the high ecological amplitude (adaptability) to varying environments (Kocurik et al. 1979) of different populations of chamomile, but it has also been found that the various chamomile population show specific heritable responses to specific environmental conditions. The populations have given high yields in certain specific environments only (Gasic et al. 1989; Massoud and Franz 1990a). Cultivars possessing high ecological amplitude, combined with desirable agronomic traits, are not yet available and, therefore, need to be developed.

5.7 MICROPROPAGATION OF CHAMOMILE

The various organs and tissues of chamomile plant have been brought under tissue culture primarily to study the ability of the callus tissues to produce secondary metabolites (Szoke et al. 1977a, 1977b; Reichling et al. 1980; Kintzios and Michaelakis 1999). Various phytohormonal combinations and coconut milk were used to culture tissues from stems, roots, and capitula. Shoot regeneration and rhizogenesis occurred in the callus cultures growing on Murashige and Skoog (MS) medium having 0.1 mg/L NAA (naphthalene acetic acid) (Cellarova et al. 1982). Rapid regeneration of plantlets from the callus was obtained in 1/4 MS liquid medium supplemented with 2.0 mg/L BAP (benzylamino purine) (Tokano et al. 1991). High-frequency plant regeneration by direct organogenesis has also been reported (Menghini et al. 1994; Hirata et al. 1996).

Tavoletti and Veronesi (1994) identified an appropriate medium for the micropropagation of the tetraploid genotypes. Passamonti et al. (1998) cultured Italia Minardi, BK-2, and Lutea in an MS medium with four different compositions of the growth regulators. The best medium contained 0.5 mg/L kinetin.

Magiatis et al. (2001) established callus cultures of chamomile rich in α-bisabolol. They distilled the oil from the callus. They found that a mixture of 42.9 µM of α-naphthylacetic acid and 18.6 µM of kinetin provided a higher yield of essential oil and higher α-bisabolol content.

5.7.1 Essential Oil in Cultures

The quantity and quality of the oil in the cultured tissues differed with levels of differentiation and organization achieved (Szoke et al. 1979). The surface of callus tissue from the callus cultures raised from stem and capitula explants produced essential oil similar to that of root (Szoke et al. 1979). The schizogenous oil passages and oil cells of the root contained certain specific compounds, namely, chamomillol, caryophyllene, caryophyllene epoxide, and the polyenes chamomillaester I and II (Reichling et al. 1980, 1984). The shoot lacked these compounds but had chamazulene, which was not reported in the root, thus showing organ-specific distribution of the chemical constituents. In the oil from callus tissues derived from inflorescence, the concentration of the sesquiterpenes was directly proportional to the kinetin concentration in the medium and inversely proportional to the light intensity. Saturated hydrocarbons increased with an increase in light intensity and 2,4-D concentration in the medium (Szoke et al. 1978).

Isolated, intact cells were cultured in a two-phase liquid medium to study the production of essential oil by cells in culture. The medium was prepared by dissolving the nutrients in water, which served as the hydrophilic phase. To this, a nontoxic triglyceride was added to serve as a lipophilic phase. The analysis of both the phases of cell suspension after 1 week showed the accumulation of several lipophilic substances in the triglyceride phase and one of the lipophilic substances was identified as α-bisabolol (Bisson et al. 1983).

Synthetic growth regulators (hydroquinone, N-isopropyl benzimidazolium chloride, and 1,4-hydropthalazine biquinidine) in the medium influenced the morphology, biomass, secondary metabolism and oil content and composition of the growing cells in tissue cultures (Petri et al. 1989).

Studies on response to salinity stress showed that the callus cultures accumulated Na^+, K^+, Ca^{2+}, and Mg^{2+} when the concentrations of their salts were increased in the medium. The cells tolerated salinity up to 85 µM (Cellarova et al. 1986; El-Bahr 1993).

Genetically transformed hairy root cultures (Szoke et al. 1997) induced using *Agrobacterium rhizogenes* were used to study the effect of magnesium ions. At a concentration of 740 mg/L $MgSO_4$, the essential oil content, the oil components, and the biomass increased (Szoke et al. 1997). Szöke et al. (2004) grew hairy root cultures of chamomile in B5 and MS medium and determined the effect of magnesium on the essential oil formation. They reported that addition of magnesium ($MgSO_4$) to the medium favorably affected the biomass and essential oil production.

A study on the mitotic activity in cells of long-term callus cultures (2–3 years) showed a prevalence of cells in prophase. The metaphases were similar to colchicine metaphases and normal bipolar anaphases were extremely rare (Cellarova et al. 1990). These observations confirmed that the cultures had reached a stationary level of growth.

Cytological instability, including the generation of polyploid cells, was observed in the long-term callus cultures when the concentration of kinetin in the growth medium was 0.1 mg/L. The cells showed subhaploid, haploid, tetraploid, and aneuploid chromosome complements, besides the usual diploids. Karyological irregularities in interphase cells, presence of multinucleate cells, agglutinated nuclei in multinucleate cells, fragmented nuclei, micronuclei, and chromosomal aberrations were also reported in such cultures (Cellarova et al. 1984).

5.7.2 Cryopreservation

The technique of tissue culture has been used to generate and maintain elite clones, and this has enabled their cryopreservation with a cryoprotectant mixture of dimethyl sulfoxide, glycerol, and sucrose. Ten weeks were required for successful regrowth after cryopreservation (Cellarova et al. 1992).

REFERENCES

Adamović, D., Borojevic, K., Joksimovic, J., Gašić, O., and Lukić, V. 1982. Genetska varijabilnost prinosa i kvaliteta sorti i populacija kamilice (*Matricaria chamomilla* L.). [Genetic variability of yield and quality in chamomile varieties and populations (*Matricaria chamomilla* L.)]. *Plant Genetics and Breeding* 14: 35–43.

Adamović, D. S. 2000. The most important results on selection of medicinal and aromatic plants in Yugoslavia. In: *First Conference on Medicinal and Aromatic Plants of Southeast European Countries &VI Meeting "Days of Medicinal Plants,"* Arandelovac (FR Yugoslavia), May 29–June 3, 2000.

Aiello, N., Scarteezzini, F., Vender, C., and Brunelli, C. 2004. Morpho-productive and qualitative characteristics of a chamomile selection compared with other cultivars. In: *Atti del convegno: Prospettive di produzione e di impiego delle piante officinali: la camomilla*, San Sepolcro (AR), Pistrino (PG), Italia, maggio 19–20, 2003, pp. 45–50.

Azizi, M., Bos, R., Woerdenbag, H. J., and Kayser, O. 2007. A comparative study of four chamomile cultivars cultivated in Iran. ISHS *Acta Horticulturae* 749: 93–96.

Bernáth, J. and Németh, E. 2005. Production of chamomilla (*Matricaria recutita* L.) in East and South European Countries. In: R. Franke and H. Schilcher (eds.) *Chamomile: Industrial Profiles*. Florida: CRC Press, p. 119.

Bettray, G. and Vömel, A. 1992. Influence of temperature on yield and active principles of *Chamomilla recutita* (L.) Rausch. Under controlled conditions. ISHS *Acta Horticulturae* 306: 83–87.

Bisson, W., Beiderbeck, R., and Reichling, J. 1983. [Production of essential oils by cell-suspensions of *Matricaria chamomilla* in a two phase system]. *Planta Medica* 47(3): 164–168.

Bundessortenamt. 2002. *Kamille. Beschreibende Sortenliste Arznei und Gewürzpflanzen.* Hannover: Deutscher Landwirtschaftsverlag GmbH, pp. 8–88. http://www.bundessortenamt.de/internet30/fileadmin/Files/PDF/bsl_arznei_2002.pdf (Accessed May 16, 2013).

Cellarova, E., Cernicka, T., Vranova, E., Brotovska, R., and Lapar, M. 1992. Variability of *Chamomilla recutita* (L.) Rausch cells after cryopreservation. *Cryoletters* 13: 37–42.

Cellarova, E., Grelakova, K., Repcak, M., and Honcariv, R. 1982. Morphogenesis in callus tissue cultures of some *Matricaria* and *Achillea* species. *Biologia Plantarum* 24(6): 430–433.

Cellarova, E., Repcakova, K., and Honcriv, R. 1986. Salt tolerance of *Chamomilla recutita* (L.) raushchert tissue cultures. *Biologia Plantarum* 28(4): 275–279.

Cellarova, E., Rychlova, M., and Honcriv, R. 1984. Cytological instability in *Matricaria chamomilla* L. tissue cultures. *Herba Hungarica* 23(3): 37–51.

Cellarova, E., Rychlova, M., Seidelova, A., and Honcriv, A. 1990. Comparison of mitotic activity and growth in two long term callus cultures of *Matricaria recutita* L. *Acta Biotechnologica* 10(3): 245–251.

Chetvernya, S. A., Lebeda, A. F., Bezmenov, A. Y., Gorban, A. T., and Perepelova, O. M. 1987. Variability of biologically active substances and the comparative evaluation of wild chamomile cultivars of different origin. *Khimiko Farmatsevticheskii Zhurnal* 21: 595–598.

Corrêa Júnior, C. 1995. 'Mandirituba': Nova cultivar brasileira de camomila. *Horticultura Brasileira* 13(1): 61.

Czabajska, W. 1960. Breeding large-flowered common chamomile (M. *chamomilla* L). I. Polyploidization of chamomile. *Biuletyn Instytut Ochrony Roslin Leczniczych* 6: 71–82.

Dai, S. and Huang, H. 2012. Tubulin gene in chamomile and use thereof. CN 102304523 A 20120104.

Das, M. 1999. Analysis of genetic variation in chamomile (*Chamomilla recutita*) for the selection of improved breeding types. PhD Thesis, University of Lucknow, Lucknow, India.

Das, M., Kumar, S., Mallavarapu, G. R., and Ramesh, S. 1999. Composition of the essential oils of the flowers of three accessions of *Chamomilla recutita* (L.) Rausch. *Journal of Essential Oil Research* 11: 615–618.

Dellacecca, V. 1994. Five years' research into chamomile [*Chamomilla recutita* (L.)]. In: *Atti del Convegno Internazionale "Coltivazione e Miglioramento di Piante Officnalis,"* Trento, giugno2–3, 1994.

Dragland, S., Paulsen, B. S., Wold, J. K., and Aslaksen, T. H. 1996. Flower yield and the content and quality of the essential oil of chamomile, *Chamomilla recutita* (L.) Rauschert, grown in Norway. *Norwegian Journal of Agricultural Sciences* 10(4): 363–370.

Dražić, S. 2000. Influence of genetic and phenotypic variability on productive traits of camomile [*Chamomilla recutita* (L.) Rausch.]. In: *Paper Presented at the First Conference on Medicinal and Aromatic Plants of Southeast European Countries and VI Meeting "Days of Medicinal Plants,"* Arandelovac (FR Yugoslavia), May 29–June 3, 2000. http://www.amapseec.org/cmapseec.1/papers/pap_p017.htm (Accessed May 16, 2013).

Drazic, S. and Drazic, M. 2011. Mina-New chamomile cultivar. In: *V symposium with International Participation. Innovations in Crop and Vegetable Production*, October 20–22, 2011. Belgrade, Serbia. http://www.agrif.bg.ac.rs/files/publications/247/V%20Inovacije%20u%20ratarskoj%20i%20povrtarskoj%20proizvodnji%20Zbornik%20izvoda.pdf (Accessed March 23, 2014)

El-Bahr, M. K. 1993. Responses of chamomile callus cultures to potassium and sodium sulfate salinity. *Egyptian Journal of Horticulture* 20(1): 15–22.

Fogola, N. 2005. Experiences with the cultivation of chamomile in Argentina. In: R. Franke and H. Schilcher (eds.) *Chamomile: Industrial Profiles*. Florida: CRC Press, p 145.

Franke, R. and Hannig, H. 2005. Cultivation in Germany. In: R. Franke and H. Schilcher (eds.) *Chamomile: Industrial Profiles*. Florida: CRC Press, pp. 162–163.

Franke, R. and Schilcher, H. 2007. Relevance and use of chamomile (*Matricaria recutita* L.). ISHS *Acta Horticulturae* 749: 29–43.

Franke, R. et al. 2005. Cultivation. In: R. Franke and H. Schilcher (eds.) *Chamomile: Industrial Profiles*. Boca Raton, FL: CRC Press, pp. 103–104.

Franz, C. 1980. Content and composition of the essential oil in flower heads of Matricaria chamomilla L. during its ontogenetical development. ISHS *Acta Horticulturae* 96: 317–322.

Franz, C. 1990. Biochemical genetics of essential oil compounds. In: *Proceedings of the 11th International Congress of Essential Oils, Fragrances and Flavours*. November 12–16. New Delhi, India, Vol. 3, pp. 17–25.

Franz, C., Hölzl, J., Máthé., and Winkhofer, A. 1985. Recent results on cultivation, harvest time and breeding of chamomile. *Chamomile in Industrial and Pharmaceutical Use*. Trieste, Italy, p. 6–17.

Franz, C., Vömel, A., and Hölzl, J. 1978a. Preliminary morphological and chemical characterization of some populations and varieties of *Matricaria chamomilla* L. ISHS *Acta Horticulturae* 73: 109–114.

Franz, C., Vömel, A., and Hölzl, J. 1978b. Variation in the essential oil of *Matricaria chamomilla* L. depending on plant age and stage of development. ISHS *Acta Horticulturae* 73: 229–238.

Franz, Ch. 1980. Content and composition of the essential oil in flower heads of *Matricaria chamomilla* L. during its ontogenetical development. ISHS *Acta Horticulturae* 96: 317–322.

Franz, Ch. and Isaac, O. 1985. Procede pour la production d'une nouvelle espece de camomille (denomination Manzana) FR2547982 (A1). http://worldwide.espacenet.com/publicationDetails/biblio?FT=D&date=19850104&DB=&&CC=FR&NR=2547982A1&KC=A1&ND=1&locale=en_ (Accessed March 25, 2014).

Franz, Ch. and Isaac, O. 1986. Tetraploid, bisabolol rich chamomile. (Degussa A.-G) Ger. Offen. DE 3,542,756 (Cl .A01H1/08), June 26, 1986. DE Appl. 3,446,216, Dec 19, 1984, p. 64.

Franz, Ch., Müller, E., Pelzmann, H., Hårdh, K., Hälvä, S., and Ceylan, A. 1986. Influence of ecological factors on yield and essential oil of camomile (*Chamomilla recutita* (L.) Rauschert syn, *Matricaria chamomilla* L.). ISHS *Acta Horticulturae* 188: 157–162.

Galambosi, B. and Galambosi-Szebeni, Z. 1992. Experiments on elaborating growing technics for chamomile in Finland. ISHS *Acta Horticulturae* 306: 408–420.

Galambosi, B., Szeheri-Galambosi, Z., Repcak, M., and Cernaj, P. 1991. Variation in the yield and essential oil of four chamomile varieties grown in Finland in 1985–1988. *Journal of Agricultural Science in Finland* 63: 403–410.

Gasic, O. and Lukic, V. 1990. The influence of sowing and harvest time on the essential oils of *Chamomilla recutita*. *Planta Medica* 56(Posters): 638–639.

Gašić, O., Lukić, V., Adamović, R., and Durković, R. 1989. Variability of content and composition of essential oil in various chamomile cultivars. *Herba Hung* 28: 21–28.

Golparvar, A. R., Carlen, C., Baroffio, C. A., and Vouillamoz, J. F. 2012. Genetic improvement of essence percent and dry flower yield in German chamomile (*Matricaria chamomilla*) populations. *Acta Horticulturae* 955: 203–208.

Golparvar, A. R., Pirbalouti, A. G., and Karimi, M. 2011. Determination of the effective traits on essence percent and dry flower yield in German chamomile (*Matricaria chamomilla* L.) populations. *Journal of Medicinal Plants Research* 5(14): 3242–3246.

Gosztola, B. 2012. Morphological and chemical diversity of different chamomile (Matricaria recutita L.) Populations of the great Hungarian plain. PhD Thesis, Corvinus University of Budapest, Hungary, p. 30.

Gosztola, B., Jaszberenyi, C., Németh, E., and Szabó, S. 2008a. Effect of vegetation year on the morphological characteristics and essential oil content of chamomile (*Matricaria recutita* L.). *Kurt Gazdasag Horticulture* 40(2): 69–77.

Gosztola, B., Németh, E., Kozak, A., Sárosi, S., and Szabó, S. 2007. Comparative evaluation of Hungarian chamomile (*Matricaria recutita* L.) populations. ISHS *Acta Horticulturae* 749: 157–162.

Gosztola, B., Németh, E., Sárosi, Sz., Szabó, S., and Kozak, A. 2006. Comparative evaluation of chamomile (*Matricaria recutita* L.). *International Journal of Horticultural Science* 12(1): 91–95.

Gosztola, B., Sárosi, Sz., and Németh, E. 2010. Variability of the essential oil content and composition of chamomile (*Matricaria recutita* L.) affected by weather conditions. *Natural Product Communications* 5(3): 465–470.

Gosztola, B., Varga, L., Németh, E., Bodor, Z., Sárosi, S., and Szabó, S. 2008b. Morphological characteristics of the selected I_2 generation of chamomile (*Matricaria recutita* L.) *Kertgazdasag Horticulture* 40(4): 72–78.

Hirata, T., Izumi, S., Akita, K., Fukuda, N., Katayama, S., Taniguchi, K., Dyas, L., and Goad, L. J. 1996. Lipid constituents of oil bodies in cultured shoot primordia of *Matricaria chamomilla*. *Phytochemistry* 41(5): 1275–1279.

Höelzl, J., Berndt, D., and Hempel, B. 1989. EP 330, 240 (CI A 61 K 35/78) European Patent Office.

Holm, L. 1997. *Matricaria chamomilla* L.: Asteraceae (Compositae), Aster family. In: L. Holm, J. Doll, E. Holm, J. Pancho, and J. Herberger (eds.) *World Weeds: Natural Histories and Distribution*. New York, NY: John Wiley & Sons, pp. 462–468.

Holubář, J. 2005. Growing varieties of chamomile in the Czech Republic. In: R. Franke and H. Schilcher (eds.) *Chamomile: Industrial Profiles*. Florida: CRC Press, p. 139.

Holz, J. 1979. Investigations on the synthesis of sesquiterpenes and spiroethers of Matricaria chamomilla L. Abstract of paper presented at the 27th Annual Meeting of the Society for Medicianl Plant Research, Budapest, Hungary, July 16–22.

Hörn, W., Franz, C., and Wickel, I. 1988. The genetics of bisaboloide in chamomile plant. *Plant Breeding* 101: 307–312.
Houshmand, S., Abasalipour, H., Tadayyon, A., and Zinali, H. 2011. Evaluation of four chamomile species under late season drought stress. *International Journal of Plant Production* 5(1): 9–24.
Irmisch, S., Krause, S. T., Kunert, G., Gershenzon, J., Degenhardt, J., and Köllner, T. G. 2012. The organ-specific expression of terpene synthase genes contributes to the terpene hydrocarbon composition of chamomile essential oils. *BMC Plant Biology* 12: 84.
Jalalil, Z., Sefidkon, F., Assareh, M. H., and Attar, F. 2008. Comparison of sesquiterpens in the essential oils of *Anthemis hyalina* DC., *Matricaria recutita* L. and *Matricaria aurea* (Loefl.) Schultz-Bip. *Iranian Journal of Medicinal and Aromatic Plants* 24(1): 31–37.
Kintzios, S. and Michaelakis, A. 1999. Induction of somatic embryogenesis and in vitro flowering from inflorescences of chamomile (*Chamomilla recutita* L.). *Plant Cell Reports* 18: 684–690.
Kocurik, S. et al. 1979. Content variability of the essential oil and chamazulene in wild chamomile (*Matricaria chamomilla* L.) *Pol'nohospodarstov* 25: 67–75.
Kumar, S., Gopal Rao, M., Khanuja, S. P. S., Gupta, S. K., Das, M., Shasany, A. K., Darokar, M. P. et al. 1999. The chamazulene-rich chamomile variety—Prashant (a new genotype). *Journal of Medicinal and Aromatic Plant Sciences* 21(4): 1096–1098.
Lal, R. K. and Khanuja, S. P. S. 2007. Induced genetic variability and their exploitation in chamomile (*Chamomilla recutita* [L.] Rauschert). ISHS *Acta Horticulturae* 749: 103–109.
Lal, R. K., Sharma, J. R., and Mishra, H. O. 1996. Vallary: An improved variety of German chamomile. *Pafai Journal* 18: 17–20.
Lal, R. K., Sharma, J. R., Mishra, H. O., and Singh, S. P. 1993. Induced floral mutants and their productivity in German chamomile (*Matricaria recutita*). *Indian Journal of Agricultural Science* 63: 27–33.
Lal, R. K., Sharma, J. R., and Sharma, S. 2000. Influence of variability and association on essential oil content of German chamomile (*Chamomilla recutita* (L.) Rauschert). *Journal of Spices and Aromatic Crops* 9(2): 123–128.
Lawrence, B. M. 1987. Progress in essential oils. Chamomile oil. *Perfumer and Flavour* 12: 35–52.
Letchamo, W., Marquard, R., and Friedt, W. 1995. Alternative methods for determination of ploidy level in chamomile (*Chamomilla recutita* (L.) Rausch.) breeding. *Journal of Herbs, Spices and Medicinal Plants* 2(4): 19–26.
Letchamo, W. and Vömel, A. 1990. The pattern of accumulation of Apigenin in the ligulate flowers of *Chamomilla recutita* in extremely varying ecological conditions. *Planta Medica* 56: 638.
Mader, E., Bein-Lobmaier, B., Novak, J., and Franz, C. 2007. The effect of contaminating an/-/-α-bisabolol with an/-/-α-bisabololoxide (A,B) chemotype of tetraploid chamomile. ISHS *Acta Horticulturae* 749: 97–102.
Magiatis, P., Michaelakis, A., Skaltsounis, A., and Haroutounian, S. A. 2001. Volatile secondary metabolite pattern of callus cultures of *Chamomilla recutita*. *Natural Product Letters* 15(2): 125–130.
Marczal, G. and Verzár-Petri, G. 1980. Essential oil production and composition during the ontogeny in *Matricaria chamomilla* L. ISHS *Acta Horticulturae* 96: 325–330.
Massoud, H. and Franz, C. 1990a. Quantitative genetical aspects of *Chamomilla recutita* (L.) Rauschert. *Journal of Essential Oil Research* 2: 15–20.
Massoud, H. and Franz, C. 1990b. Quantitative genetical aspects of *Chamomilla recutita* (L.) Rauschert II. Genotype-Environment interactions and proposed breeding methods. *Journal of Essential Oil Research* 2: 299–305.

Menghini, A., Standardi, A., Tavoletti, S., and Veronesi, F. 1994. [In vitro culture technique for chamomile propagation] Possibilita applicative della propagazione in vitro della camomilla commune. In: *Atti convegno internazionale Coltivazionale. Coltivazionale e miglioramento di piante officinali.* giugno, 2–3, Trento, Italy.

Meriçli, A. H. 1985. Chamazulene content of *Matricaria chamomilla* specimens from Turkey. *Instabul Universitesi Eczacilik Fakultesi Mecmuasi* 21: 63–68.

Mohammadi, R., Dehghani, H., and Zeinali, H. 2012. Interrelationships among flower yield and related characters in chamomile populations (*Matricaria chamomilla* L.). *Journal of Medicinal Plants Research* 6(19): 3549–3554.

Mötl O, Felklová, M., Jasicova M, Lukes V. 1977. Zur GC Analyse and zu Chemischen Typen von Kamillenol(GLC analysis and chemical types of chamomile essential oil [author's transl]). *Arch Pharm* (Weinheim), 310(3): 210–5

National List. 2010. http://photos.state.gov/libraries/sarajevo/30982/pdfs/national-list-of-recognized-varieties.pdf (Accessed May 16, 2013).

Nedkov, N. 2006. The scientific and applied activity of the institute of rose and essential oil cultures in Kazanlak, Bulgaria. *BJAS Archive* 12(5). http://www.agrojournal.org/12/05-00.htm (Accessed May 17, 2013).

Nurzyńska-Weirdak, R. 2011. The essential oil of *Chamomilla recutita* (L.) Rausch. Cultivated and wild growing in Poland. *Annales Universitatis Mariae Curie-Sklodowska Lublin-Polonia* 24(2): 25, 197–206.

Ohe, C., Minami, M., Hasegawa, C., Ashida, K., Sugino, M., and Kanamori, H. 1995. Seasonal variation in production of the head and accumulation of glycosides in the head of *Matricaria chamomilla*. ISHS *Acta Horticulturae* 390: 75–82.

Okón, S. and Surmacz-Magdziak, A. 2011. The use of RAPD markers for detecting genetic similarity and molecular identification of chamomile (*Chamomilla recutita* (L.) Rausch.) genotypes. *Herba Polonica* 57: 38–47.

Okoń, S., Surmacz-Magdziak, A., and Paczos-Grzeda, E. 2013. Genetic diversity among cultivated and wild chamomile germplasm based on ISSR analysis. *Acta Scientiarum Polonorum-Hortorum Cultus* 12(2): 43–50.

Orav, A., Raal, A., and Arak, E. 2010. Content and composition of the essential oil of *Chamomilla recutita* (L.) Rauschert from some European countries. *Natural Product Research* 24(1): 48–55.

Oravec Sr., V., Oravec Jr., V., and Gaia-Mgr. Oravec, V. 2007. Breeding of bisabolol diploid and tetraploid varieties of chamomile in Slovakia. ISHS *Acta Horticulturae* 749: 115–120.

Passamonti, F., Piccioni, E., Standardi, A., and Veronesi, F. 1998. Micropropagation of *Chamomilla recutita* (L.) Rauschert. ISHS *Acta Horticulturae* 457: 303–310.

Peneva P. T., Ivancheva, S., and Terzieva, L. 1989b. Essential oil and flavonoids in the racemes of the wild chamomile (*Matricaria recutita*). *Plant Science* 26: 25–33.

Peneva, P., Popova, E., and Kuzmanov, B. 1989a. Morphological variability of the chamomile *Matricaria recutita* L.: I. Cluster Analysis. *Fitologiya* 36: 15–25.

Petri, G., Kursinszki, L., and Szoke, E. 1989. Essential oil production in *Matricaria* tissue culture influenced by different chemicals. In: S. C. Bhattacharya, N. Sen, K. L. Sethi (eds.) *Proceedings of the 11th International Congress of Essential Oils, Fragrances and Flavours*. New Delhi, India: Oxford and IBH, Vol. 3, pp. 35–41.

Petronilho, S., Maraschin, M., Delgadillo, I., Coimbra, M. A., and Rocha, S. M. 2011. Sesquiterpenic composition of the inflorescences of Brazilian chamomile (*Matricaria recutita* L.): Impact of the agricultural practices. *Industrial Crops and Products* 34(3): 1482–1490.

Piccaglia, R., Marotti, M., Giovanelli, E., Deans, S. G., and Eaglesham, E. 1993. Antibacterial and antioxidant properties of Mediterranean aromatic plants. *Industrial Crops and Products* 2: 47–50.

Pirkhezri, M., Hassani, M. E., and Fakhre Tabatabai, M. 2008. Evaluation of genetic diversity of some German chamomile populations (*Matricaria chamomilla* L.) using some morphological and agronomical characteristics. Contribution from College of Agriculture University of Tehran. http://www.sid.ir/en/ViewPaper.asp?ID = 141486&varStr = 3.14159265358979;PIRKHEZRI%20M.A.D.,HASANI%20 M.E.,FAKHR%20TABATABAEI%20M.;HORTICULTURE%20 SCIENCE%20%28AGRICULTURAL%20SCIENCES%20AND% 20TECHNOLOGY%29;22;2;87;99 (Accessed May 18, 2013)

Pirkhejri, M., Hassani, M. E., and Hadian, J. 2010. Genetic diversity in different populations of *Matricaria chamomilla* L. growing in Southwest of Iran, based on morphological and RAPD markers. *Research Journal of Medicinal Plant* 4(1): 1–13.

Popova, E. and Peneva, P. 1987. Morphological variation in chamomile (*Matricaria recutita* L.). II. Discriminant analysis. *Genetika i Selektsiya* 20(4): 319–326.

Raal, A., Arak, E., Orav, A., and Ivask, K. 2003. Comparison of essential oil content of *Matricaria recutita* L. from different origins. *Ars Pharmaceutica* 44(2): 159–165.

Reichling, J., Beiderbeck, R., and Becker, H. 1980. Comparative studies on the secondary products from tumours, flowers, herb and roots of *Matricaria chamomilla*. *Planta Medica* 36: 322–332.

Reichling, J., Bisson, W., and Becker, H. 1984. Comparative study on the production and accumulation of essential oil in the whole plant and in the callus culture of *Matricaria chamomilla*. *Planta Medica* 50(4): 334–337.

Repcak, M., Cernaj, P., and Martonfi, P. 1993. The essential oil content and composition of diploid and tetraploid *Chamomilla recutita* during the ontogenesis of anthodia. *Journal of Essential Oil Research* 5: 297–300.

Repcak, M., Martonfi, P., and Oravec, V. 1994. Variability of apigenin glucosides in diploid and tetraploid chamomile. In: *Atti del convegno internazionale: Coltivazione miglioramento di piante officinali*, giugno 2, 1994. Trento, Italy: Institute Sperimentale per l' Assessmer Forestale e per l' alpicoltura, pp. 413–416.

Repcak, M., Smajda, P., Cernaj, P., Honcariv, R., and Podhradsky, D. 1980. Diurnal rhythms of certain sesquiterpenes in wild camomile (*Matricaria chamomilla* L.). *Biologia Plantarum* 22: 420–427.

Repcak, M., Smajda, B., Kovacik J., and Eliasova, A. 2009. Circadian rhythm of (Z)- and (E)-2-β-D-glucopyranosyloxy-4-methoxy cinnamic acids and herniarin in leaves of *Matricaria chamomilla*. *Plant Cell Reports* 28: 1137–1143.

Repčak, M. and Martonfi, P. 1995. The variability pattern of apigenin gucosides in *Chamomilla recutita* diploid and tetraploid cultivars. *Pharmazie* 50(H 10): 696–699.

Repčák, M., Pastírová, A., Imrich, J., Švehlíková, V., and Mártonfi, P. 2001. The variability of (Z)- and (E)-2-β-D-glucopyranosyloxy-4-methoxycinnamic acids and apigenin glucosides in diploid and tetraploid *Chamomilla recutita*. *Plant Breeding* 120(2): 188–190.

Repčák, M., Švehliková, V., Imrich, J., and Pihlaja, K. 1999. Jaceidin and chrysosplenetin chemotypes of *Chamomilla recutita* (L.) Rauschert. *Biochemical Systematics and Ecology* 27(7): 727–732.

Ristić, M. S., Đorđević, S. M., Đoković, D. D., and Tasič, S. R. 2007. Setting a standard for the essential oil of chamomile originating from Banat. ISHS *Acta Horticulturae* 749: 127–140.

Salamon, I. 1993. Production of chamomile, Chamomilla recutita (L.) Rauschcrt and its production ecology. PhD Thesis, Comenius University in Bratislava, Slovakia.

Salamon, I. 1994. Growing conditions and the essential oil of chamomile, *Chamomilla recutita* (L.) Rauschert. *Journal of Herbs, Spices and Medicinal Plants* 2(2): 31–37.

Salamon, I. 2004. The Slovak gene pool of German chamomile (*Matricaria recutita* L.) and comparison in its parameters. *Zahradnictví (Horticultural Science)* 31(2): 70–75.

Salamon, I. 2009. Chamomile biodiversity of essential oil qualitative-quantitative characteristics. In : B. Sęner (ed.) *Innovations in Chemical Biology.* Dordrecht, the Netherlands: Springer Science+Business Media B.V, pp. 83–90.

Salamon, I., Ghanavati, M., and Khazaei, H. 2010. Chamomile biodiversity and essential oil qualitative-quantitative characteristics in Egyptian production and Iranian landraces. *Emirates Journal of Food and Agriculture* 22(1): 59–64.

Saleh, M. 1971. The effect of air temperature and thermoperiod on the quality and quantity of *Matricaria chamomilla* L. oil. *Mededelingen Landbouwhogeschool Wageningen* 70: 1–17.

Saleh, M. 1973. Effects of light upon quantity and quality of *Matricaria chamomilla* oil. III. Preliminary study of light intensity effects under controlled conditions. *Planta Medica* 24: 337–340.

Sashidhara, K. V., Verma, R. S., and Ram, P. 2006. Essential oil composition of *Matricaria recutita* L. from the lower region of the Himalayas. *Flavour and Fragrance Journal* 21(2): 274–276.

Schilcher, H. 1973. Neuere erkenntnisse bei der qualitätsbeurteilung von kamillenblüten bzw. Kamillenöl: Qualitative Beurteilung des ätherischen Öles in Flores Chamomillae. Aufteilung *der H*andelskammillen in vier bzw. fünf chemische Typen [Recent findings for the quality assessment of flowers chamomile or chamomile oil: Qualitative assessment of the essential oil in Chamomile flowers. Distribution of commercial chamomile in four or five chemical types]. *Planta Medica* 23(2): 132–144.

Schilcher, H. 1974. Newly acquired information on evaluating the quality of chamomile flowers or chamomile oil. *Planta Medica* 23: 132–144.

Schmiderer, C., Lukas, B., and Novak, J. 2013. Effect of different DNA extraction methods and DNA dilutions on the amplification success in the PCR of different medicinal and aromatic plants. *Zeitschrift für Arznei- & Gewürzpflanzen* 18(2): 65–72.

Seidler-Łożykowska, K. 2003. Determination of the ploidy level in chamomile (*Chamomilla recutita* (L.) Rausch.) strains rich in α-bisabolol. *Journal of Applied Genetics* 44(2): 151–155.

Seidler-Lozykowska, K. 2007. Chamomile cultivars and their cultivation in Poland. ISHS *Acta Horticulturae* 749: 111–114.

Solouki, M., Mehdikhani, H., Zeinali, H., and Emamjomeh, A. A. 2008. Study of genetic diversity in chamomile (*Matricaria chamomilla*) based on morphological traits and molecular markers. *Scientia Horticulturae* 117(3): 281–287.

Stanev, S., Zheljazkoiv, V., and Janculoff, Y. 1996. Variation of chemical compounds in the essential oil from some native forms of chamomile (*Chamomilla recutita* L.) In: F. Pank (ed.) *Proceedings of International Symposium. Breeding Research on Medicinal and Aromatic Plants,* June 30–July 4, 1996. Quedlinburg, Germany: Beitrage zur Zuctungsforschung—Bundesanstalt fur Zuchtingsforschung and kulturpfanzen, Vol. 2, pp. 214–217.

Sváb, J. 1983. Results of chamomile cultivation in large-scale production. ISHS *Acta Horticulturae* 132: 43–48.

Sváb, J., Marczal, G., Verzár-Petri, G., and Rajki, E. 1980. Cold-treatment effect on the flower and volatile oil building of camomile (*Matricaria chamomilla* L.). ISHS *Acta Horticulturae* 96: 235–244.

Švehlíková, V. and Repčák, M. 2000. Variation of apigenin quantity in diploid and tetraploid *Chamomilla recutita* (L.) Rauschert. *Plant Biology* 2(4): 403–407.

Szabó K., Németh, 'E., Sárosi Sz., and Czirbus, Z. 2010. Essential oil content of Hungarian wild chamomile (*Chamomilla recutita* L.) and its composition during primary processing-survey of practice. *Z Arznei-Gewurzpfla* 15(2): 63–68.

Szoke, E., Kuzovkina, I. N., Verzar, G., and Smirnov, A. M. 1977a. Cultivation of wild chamomile tissues. *Fitologiya Rastenii* 24(4): 832–840.

Sžoke, E., Máday, E., Lemberkovics, É., Kiss, A. S., and Muskáth, Zs. 1997. Influence of magnesium on the essential oil production in chamomile cultures. In T. Theophanides and J. Anastassopoulou (eds.) *Magnesium: Current Status and New Developments*. Dordrecht, the Netherlands: Kluwer Academic Publishers, pp. 407–411.

Szöke, E., Máday, E., Kiss, S., Sonnewend, L., and Lemberkovics, E. 2004. Effect of magnesium on essential oil formation of genetically transformed and non-transformed chamomile cultures. *Journal of American College of Nutrition* 23(6): 763S–767S.

Szoke, E., Shavarda, A. L., and Kuzovkina, I. N. 1978. The effect of conditions of growing wild chamomile inflorescence callus tissue on the production of essential oil. *Fiziologiya Rastenii* 25(4): 743–750.

Szoke, E., Verzar, G., Kuzovkina, I. N., Lemberkovics, E., and Kery, A. 1977b. Phytochemical analysis of tissue cultures of chamomile (*Matricaria chamomilla* L.) grown in dark and light. Use of tissue cultures in Plant Breeding. In: J. N. Frantisek (ed.) *Proceedings of the International Symposium*. September 6–11, 1976. Olomouc, Czechoslovakia, pp. 593–605.

Szoke, E., Verzar, G., Shavarda, A. L., Kuzovkina, I. N., and Smirnov, A. M. 1979. Differences in essential oil component, composition of isolated root callus tissue and cell suspension of chamomile. *Izvestiia Akademii Nauk SSSR Seriia Biologicheskaia* 6: 943–949.

Tavoletti, S. and Veronesi, F. 1994. Response to in vitro culture of two chamomile (*Matricaria chamomilla* L) populations with different ploidy levels. *Journal of Genetics and Breeding* 48(2): 125–130.

Tokano, H., Hirana, M., Taniguchi, K., Tanaka, R., and Kondo, K. 1991. Rapid clonal propagation of *Matricaria chamomilla* by tissue cultures shoot primordial. *Japanese Journal of Breeding* 41(3): 421–426.

Wagner, T. 1993. Chamomile production in Slovenia. ISHS *Acta Horticulturae* 344: 476–478.

Wagner, C., Marquard, R. A., Friedt, W., and Ordon, F. 2004. Implementation of molecular techniques (RAPDs, AFLPs) on camomile (*Chamomilla recutita* (L.) Rausch.) for genotyping and marker development. ISHS *Acta Horticulturae* 629: 509–516.

Wagner, C., Marquard, R. A., Friedt, W., and Ordon, F. 2005a. Gentic analysis of (–)-α-bisabolol and chamazulene content in tetraploid chamomile (*Chamomilla recutita* (L.) Rausch.) and identification of molecular markers. ISHS *Acta Horticulturae* 676: 185–191.

Wagner, C., Marquard, R. A., Friedt, W., and Ordon, F. 2005b. Molecular analyses on the genetic diversity and inheritance of (–)-α-bisabolol and chamazulene content in tetraploid chamomile (*Chamomilla recutita* (L.) Rausch.). *Plant Science* 169(5): 917–927.

Yuan, Y., Tai, Y., and Yang, X. 2012. Chamomile farnesyl diphosphatesynthase gene clone and prokaryotic expression. CN 102703397 A 20121003.

6 Cultivation of Chamomile

6.1 INTRODUCTION

The yield and quality of the compounds as well as the essential oil depend on many factors, such as the genotype of the cultivated plants (Zalecki 1972a,b; Adamovic et al. 1982), environments in which the crop is cultivated, time of sowing (Galambosi et al. 1991) and harvesting (Ohe et al. 1995), and the conditions of the hydrodistillation of the capitula. Therefore, it becomes imperative to deploy the above determinants of yield and quality at the optimum levels during cultivation and oil extraction.

In Europe, till 1955, the chamomile consumed was obtained from the wild. Before cultivation began, in Slovakia the chamomile came from the wild flowers collected by the gypsies in the East-Slovakian lowlands. These were sold to the pharmacological companies (Salamon 2007).

Breeding of chamomile began in the late 1990s in Italy. Aiello et al. (1998) reported that efforts were initiated to breed Italian varieties with 12 populations of chamomile collected from the wild. On the basis of the quantity and quality of the oil, one population (Population 11) was identified.

The appropriateness of the mountainous regions of Yugoslavia for cultivating chamomile was investigated by Vukmanovic and Stepanovic (1987). These areas were reported to be satisfactory for growing chamomile, especially because the plants yielded better quality essential oils in these mountain regions. Dellacecca et al. (1992) conducted a 2-year research into the feasibility of cultivating chamomile in northern Italy. On the basis of the results, they recommended that autumn sowing would provide optimum flower yields.

Section 6.2 is a brief account of the climatic conditions that support cultivation of chamomile. The subsequent sections describe the cultivation practices of chamomile.

6.2 HABITAT AND CLIMATIC CONDITIONS

The occurrence of chamomile has been described in detail in Section 2.3. The herbaceous annual chamomile plant grows wild in the temperate climate. Chamomile is found both as a weed and in the cultivated form in many countries of the north temperate zone (Holm et al. 1997). It grows equally well in moderate and tropical zones. Other than the countries where it is traditionally grown, introductory trials of chamomile crop have given encouraging results in several countries that have varying altitudinal, ecological, and edaphic conditions.

At present, chamomile can be found growing all over Europe, Western Siberia, Asia Minor, Iran, Afghanistan, and India (Franke 2005).

Chamomile can be grown successfully at different altitudes ranging from 1000 to 5000 ft. above sea level. The plant can withstand considerable cold conditions ranging from 2°C to 20°C (Singh 1970).

Chamomile is essentially a long day plant, but can grow in diverse seasons. The yields of flowers and essential oil are significantly affected by the sowing seasons (Letchamo 1992). However, the maximum yields of the flowers, oil, and chemical compounds are obtained under long day conditions (Franz et al. 1986). Franz et al. (1986) mentioned that studies in Hungary and Egypt by Sváb and colleagues in 1967 showed that the weight of flowers decreased with temperature. However, the oil composition was not affected. Saleh in 1968, however, found that chamomile grown in 18-hour daylight conditions produced higher levels of chamazulene compared with those grown in 14-hour daylight conditions. The content of the bisaboloids was found to be affected by the day length under phytotron conditions (Franz et al. 1986).

The autumn-sown crops were found to have higher yields than the spring-sown crops (Letchamo and Marquard 1993). In Italy, a study in 1992 indicated that chamomile could be grown under different environmental conditions. However, autumn sowing was advised for northern Italy. It was advised that the best time for sowing in this region was the month of October. In southern Italy, November was recommended to be the best time for sowing (Dellacecca et al. 1992).

The impact of the sowing season on the yield was also reported by Dražić (2000) in Yugoslavia. He found that the yield of the flowers and essential oil was significantly higher in the autumn-sown plants as compared to the spring-sown plants.

A study in Iran by Hadi et al. (2004) also indicated that the sowing time significantly influenced the yield of flowers and the essential oil. In this study, it was found that the plants sown in early March yielded higher quantity of flowers and essential oil compared with those sown in late march after a gap of 20 days. The chamazulene content was also reported to decrease with late planting (Hadi et al. 2004). The researchers had sown the seeds on March 5, 15, and 25. They found that the highest fresh flower yield (2132.9 kg/ha), dry flower yield (389.8 kg/ha), essential oil yield (2.47 L/ha), and chamazulene percentage (6.4%) and content (152 mL/ha) were shown when the seeds were sown on March 5.

In another study in Iran, it was found that the height of the plants and flower yield (dry weight) were more when the plants were sown in the fall compared with the spring-sown plants. However, it was observed that the harvest index was higher in the spring-sown plants (Shams et al. 2012). Chamomile is now widely grown in Iran (Fanid et al. 2012).

In northern India, chamomile starts flowering from the second fortnight of February and continues till April (Singh 1970).

High flower yield in autumn-grown crops (Falistocco et al. 1994) is reported from north European countries, whereas, the middle European countries report high yield from spring-sown crops (Gasic et al. 1991).

Chamomile can be grown in the rabi (winter) season in the semitropical environment of north India, south of Himalayas. The crop can be raised twice a year in the north and south Indian hills as a rabi and kharif (summer) crop. In the high hills, the crop requires planting in June–July, harvesting in October–November, replanting in October–December, and harvesting in March–April.

Cultivation of Chamomile 245

Increased flower yield is reported at 25°C day temperature and 15°C night temperature (Saleh 1971), high humidity, and rainfall (Gasic et al. 1991). In the south Indian hills, where chamomile can be grown, the day temperature is 20°C, night temperature 10°C, rainfall 150 mm, and humidity 80%. The areas of north Indian plains suitable for chamomile cultivation show, during the cropping season, a day temperature of 30°C and a night temperature of 15°C, relative humidity 60%, and rainfall 56 mm.

Gosztola et al. (2010) studied the effect of weather conditions on the essential oil content and its composition for 3 years. They found that when the conditions were moderately warm and wet, the contents of essential oil were high. Further, under these conditions, the α-bisabolol content was also increased.

A correlation between the seasons and effect of fertilizers on yield has also been observed. Bagheri et al. (2008), in Iran, investigated the effect of planting date and nitrogen fertilizer on the flower and essential oil yield. The plants were grown on March 6, March 20, and April 5. They applied different doses of nitrogen (urea) (75, 150, and 225 kg/ha). They found a significant interaction between the planting date and the level of nitrogen on the yield of essential oil. The highest number of flowers were obtained on first planting date (March 6) and on a dose of 75 kg urea/ha. The highest percentage of essential oil was obtained from the third planting, that is, April 5, and on a dose of 150 kg urea/ha.

6.3 SOIL CONDITIONS

The chamomile plant is reported to thrive in humus-rich soils. However, it can grow in any type of soil. It is reported to grow equally well in both acidic and alkaline soils (Franke 2005). Chamomile can grow in acidic (4.5 pH) as well as in alkaline (9.5 pH) (Patra and Singh 1995; Prasad et al. 1997) soils. The plant is recommended for reclamation of usar soils of north India, which are highly alkaline (~10 pH). The optimum pH measured for chamomile is 7 (Bernáth and Németh 2005).

It grows well in Hungary in the clayey lime soils that are barren and other crops cannot possibly grow. In Yugoslavia, this plant is reported to grow in a huge area of saline-soda soils. The optimum pH for chamomile is estimated at 8, but it can very well tolerate a pH of 9–9.2 (Singh 1970).

It was found that in India, chamomile could grow well in the saline-alkaline soils of Banthra, Uttar Pradesh. The plant was found to show high uptake of salt, thereby decreasing the existing high content of salt in the top soil. It was found to absorb high amounts of sodium (66 mEq/100 g dry plant material or 28 kg/ha) (Singh 1970). The high salt uptake is believed to facilitate water uptake, even when the level of water in the oil is very low (Bérnath and Németh 2005).

In 1987, a study in Yugoslavia indicated that the mountainous regions possessed optimum conditions for the cultivation of chamomile (Vukmanovic and Stepanovic 1987).

In Slovenia, Wegner (1993) reported that chamomile was cultivated in the heavy clayey soil and light sandy soil. However, the plants thrived in the light sandy soil as compared with the heavy clayey soil. Apparently the heavy clayey soil caused unequal development of the plants and also favored the development of the vegetative parts instead of the flowers.

A study by Filipović et al. (2007) indicated that chamomile grew well in chernozem (a dark top soil rich in humus), and also in loess (fine calcareous silt or clay) soils containing gley (bluish clay found in subsurface). The plant did not grow well in dark soils of the marshes.

Today, the possibility of identifying the suitable areas for chamomile cultivation has been enhanced using geographic information system (GIS). This technique allows identifying the ecological conditions, such as topography, climate (temperature and precipitation), and soil (pH, electrical conductivity, and organic matter) that are suitable for the crop. A study in Kuzhestan, Iran using the GIS technology indicated that 1.5% of the land was moderately suitable, 32.7% of the land was marginally suitable, and 65.8% of the land was unsuitable for chamomile cultivation (Pirbalouti et al. 2011).

6.4 SEEDS

The mature achenes are harvested while they are still attached to the receptacle of the flower. This it to avoid loss of seeds due to shedding. To obtain seeds, the flowers that are mature and not yet dry are harvested. These are then fully dried and stored in dry and cool environment until used. The seeds can be disinfested by fumigation with methyl bromide (25 g/m^2, 24 hours) or ethylene oxide (90% of EO/10% CO$_2$). Just before use, the flowers are threshed and the liberated achenes are cleared to separate the healthy seeds. On sowing, an achene germinates into a seedling. It has been reported that a minimum of 200 days of dormancy period is required for optimum germination rate of the seeds (Reinhold et al. 1991).

High-quality chamomile seeds should have at least 70% germinability. A good germination rate is important to obtain optimum yields of chamomile. Several factors have been found to hamper the germination rate, which include temperature, humidity, seed age, seed size, water content in seed, and storage methods. It was found by Carle et al. (1991) in Germany that the optimum germination of 70% is reached 200–300 days after the mature seeds are harvested. A decrease in the germination rate was observed when the seeds were stored for 5 years under the controlled conditions of 10°C and 30% relative humidity and airtight seal. The seeds showed ~50% germination in the fifth year. Chamomile seeds have also been found to germinate after a prolonged dormancy. To delay the dormancy, fresh seeds are stored in deep freeze. Cold temperatures and frost have been found to increase chamomile seed germination. The cold temperature acts on the swollen seeds, which probably reinforces their need for cold stratification (Carle et al. 1991).

It was found that high temperatures of up to 90°C did not affect the germination rate of mature chamomile seeds even when the seeds were exposed for up to 6 hours (Carle et al. 1991). It was found that even after 65 days of the heat treatment, the germination rate was as high as 90%.

Small seeds have been found to have a lower germination percentage and slower germination time than the large seeds. Larger seeds have larger embryos and more nutrients, which result in the growth of a vigorous hypocotyls, larger cotyledons and more intensive root formation (Carle et al. 1991).

Carle et al. (1991) also found that the disinfectants ethylene oxide and methyl bromide did not have any adverse effects on the germination rates. The rates were 80%–85% with ethylene oxide.

Tilebeni and Sadeghi (2011) investigated the mechanism of seed aging and deterioration and the effect of hydropriming on their regeneration. Seeds were artificially aged for 24, 48, 72, and 96 hours. Thereafter, they were primed through hydropriming for 2 hours and acid priming for 12 hours. With progressive time of artificial aging, the germinability was found to decrease. The deterioration of the cell membrane was observed with progressive aging as assayed by electric conductivity of the seed leachates. In addition, there was an increase in the level of total peroxide and malondialdehyde content. The decrease in the activities of the antioxidant enzymes, peroxidase, catalase, ascorbate peroxidase, glutathione reductase, and superoxide dismutase was observed. When the aging seeds were hydroprimed and primed with ascorbic acid, the germination was partially maintained. A decrease in the content of the antioxidant enzymes and malondialdehyde was also observed.

6.5 SOWING

Spring sowing is carried out in Europe in April/May when the rainfall is less than 50 mm and the day temperatures are more than 15°C (Oravec et al. 2005). The possible disadvantage of growing chamomile in summer months is that the plants become more susceptible to parasites and diseases (Franke 2005).

Autumn sowing begins in September under conditions of irregular precipitation (rain and frosts). Sowing is continued at intervals of 8–14 days. This ensures that the plant enters the winter months at the stage of 6–8 leaves, which is optimum for resisting the cold. Chamomile requires a day length of 17 hours. The plants sown in autumn start flowering in the month of May and continue to flower in several flushes until June (Franke 2005).

Although slightly humid conditions are favorable for the growth of the plant, excessive rains are not desirable as it leads to the higher vegetative growth (Franke 2005).

The seeds are either directly sown in the field or a nursery is raised and the seedlings are transplanted into the field.

In Argentina, sowing is carried out in autumn (March–June) because autumn sowing ensures good germination of seeds (Fogola 2005).

The seeds are sown in a raised nursery bed. The nursery bed is prepared by tilling the land and raising the soil to form beds. In India, farmyard manure (FYM) is added to the soil and mixed before tilling.

The seeds are sown in the bed, covered lightly with soil, and lightly sprinkled with water. For sowing, some researchers recommend ~25 kg/ha of seeds mixed with the flower parts (called pulvis). Pulvis is 20%–30% seeds and 70%–80% flower parts (Bernáth and Németh 2005). Others recommend 1.5–2.5 kg/ha of seeds depending on their quality (Oravec et al. 2005). In Chile, 6–8 kg/ha of seeds are sown mixed with fertilizer (Weldt 2005).

In Egypt, the nursery beds are prepared by plowing the land after adding compost and then irrigated. Afterward, the area is plowed, softened, and divided into nursery beds. One part of seed is mixed with three parts of sand and sown in the nursery beds. The seeds are covered with soil, and water is sprayed.

Dražić (2000) in Yugoslavia investigated the effects of planting mode (freehand and in rows) and seed quantity (2 and 4 kg/ha) on the flower yield and essential oil yield. He reported that he did not observe any significant effect of the planting mode or seed quantity on the flower and oil yield.

Hadi et al. (2004) in Iran tested the effects of three types of plant spacing, 50 cm × 20 cm, 50 cm × 30 cm, and 50 cm × 40 cm, on the plant height, flower diameter, fresh flower yield, dry flower yield, essential oil yield, essential oil per unit dry flower weight, chamazulene content, and chamazulene percentage.

The traits that were influenced by the plant spacing were the fresh flower yield, dry flower yield, essential oil yield, and the chamazulene content. The highest yields of fresh flowers (2233.6 kg/ha), dry flowers (412.4 kg/ha), essential oil (2.44 L/ha), and chamazulene (150 mL/ha) were obtained at plant spacing of 50 cm × 20 cm. The spacing of 50 cm × 40 cm yielded more number of flowers and chamazulene percentage. The other traits, such as plant height, flower diameter, essential oil per unit dry flower weight, and chamazulene percentage, were not influenced by plant density.

Pirzad et al. (2011) in Iran studied the effect of plant density on the yield of chamomile. They conducted experiments with the following plant spacings: 5 cm × 30 cm, 10 cm × 30 cm, 15 cm × 30 cm, 20 cm × 30 cm, and 25 cm × 30 cm. It was found that the highest flower and essential oil yield and biomass were obtained with a plant spacing of 10 cm × 30 cm.

6.5.1 Direct Sowing

This method is practiced in Europe. Optimum yields are obtained with a spacing of 20 cm and using 1.5–3.0 kg seed per hectare (Zalecki 1972b). Higher yields are recorded when the seeds are broadcast as compared to sowing in rows. Sowing in rows is, however, required to facilitate herbicide application (Galambosi et al. 1991) and mechanical harvesting (Zalecki 1972b). Sowing time is known to affect the vegetative growth and the flower yield.

6.5.2 Indirect Sowing

The nursery is prepared in the field by adding FYM (Hussain et al. 1988) to the soil. Small raised beds are made to facilitate irrigation and drainage. Healthy seeds are sown superficially by mixing with the top soil. One kg of seed is sufficient for one hectare. In the first week, watering is done with a manual sprinkler. Subsequent 2–3 irrigations are done by flooding.

6.5.3 Self-Sowing

Self-sowing is another method of chamomile cultivation in addition to spring sowing and autumn sowing. In this method, the chamomile plants are allowed to grow continuously as a monoculture in the fields. There is no rotation with other crops. Each season, chamomile plants grow through the seeds that had been dropped from the flowers in the previous season. In some cases, new seeds are also sown. When the optimum collection has been made from the crop, the plants are mulched. This is followed by a field preparation where the field is cleared, but the soil is not turned. The self-sown seeds germinate in the next season. This form of cultivation is not recommended for the breeding lines (Franke 2005).

6.5.4 Field Preparation

In India, the land is prepared by adding FYM (Hussain et al. 1988) at the rate of 15 t/ha. Phosphogypsum along with nitrogen in the form of urea (Marczal and Petri 1988) and a herbicide are also added. Deep plowing once or twice using a disc plow is followed by leveling. Plots with raised bunds, as per layout specification, and channels are prepared. Channels alternate with plots and are used for irrigation. The field is flood irrigated and transplanting is carried out after 3 days.

For monocultures, it is recommended that the soil be made free of the previous years' stubble and then turned over using a disc. Thereafter, the soil should be made compact through the use of rollers. This ensures proper germination of the seeds (Bernáth and Németh 2005).

6.5.5 Germination

Germination takes place within 1–2 weeks of sowing (Franke and Hannig 2005). Germination requires the presence of light. Further, the humidity is helpful in germination (Fogola 2005).

6.5.6 Transplantation

The transplantation is carried out in rows. The recommended spacing from one row to another is 30–45 cm and from one plant to another in a row is about 35–45 cm.

No sowing is necessary in monocultures. In Yugoslavia, a distance of 12–30 cm is recommended, and in Romania, a distance of 40 cm is recommended (Bernáth and Németh 2005). A distance of 45 cm between two rows is recommended in Slovakia (Oravec et al. 2005), whereas a distance of 45–50 cm is recommended in the Czech Republic (Holubář 2005) and 40–60 cm in Argentina (Fogola 2005).

In India, the 2-month-old seedlings are transplanted directly from the nursery to the field while the soil is still moist. Spacing of 30 cm × 30 cm (Johri et al. 1991) or 30 cm × 20 cm (Gowda et al. 1991) has been found to give high yield of flowers.

6.6 FERTILIZATION

6.6.1 Inorganic Fertilizers

The fertilization containing potassium (K), phosphorous (P), and nitrogen (N) (Yaskonis 1978; Emongor and Chweya 1992) enhances the crop growth. It is reported that the increase in the vegetative growth results in more branching, which in turn gives larger yield of flower. Their rates of application are N (150–180 kg/ha),

P as P_2O_5 (60 kg/ha), and K as K_2O (60–80 kg/ha). P and K are applied as basal dose, whereas N is applied in 2–3 splits with one-fourth quantity as basal dose (Ram et al. 1997). N increases chamazulene and bisabolol contents (Emongor and Chweya 1992) and K increases the bisabolol oxide content (Felkova et al. 1981) in the oil.

The recommended fertilizers for a chamomile crop are nitrogen (urea), phosphorus (P_2O_5), and potassium (K_2O). It has been estimated in Hungary that to produce 1 t of chamomile flowers, 53 kg of N, 85 kg K_2O, and 21 kg of P_2O_5 are required (Bernáth and Németh 2005).

Nitrogen promotes vegetative growth. Potassium is important for the development of the active secondary metabolites and flowering. Phosphorus is important for metabolism but is detrimental in excess quantities. Organic manure is not recommended by several researchers owing to possible microbial contamination of the plants (Franke 2005).

The fertilizer application regime could be split into two doses: basal dose and top dressing. The basal dose may be applied initially after planting and the top dressing after a few weeks.

The recommended doses of fertilizers vary from region to region. In Argentina, the recommended dose is 40–60 kg N, 36–45 kg P_2O_5, and 80–120 kg K_2O per hectare (Fogola 2005).

In Slovakia, for the autumn-sown plants, it has been reported that the fertilizers are recommended in the ratio of 60 kg N, 10 kg P, and 142 kg K. Only one-third of the nitrogen is applied before sowing. The remaining two-thirds are applied in two doses after overwintering and the second weeding (Oravec et al. 2005). In the Czech Republic, the recommended doses of the fertilizers are 20–40 kg N, 20 kg P, and 66–100 kg K per hectare (Holubář 2005).

In Chile, the recommended doses of NPK are 50–80 kg N, 100–200 kg P, and 50–70 kg K per hectare (Weldt 2005). In Egypt, the recommended dose is 40 kg N, 50 kg P_2O_5, and 100 kg K_2O per hectare (Fahmi 2005).

In Kenya, Mohammadreza et al. (2012) found that the highest yield of flowers was obtained by the application of 100 kg N and 50 kg P per hectare. They also found that the highest amount of chamazulene (28.5%) was obtained when the soil was fertilized with 50 kg N and 20 kg P per hectare and without any potassium fertilizer. From this finding, it appears that potassium does not play a direct role in the yield of the flowers and the quantity of the active secondary metabolites.

Krstic-Pavlovic and Dzamic (1984) reported that in Yugoslavia the highest yields of flower and essential oil were obtained using NPK in the ratio of 60:30:30. A high yield was also obtained using the NPK ratio of 40:20:20.

Alijani et al. (2010) recommended that 40 kg N and 60 kg P for every hectare was optimum for Iranian conditions. They did not find any significant increase or decrease in the chamazulene content of the oil in the plants treated with different levels of nitrogen and phosphorus fertilizers.

Tamizkar and Khoshouei (2011) reported from Iran that the maximum yield of essential oil was obtained with the application of 150 kg N and 225 kg P per hectare.

Nitrogen has been reported to increase the essential oil yield. Phosphorus at low doses (17.47 kg/ha) increased the essential oil yield, but at higher doses decreased the essential oil yield. Application of 17.47 kg P per hectare at transplanting and top dressing later with 50 kg N per hectare gave the best results (Emongor et al. 1990).

Nitrogen and phosphorus have been reported to increase the content of chamazulene and α-bisabolol in the essential oil (Emongor and Chweya 1992). However, nitrogen in high doses has been reported to decrease the chamazulene content (Emongor et al. 1991). It was also reported that nitrogen decreased the levels of bisabolol oxide A and bisabolol oxide B (Emongor and Chweya 1992). Emongor et al. (1991) found that the essential oil yield increased doses of nitrogen. When the plants were treated with 100 kg N per hectare there was a 65% increase in the essential oil yield. Similarly, there was a 185% increase in the weight of the dry flowers when the dose of nitrogen fertilizer was increased from 0 to 100 kg/ha.

The increase in the essential oil content due to nitrogen may be attributed to the role of nitrogen in development and division of new essential oil cells, cavities, secretory ducts, and glandular hairs. It also might have increased the contents of carbohydrate, gibberellins, and auxins, which were used to form more essential oil cells, cavities, secretory ducts, and glandular hairs (Emongor et al. 1991).

The composition of the essential oil was also bound to be influenced by nitrogen application. With the increasing doses of nitrogen from 0 to 50 kg/ha, the chamazulene, α-bisabolol, farnesene, and *cis*-spiroether contents increased by 25%, 13%, 11%, and 15%, respectively. It was also observed that there was a simultaneous decrease in the levels of bisabolol oxide A and bisabolol oxide B. When nitrogen was increased from 50 to 100 kg and 150 kg, there was a drop in the levels of the constituents except α-bisabolol, which continued to increase. This implied that the biosynthesis of matricin, farnesene, and α-bisabolol was at the expense of bisabolol oxide A and bisabolol oxide B. The application of phosphorus and the interaction of nitrogen with phosphorus did not significantly influence the composition of essential oil of chamomile.

Nikolova et al. (1999) found, in pot experiments, that nitrogen and potassium increased the yield of the flower, whereas phosphorus increased the yield of the essential oil. They found that in the field the best fertilizer regime was NPK in the ratio of 1:1:1 at the rate of 120 kg/ha in field conditions. They observed that the α-bisabolol levels were increased when treated with high levels of phosphorus. Chamazulene level was found to be stable. They found that the elements K, Ca, and Mg produced optimal yields when applied in the ratio of 40:38:28.

6.6.2 Vermicompost

The effects of vermicompost and amino acids on the yield of chamomile oil and flowers were determined by Hadi et al. (2011) in Kenya. They found that 20 t of vermicompost per hectare significantly increased the plant height, flower diameter, fresh and dry flower weight, and the essential oil content. They also found that spraying the plants at the budding + flowering stage with amino acids significantly enhanced the plant height and flower diameter and increased the yield of flowers and essential oils.

Hadi (2010) conducted a study in Iran to assess the effect of vermicompost on the flower yield of chamomile. The highest flower yield was observed when the plants were treated with 20 tons of vermicompost per hectare. In another experiment, the chamomile plants were sprayed with amino acids at budding, flowering, and budding + flowering stage. The highest yields were obtained when the amino acids were sprayed at the budding + flowering stage.

6.6.3 COMBINED FERTILIZERS

Alijani et al. (2011) in Iran found that the application of 40 kg triple superphosphate per hectare combined with a phosphate biofertilizer (Barvar 2) and 80 kg nitrogen significantly increased the yield of flowers and essential oil. They recommend the above doses of K and N along with the phosphate biofertilizer to enhance crop yield.

In a greenhouse pot experiment conducted by Aguilera et al. (2000) in Brazil, it was found that the best flower yield was obtained with 4 kg controlled release fertilizer (NPK in the ratio of 15:9:12) per m^3 and 4 kg vermicompost per m^3. Another combination that yielded good results was 2 kg controlled release fertilizer and 6 kg vermicompost per m^3.

Juárez-Rosete et al. (2012) in Belgium carried out an investigation to determine the effect of inorganic and organic fertilization on the flower yield, flower diameter, plant height, and the essential oil content of chamomile with two substrate systems—one a volcanic rock called *tezontle* and another a mixture of soil, compost, and perlite in the ratio of 50:20:30. The organic fertilizer consisted of humic acids and the inorganic fertilizer consisted of had biosynthetic amino acids. The organic fertilizer was the Steiner solution at 75% nutrient concentration. They found that the inorganic fertilizer had a significant impact on the morphological and yield characteristics, but the content of the essential oil was not influenced. The humic acid in organic fertilizer was found to increase the α-bisabolol levels.

Biofertilizers have been recommended by Mashkani et al. (2011) in Iran. In a study, they concluded that the biofertilizer biosulfur could be applied with good results along with the chemical fertilizers. This is expected to reduce the application of excessive chemicals, which is detrimental to the agroecosystem.

The influence of organic and inorganic fertilizers on the flower and oil yield and the essential oil composition was studied in Brazil by Corrêa Júnior et al. (1999). They used a green manure containing *Mucuna aterrima* and *Crotalaria spectabilis*, another green manure containing *plant cocktail*, and FYM. They also used N (urea); N (ammonium sulfate); and N, P, K with ammonium sulfate as the nitrogen supplement. The results revealed that there were no significant differences on the yields of flower or essential oil content in the plants treated with organic and inorganic fertilizers. Chamazulene content was, however, higher in the green manure containing *Mucuna aterrima* and *Crotalaria spectabilis*.

Shams et al. (2012) in Iran evaluated the effect of organic and chemical fertilizers on the yield of flowers and essential oil of chamomile. They used chemical fertilizers (100 kg/ha urea + 100 kg/ha ammonium phosphate), manure (15 t/ha), and mixed fertilizers (50 kg/ha urea + 50 kg/ha ammonium phosphate + 7.5 t/ha manure). The results indicated that maximum plant height was obtained when mixed fertilizers were used in spring planting. A 47.8% increase in height was reported. In fall planting, the maximum plant height was obtained when chemical fertilizers were applied. They found that there was an 18.9% increase of harvest index in spring and 25.85% increase in fall. The highest flower dry weight (an increase of 325% was reported) was obtained by using mixed fertilizer in spring as well as in fall. The essential oil content also showed maximum levels with the use of mixed fertilizer in spring and fall.

6.6.4 Physiological Treatments

The application of gibberellic acid in experimental plants is reported to enhance the plant height (Haikal and Badr 1982) and number of flowers per plant (Abou-Zeid and El-Sherbeeny 1975). Kinetin has also been reported to stimulate growth, weight of flowers, and yield of oil and flowers (Al-Badawy et al. 1984a, 1984b). Pyrimidines such as CDOP (3-cyclohexyl-2,5,6-trimethylene-2,4-dioxypyrimidine) increased the content and composition of oil (Vaverkova and Herichova 1980). A similar effect of cycocel (Abou-Zeid and El-Sherbeeny 1971) has been demonstrated. These results remain to be standardized for field applications.

Meawad et al. (1984) studied the effect of nitrogen combined with growth regulators on plant growth, flowering, essential oil yield, and chamazulene content of the essential oil of chamomile. The different growth regulators tested were gibberellin, ethephon, cycocel, and B-9. They concluded that nitrogen at 300 kg/feddan (0.42 ha) along with gibberellin (10–100 ppm), ethephon (1–10 ppm), cycocel (50–500 ppm), and B-9 (50–500 ppm) was effective for the growth and yield of flowers and oil along with the chamazulene content in the oil.

Fatma et al. (1999) reported that treating moistened seeds of chamomile at 6°C for 15 days slightly increased bisabolol oxide A and B and chamazulene content in the essential oil.

El-Shamy and Gendy (2013) assessed the effects of complete fertilizer (Kristalon) and salicylic acid on the growth and productivity of chamomile. They found that spraying the plants with 4 g/L Kristalon and 200 ppm salicylic acid provided maximum values of plant height, plant fresh weight, plant dry weight, fresh weight of flowers per plant, dry weight of flowers per plant, dry weight of flowers per feddan (0.48 ha), essential oil yield per plant, and essential oil yield per feddan. These concentrations of Kristalon and salicylic acid also gave maximum values of total chlorophylls and total percentage of carbohydrate, nitrogen, crude protein, phosphorus, and potassium.

6.6.5 Vascular Arbuscular Mycorrhiza

The yield of chamomile flowers has been found to increase when treated with vascular arbuscular mycorrhizal fungi (AMFs). Farkoosh et al. (2011) investigated the effects of five vascular AMFs on chamomile flower yield, oil yield, and the yield of the chemical components in the oil. The five species were *Glomus etunicatum*, *Glomus caledonium*, *Glomus mosseae*, *Glomus intraradices*, and *Glomus spp.* They reported that the number of flowers was increased by inoculation, but the flower diameter was decreased. The fresh and dry flower weights were also reported to be increased by inoculation. The content of essential oil in the flowers was not influenced by the mycorrhiza; however, the chamazulene and α-bisabolol content was reported to increase.

Urcoviche et al. (2012) reported that in Brazil, the soil cultivated with chamomile showed higher spore density of AMFs.

6.6.6 RHIZOBACTERIA

Dastborhan et al. (2012) investigated the effect of rhizobacteria on the growth and yield of chamomile. The chamomile plants were inoculated with *Azotobacter chrocooccum*, *Azospirillum lipoferum*, and a combination of *Azotobacter* and *Azospirillum*. Nitrogen as fertilizer was used in doses of 50, 100, and 150 kg N per hectare. The highest yield of essential oil was obtained when the plants were inoculated with *Azotobacter* and grown without nitrogen fertilizer.

6.7 IRRIGATION

The crop requires a total of three to four irrigations during the entire cropping season. Immediately after transplanting, watering is done manually using sprinklers for few days to 1 week after which flood irrigation is carried out. Flower yield and oil content in capitula increases with better water supply (Wali and Kawy 1982) to the plants.

The chamomile plant does not require much water. Pirzad (2012) conducted a class A pan experiment to evaluate the evapotranspiration and thereby schedule the irrigation regime. It was found that the highest yields of chamomile flowers and essential oil were obtained when the plants were irrigated after 60 and 120 mm evaporation from the pan. The other conditions were 30 and 90 mm, which did not produce significant yields.

Rojas-Valencia et al. (2011) advocated the use of waste water after treating it with ozone. They found that ozone, in addition to eliminating the highly pathogenic microbes, increased the rate of plant growth in chamomile.

Pirzad et al. (2011) in Iran studied the irrigation regime on the yield of chamomile flowers, biomass, and essential oil. They experimented with four irrigation regimes, namely 25, 50, 75, and 100 mm from class A pan. The highest yields of dried flowers and essential oil were obtained from 50 mm evaporation. Irrigation at 50 mm yielded the highest biomass.

6.8 WEEDING

Chamomile crop requires weeding only for the initial 2–3 months after which the vigorous growth of the plant suppresses the growth of the weeds. In European and other countries, herbicides (Vomel et al. 1977; Bouvert-Bernier and Gallote 1989), such as linuron, propyzamide, ethofumesate, and auxins (mecoprop, 2,4-dimethylphenoxiacetic acid [2,4-D], 2-methyl-4-chlorophenoxyacetic acid), are used for the removal of weed. However, the weedicides are known to reduce plant growth and oil yield and alter the quality of oil (Schilcher 1978; Reichling et al. 1980; Betti et al. 1991; Carle et al. 1992). Linuron appears to be more suited to chamomile than the other herbicides (Reichling 1980; Adamovic et al. 1988).

In Europe, herbicides are used both in spring-sown and autumn-sown crops of chamomile. Bernáth and Németh (2005) tabulated 42 weeds that were identified by Mathé in the chamomile fields in different locations of Hungary. Some of these occurring in high frequencies were *Polygonum aviculare*, *Capsella bursa-pastoris*, *Malva neglecta*, *Lepidium draba*, *Festuca pseudovina*, *Ranunculus arvensis*, and *Bromus mollis*.

The weeds occurring in the chamomile fields of Chile have been listed by Weldt (2005). These are, among others, *Chenopodium album*, *Taraxacum officinale*, *Anthemis cotula*, *Convolvulus arvensis*, *Polygonum persicaria*, and *Plantago lanceolata*. The weed control in Chile is done through crop rotation and no herbicides are used (Weldt 2005).

Sváb (1983) recommended the herbicide Maloran (chlorbromuron) to be applied in April in the doses of 3–4 kg/ha. Its optimum time of application is when the chamomile plants are in two- or three-leaf stage and the weeds have just begun to sprout. Another herbicide suggested by Wagner (1993) is Afalon (linuron) to be applied in the doses of 1–2 kg/ha.

In Argentina, the herbicide Treflan (α,α,α-trifluoro-2,6-dinitro-N,N-dipropyl-p-toluidine) is used in the fields before sowing chamomile. While applying the herbicides, care has to be taken that the chemicals do not come in contact with the seeds as this could hamper their germination (Fogola 2005).

In Egypt, the weeds are removed by hoeing at regular intervals. At least three hoeings are carried out to remove the weeds for the entire duration of the crop (Fahmi 2005).

In Germany, a combination of herbicides and mechanical hoeing is adopted to remove weeds in chamomile fields (Franke and Hannig 2005).

In India, weeding is done manually with a small handheld spade for the initial 2 months of crop growth. Two weedings, one after ~4 weeks is usually sufficient.

Stevanovič et al. (2007) reported that in Serbia, the weeds were mostly biannual and a high percentage of these weeds were related to the sowing date of chamomile.

Adamovic et al. (1989) found that the use of the herbicides Afalon and Monosan affected the oil composition. They reported that there was no variation in the oil content after herbicide treatment. They reduced the contents of bisabolol oxide A and chamazulene.

Reichling (1980) investigated the effect of the herbicides Tramat (ethofumesate), Kerb 50 (propyzamide), and Afalon (linuron) on the flower and essential oil yield. He reported that Afalon was very effective in exterminating the weeds. Kerb 50 had selective effect and Tramat was not found to be efficient. He also reported that the individual components of the essential oil were not significantly influenced by Afalon.

Schilcher (1978) studied the influence of the herbicides atrazine, prometon, prometryn, linuron, and 2,4-D on the growth of chamomile plants. He found that higher doses of atrazine had a detrimental effect on the growth of the plants and the chemical components of the essential oil. He reported that the application of linuron just after germination damaged the chamomile plants more than the weeds. The application of 2,4-D did not significantly alter the composition of the essential oil.

Stevanovič et al. (2007) reported that in Serbia, the majority of the weeds in chamomile fields were biannual and a high percentage of the winter and spring weeds were related to the sowing date of chamomile.

Budimir (1986) from Yugoslavia reported that the chamomile fields had weeds growing, such as *Convolvulus arvensis* and *Sorghum halepenses* along with other annual weeds of Poaceae and some other dicotyledon species. He investigated the effect of two herbicides penoxalin and prometrin. These herbicides were found to control the weeds efficiently, which resulted in the decrease in the number and biomass of the weeds per unit area. No phytotoxic effects of the herbicides on chamomile were reported.

6.9   HERBICIDE RESISTANCE

Mackova and Helemikova (1992) in Slovakia studied the effect of the herbicide Hedonal DP (dichlorprop) on chamomile plants. The chamomile plants were found to be resistant at 1.5 L/ha. At a higher dose of 3 L/ha, it was found that 33% of chamomile plants died. Hedonal DP had no negative effect at 1.5 L/ha on either the content or composition of the essential oil or the fertility of the chamomile seeds.

6.10   PEST CONTROL

Chamomile is susceptible to southern root-knot nematode (*Meloidogyne incognita* [Kofoid and White] Chitwood, race 3), which decreases the dry weight of the plant (Walker 1995). Insect pests reported are *Aphis fabae*, *Nysius minor*, and *Antographa chryson*. These can be controlled with 20% EC (280) or 0.05% malathion (Ram et al. 1997). No cases of insects are reported from Argentina. The losses caused by the pests and diseases in the field conditions in India remain to be estimated.

6.11   HARVESTING

Harvesting is carried out when the flowers are open. Franz (1980) had identified four stages of flowers when they would contain the maximum amount of essential oil and chemical constituents. These stages and the corresponding predominance of the chemical compounds are as follows:

1. Stage I: Flowers in bud stage, development of the ligulate flowers, while tubular flowers remain closed.
2. Stage II: Tubular flowers were partially up to 50% completely open. Ready to be harvested.
3. Stage III: More than 50% of the tubular flowers open. Ready for harvesting.
4. Stage IV: The ligulate and tubular flowers dry and withered, decomposing. Unfit for harvesting.

Harvesting the flowers at these stages when the flowers are in full bloom ensures maximum yield of the oil and the active substances. If harvesting is done early at the bud stage, the buds are unlikely to yield essential oil in high amounts. Harvesting should not be carried out late when the old flowers tend to shatter and fall off.

The harvesting of stems and leaves is also carried out from which oil could be distilled.

Harvesting is done two or three times usually after a gap of a couple of weeks depending on the climatic conditions during the entire flowering period. Oravec et al. (2005) reported that the spring-sown crop requires 2–3 pickings and the autumn-sown crop requires 5–8 pickings in Slovakia.

Three to four months after transplanting, the flowers are harvested. Harvesting of flower is done three to four times with an interval of 15–20 days. Flower yields vary greatly depending on the number and time of harvests (Galambosi et al. 1991; Salamon 1993a; Salamon 1993b; Letchamo 1996). The first harvest gives high flower and oil yields and oil of better quality (Gasic et al. 1989; Salamon 1993b). In European countries, mechanized harvesters have been developed, which use a comb roller for harvesting the flowers (Nikitushkin and Martynov 1978; Carle and Gomma 1992; Oravec et al. 1993; Bovelli 1996; Chiumenti et al. 1996). In India, harvesting is done manually by handpicking of the flowers.

In a study conducted in Japan, the mid-flowering stage was found to be the best time for harvest when the contents of the effective anti-inflammatory materials such as phenylpropanoids and the flavonoid glycoside cosmosiin were highest (Nikitushkin and Martynov 1978).

The highest apigenin content was found in the first harvest for Degumille (Letchamo 1992) and R-43 (Letchamo and Marquard 1993) and second harvest for Degumille and BK2-39, and Diploide (Letchamo and Marquard 1993).

Letchamo and Marquard (1993) found that the autumn-sown chamomile plants (diploid) produced ~60% more flowers than the spring-sown plants. The content of chemical constituents, such as α-bisabolol and chamazulene, were found to be slightly higher in the autumn-sown plants than in the spring-sown plants. However, apigenin content was found to be 21% more in the spring-sown plants. The essential oil content was found to decrease with the harvesting frequency.

The harvesting is done either manually by tearing off the flowers using bare hands or hand-held tools, or with machines. Chamomile growing in the wild is harvested by hands. Cultivated chamomile is harvested using various types of harvesting machines (harvesters) across the world. In some countries, such as India and Egypt, it is still harvested by hand. In Egypt, chamomile harvesting is carried out in large farms and is highly labor intensive (Santucci et al. 2013).

The handheld tools used are rakes and toothed shovels. The machines used are either manually pushed picking carts or harvesters. Harvesting with machines commonly involves the stripping process. Specialized harvesters have been developed in many countries, such as Italy, Argentina, Hungary, Russia, Slovakia, Czech Republic, and Germany (Fogola 2005; Franke 2005).

The process of harvesting is a special process as it involves picking of flowers as delicately as possible to maintain the quality of the picked material. The quality of the picked flowers has been divided into three categories. The first-class picked material comprises flowers with stems not longer than 20 mm. The second-class picked material comprises flowers with stems between 20 and 40 mm. The third category is called the *pulvis* where the flowers have stems longer than 40 mm. The aim is to harvest first-class picked material (Radojević et al. 2000).

The harvesters, therefore, should include working speed and efficiency to achieve better quality of the harvested product (Pajic et al. 2007a).

The chief working part of a harvester is the stripping rotor that has combs that plucks the flowers. The design and speed of the rotor are important parameters in harvesting quality material (Pajic et al. 2007b).

Several types of harvesters are available in the market and the best option is to be selected (Ivanovic et al. 2007). Combined harvesters could also be an economical option (Stričík and Salamon 2007).

High-performance harvesters are crucial for large-scale cultivation. At present, most of the harvesters do not meet the current requirements of quality flower harvesting, picking efficiency, and area capacity. New prototypes were suggested that have better efficiency than the existing ones (Ehlert and Roschow 2011).

The most efficient harvesters are those that have either linearly moving picking combs with additional stalk cutting parts, or rotating picking combs with central or outside discharge of flowers, or rotating picking drums with comb sections (Brabandt and Ehlert 2011).

The most important reason found for quality losses was the use of wiper blades for reducing stem length. As a result, the flowers get damaged (Brabandt and Ehlert 2011).

The combs are an important part of the manual or machine-based harvesters. For maintaining the quality of the picking, these combs have to be optimized. The picking quality of the comb was found to depend on the thickness of the comb blade, shape of the gap between the comb teeth, and substantially on the strength of the individual chamomile stalks (Ehlert et al. 2011).

6.12 YIELD

Wagner (1993) reported that in Slovenia, the chamomile cultivation began in 1988. The yield of the dry flowers increased from 203 kg/ha in 1988 to 563 kg/ha in 1992. Light sandy soil has been found to be optimum for chamomile cultivation in Slovenia.

Aly and Hussein (2007) reported that the net return for chamomile was 2095 LE/feddan (1 ha = 2.34 feddan) yearly in Egypt.

Dražić (2000) reported that the fresh flower yield of autumn-sown plants was 782 kg/ha and spring-sown plants was 682 kg/ha in 1998–1999. The essential oil yield (4.47 kg/ha) was also higher in the autumn-sown plants than the oil yield (3.49 kg/ha) in spring-sown plants. The yield of chamomile flowers in some countries in 2005 is presented in Table 6.1.

6.13 POSTHARVEST TREATMENT

The quality and quantity of chamomile essential oil depend on the raw material (fresh and dry weight of the herb and flower), time between harvest and distillation, and plant variety. It might also depend on various harvesting methods, size of the green parts in flowers, and drying process (Bucko and Salamon 2007).

The harvested flowers are either immediately distilled or dried in shade for a few days and stored in cool and dry conditions. Extraction of oil immediately after

TABLE 6.1
Yield of Fresh Chamomile Flowers in Some Countries

S. No.	Country	Yield of Fresh Flowers (kg/ha)	References
1.	Argentina	2000–3000	Fogola 2005
2.	Chile	200–250 (dry flowers)	Weldt 2005
3.	Egypt	3152	Fogola 2005
4.	Germany	800	Fogola 2005
5.	Hungary	500–2000	Bernáth and Németh 2005
6.	India	600	Fogola 2005
7.	Romania	600–1000	Bernáth and Németh 2005
8.	Slovenia	500	Bernáth and Németh 2005
9.	Soviet Union	300–500	Fogola 2005
10.	Yugoslavia	400–500	Bernáth and Németh 2005

harvesting or freezing and storing the harvested flowers is recommended for the production of a superior product rich in active compounds (Carle et al. 1989). Dried chamomile flowers can be stored in plastic containers, cellophane bags, or paper bags at 0°C–23°C for up to 2 years. The material stored in plastic containers at 0°C–2°C retained more than 60% of the oil and that at 20°C–23°C had ~60% of the original value at the end of the first year and 30% by the end of second year. There occurs, on storage, a marginal decrease in the quality of the oil (Dragland and Aslaksen 1996). Dried flowers and extracts are decontaminated (Katušin-Ražem et al. 1983, 1985) from bacteria, fungi, and yeasts using radiation treatment. The effective dose of γ-rays is 10 kGy for commercial chamomile tea and 15 kGy for extracts (Vaverkova and Herichova 1980; Katušin-Ražem et al. 1988).

Borsato et al. (2009) in Brazil studied the yield of essential oils during a drying process at 80°C. The harvested flowers were dried in a fixed layer where there was a progressive reduction of water content. This water was condensed and analyzed. It was found that 6.6% of the volatile substances could be recovered from the condensed water. This water contained artemisia ketone, bisabolol oxide B, α-bisabolol, bisabolone oxide, and bisabolol oxide A, but a significant reduction in the level of chamazulene was observed. The researchers recommended that the losses in the volatile constituents could be minimized by shortening of the condensation period coupled with increasing the drying temperature.

Plasma energy has been used for food preservation. Solís-Pacheco et al. (2013) studied the effect of plasma energy on the antioxidant activity, total polyphenols, and fungal viability in chamomile. The plasma energy at 850 V for 10 minutes significantly reduced the yeasts (0.68 ± 0.19 log CFU/g) and molds (<1.0 log CFU/g). There was a 55% reduction in the antioxidant activity in chamomile. The highest total polyphenol content in chamomile was observed after a plasma treatment at 650 V and 750 V for 10 minutes. They concluded that exposure of chamomile at 750 V for 10 minutes was the best treatment for significantly reducing the yeasts and molds without affecting the total polyphenol content.

After collection and sorting, the flowers are dried in the shade for 1 or 2 days. In some places like Slovakia, sorting of the flowers is carried out before drying to remove foreign materials such as plants and insects, whereas in other countries such as Hungary sorting is carried out just after drying. In some cases, mechanized sorters are used (Oravec et al. 2005). Drying is carried out naturally in the shade in countries such as Argentina (Fogola 2005) and Egypt (Fahmi 2005). It is recommended that during drying, the flowers should be uniformly spread so that the layer should not exceed 2 cm (Fahmi 2005).

In most European countries, drying is carried out under controlled conditions in hot air chambers under temperatures of 40°C–45°C (Oravec et al. 2005; Weldt 2005). The drying techniques such as counter flow drier comprising a belt conveyor or tunnel are also used, which operate at 40°C–70°C.

Chamomile flowers have been reported to undergo postharvest period respiration, senescence, transpiration, ripening, and changes in the chemical constituents due to secondary metabolism. The respiration is reported to lead to heating of the stacked crop, extreme heating up to spontaneous combustion and fermentation. The senescence process included chlorophyll degradation, drying of leaf, and changes in the quality of the secondary metabolites. Transpiration resulted in wilting and shriveling. Because of the high respiration rate, the researchers recommended technological treatments, such as ventilation with fresh air and cooling. The researchers found that when the harvested flowers were stored at 10°C, they remained fresh up to 70 hours. However, when the flowers were stored at 30°C, they remained fresh for only 15–20 hours. The researchers recommended that the flowers be stored at 20°C for 25–30 hours for optimum quality of essential oil and chemical constituents (Böttcher and Günther 2005).

However, Szabó et al. (2010) opined that the existing data were for belt dryers at industrial scale. They reported that when the chamomile flowers were dried using various methods, such as drying traditionally in a drying room at 30°C–35°C for 1–2 days, or in a tray dryer at 50°C for 12 hours or on a belt drier at 50°C–55°C, there was no drastic change in the levels of the chemical constituents. The quality of the essential oil was also found to be excellent.

There is a possibility of an increase in the microbial contamination and spoilage because of lesions and bruises, partial heating, and rotting (Böttcher and Günther 2005). Predrying in natural conditions could also aggravate microbial growth (Szabó et al. 2010).

The complete drying process usually takes 10–15 days (Fahmi 2005; Oravec et al. 2005). It is recommended that during this time the flowers should be kept in open packages (Oravec et al. 2005). The flowers are packed in paper bags or cartons (Fahmi 2005; Oravec et al. 2005).

The spray drying method of chamomile extracts has been explored by López Hernández et al. (2010) to quickly and efficiently dry chamomile flower extracts. This method could provide various advantages of the powder form, such as the absence of ethyl alcohol, easy storage, manipulation, transportation, and stability.

Another drying method using solar power has been developed by Amer and Gottschalk (2012) in Germany. They developed a drier that used solar power to dry the chamomile flowers and also a water tank to store excess solar power.

The capacity of the dryer is 32–35 kg fresh chamomile flowers. The dryer could reduce the moisture from 72%–78% to 7.2% in about 18–30 hours. During night, the heat stored in the water tank is transferred to the dryer, thus controlling the drying process at night. The appearance and quality of the oil were reported to be better than that from the sun-dried flowers.

6.14 FITNESS FOR ROTATION WITH OTHER CROPS

Chamomile can be grown in rotation with other crops. The economic returns from a unit area of land in north Indian plain conditions using the cropping sequence of rice, chamomile, and menthol mint were 230% of that of rice and wheat cropping sequence (Ram and Kumar 1996).

Darbaghshahi et al. (2012) evaluated the yield of chamomile when grown mixed with saffron. The yield of chamomile flowers when grown in rotation with other crops was found to be not significantly different than when grown as a single crop. There was no competition when chamomile was grown with saffron. They concluded that chamomile could be cultivated immediately after the harvest of saffron in late November or February.

6.15 HYDROPONIC CULTURE

Chamomile can be potentially grown in hydroponic culture. Ghasemi et al. (2013) evaluated the composition of the essential oil of chamomile growing in hydroponic culture. The plants were grown in Hoagland's solution. The flowers were harvested when the plants were 3 months old at the end of the growing season. The essential oil was obtained by distillation and analyzed by gas chromatography mass spectrometry and thin layer chromatography. They found the following composition: bisabolol oxide A (59.53%), bisabolone oxide B (29.86%), bisabolol oxide B (6.57%), chamazulene (2.27%), spathulenol (1.32%), and farnesene (1.23%).

REFERENCES

Abou-Zeid, E. N. and EL-Sherbeeny, S. S. 1971. Effect of cycocel on flower production and volatile oil of *Matricaria chamomilla* L. *Zeitschrift für Pflanzenphysiologie* 65: 35–38.

Abou-Zeid, E. N. and EL-Sherbeeny, S. S. 1975. A preliminary study on quality and quantity of volatile oil of *Chamomilla Egypt. Journal of Physiological Sciences* 1: 63–70.

Adamovic, D., Borojevic, K., Joksimovik, J., Gasic, O., and Lukic, V. 1982. Genetic variability of yield and quality in chamomile varieties and populations. *Bilt Hmelj Sirak Lek Bilge* 39: 35–44.

Adamovic, D. S., Kisgeci, J., Lukic, V., and Gasic, O. 1988. Variability of herbicide efficiency and their influence on yield and quality of chamomile. ISHS *Acta Horticulturae* 249: 63–67.

Adamovic, D. S., Kišgechi, J., Lukic, V., and Gašic, O. 1989. Variability of herbicide efficiency and their influence upon yield and quality of chamomile. ISHS *Acta Horticulturae* 249: 61–66.

Aguilera, D. B., de Souza, J. R. P., and Miglioranza, E. 2000. Effect of controlled release fertilizer and vermicompost on chamomile (*Matricaria chamomilla* L.) yield. *Revista Brasileira de Plantas Medicinais* 3(1): 61–65.

Aiello, N., D'Andrea, L., and Scartezzeni, F. 1998. Traits of seeds and essential oil of some spontaneous populations of common chamomile of northern Italy [*Chamomilla recutita* Rauschert]. *Agricoltura Ricerca* 20(176): 3–7.

Al-Badawy, A. A., Abdalla, N. M., Rizk, G. A., and Ahmad, S. K. 1984a. Growth and volatile oil content of chamomile plants as influenced by kinetin treatments. In: *Proceedings of the 11th Plant Growth Regulator Society of America*, Boston, MA, pp. 215–219.

Al-Badawy, A. A., Abdalla, N. M., Rizk, G. A., and Ahmad, S. K. 1984b. Influence of Atonik and Atonik-G treatment on growth and volatile oil content of *Matricaria chamomile*. In: *Proceedings of the 11th Plant Growth Regulator Society of America*, Boston, MA, pp. 220–223.

Alijani, M., Dehaghi, M. A., Malboobi, M. A., Zahedi, M., and Sanavi, S. A. M. M. 2011. The effect of different levels of phosphorus fertilizer together with phosphate bio-fertilizer (Barvar 2) on yield, essential oil amount and chamazulene percentage of *Matricaria recutita* L. *Iranian Journal of Medicinal and Aromatic Plants* 27(3): 450–459.

Alijani, M., Dehaghi, M. A., Sanavi, S. A. M. M., and Rezaye, S. M. 2010. The effects of phosphorous and nitrogen rates on yield, yield components and essential oil percentage of *Matricaria recutita* L. *Iranian Journal of Medicinal and Aromatic Plants* 26(1): 101–113.

Aly, M. S. and Hussien, M. S. 2007. Egyptian chamomile—cultivation & industrial processing. *ISHS Acta Horticulturae* 749:81–91. http://www.actahort.org/books/749/749_6.htm (Accessed July 08, 2013).

Amer, B. M. and Gottschalk, K. 2012. Drying of chamomile using a hybrid solar dryer. In: *International Conference of Agricultural Engineering—CIGR-AgEng 2012: Agriculture and Engineering for a Healthier Life*. Valencia, Spain, July 8–12, 2012, p. C-0347. http://www.researchgate.net/publication/256457594_Drying_of_Chamomile_Using_a_Hybrid_Solar_Dryer (Accessed March 26, 2014).

Bagheri, M., Golparvar, A. R., Rad, A. H. S., Zeinali, H., and Jafarpour, M. 2008. Assessment effect of planting date and N_fertilizer on quantitative and qualitative attributes of medicinal *Matricaria chamomilla* in Esfahan. *Journal of Research in Agricultural Science* 4(1): Pe29–Pe40.

Bernáth, J., and Németh, E. 2005. Production of chamomile (*Matricaria recutita* L.) in East and South European countries. In: R. Franke and H. Schilcher (eds.) *Chamomile: Industrial Profiles*. Boca Raton, FL: CRC Press, pp. 108–121.

Betti, A., Lodi, G., Fuzzati, N., Coppi, S., and Benedetti, S. 1991. On the role of planar multiple development in a multidimensional approach to TLC-GC. *Journal of Planar Chromatography Modern TLC* 4: 360–364.

Borsato, A. V., Doni-Filho, L., Rakocevic, M., Côcco, L. C., and Paglia, E. C. 2009. Chamomile essential oils extracted from flower heads and recovered water during drying process. *Journal of Food Processing and Preservation* 33(4): 500–512.

Böttcher, H. and Günther, I. 2005. Raw plant material and postharvest technology. In: R. Franke and H. Schilcher (eds.) *Chamomile: Industrial Profiles*. Boca Raton, FL: CRC Press, pp. 173–185.

Bouvert-Bernier, J. P. and Gallote, P. 1989. Chemical herbicides of *Matricaria chamomilla* L. *Herba Gallica* 1: 17–23.

Bovelli, R. 1996. Meccanizzacional: Coltivazione e miglioramento di painte officinali (Mechanizing the harvesting of chamomile). In *Atti Convegno Internazionale: Coltivazione e miglioramento di piante officinali*, Trento, Italy, 2–3 Giugno 1994. Trento Italy: Istituto Sperimentale per L'Assestamento Forestale e per L' Apicoltura 465–468.

Brabandt, H. and Ehlert, D. 2011. Chamomile harvesters: A review. *Industrial Crops and Products* 34(1): 818–824.

Bucko, D. and Salamon, I. 2007. The essential oil quality of chamomile, *Matricaria recutita* L., after its large-scale distillation. *ISHS Acta Horticulturae* 749: 269–273.

Budimir, M. 1986. Weed control with herbicides in chamomile. *Pesticidi* 1(3): 115–117.

Carle, R., Beyer, J., Cheminat, A., and Krempp, E. 1992. Proton NMR determination of site specific natural isotope fractionation in levo-a-bisaboloids. *Phytochemistry* 31: 171–174.
Carle, R., Dolle, B., and Reinhard, E. 1989. A new approach to the production of chamomile extracts. *Planta Medica* 55: 540–543.
Carle, R. and Gomma, K. 1992. Effect of harvest technology on the quality of chamomile flower and essential oil. *P2 (Pharm Ztg) Wiss* 137: 71–77.
Carle, R., Seidel, F., and Franz, Ch. 1991. Investigation into seed germination of *Chamomilla recutita* (L.) Rauschert. *Angewandte Botanik* 65(1–2): 1–8.
Chiumenti, R., Borso, F., and Da Bizzotto, A. 1996. Una macchina selezionatric per i capolinidi chamomilla [A sorting machine for chamomile flowers]. In: *Atti convegno internazionale: Coltivazione e miglioramento de piante officinali*. Trento, Italy: Istituto Sperimentale per L' Assestamento Forestale e per L'Apicoltura, pp. 469–474.
Corrêa Júnior, C., Castellane, P. D., and Jorge Neto, J. 1999. Influence of organic and chemical fertilization on the yield of flowers, contents and composition of essential oil of (*Chamomilla recutita* (L.) Rauschert). *ISHS Acta Horticulturae* 502: 195–202.
Darbaghshahi, M. N., Banitaba, A., and Bahari, B. 2012. Evaluating the possibility of saffron and chamomile mixed culture. *African Journal of Agricultural Research* 7(20): 3060–3065.
Dastborhan, S., Zehtab-Salmasi, S., Nasrollahzadeh, S., Tavassoli, A. R., Ghaemghami, J., Khosh-Khui, M., and Omidbaigi, R. 2012. Effect of plant growth-promoting rhizobacteria and nitrogen fertilizer on yield and essential oil of German chamomile (*Matricaria chamomilla* L.). *ISHS Acta Horticulturae* 964: 121–128.
Dellacecca, V., Bezzi, A., Chiumenti, R., Galigani, P. F., Leto, C., Marzi, V., Nano, G. M., and Zanzucchi, C. 1992. Chamomile (*Chamomilla recutita* (L.) Rausch.). First results obtained with the program "Cultivation and improvement of medicinal plants." *Agricoltura Ricerca* 14(131): 77–86.
Dragland, S. and Aslaksen, T. H. 1996. Storing chamomile flowers at different temperatures and with different packagings. *Norsk Landbruksforsikring* 10: 167–174.
Dražić, S. 2000. Effects of planting date, planting mode and seed quantity on chamomile yield and quality. In: *First Conference on Medicinal and Aromatic Plants of Southeast European Countries & VI Meeting "Days of Medicinal Plants."* Arandjelovac (Federal Republic of Yugoslavia), May 29–June 3, 2000. http://www.amapseec.org/cmapseec.1/abstracts/abs_p028.htm (Accessed July 05, 2013).
Ehlert, D., Adamek, R., Giebel, A., and Horn, H. J. 2011. Influence of comb parameters on picking properties of chamomile flowers (*Matricaria recutita*). *Industrial Crops and Products* 33(1): 242–247.
Ehlert, D. and Roschow, K. 2011. Current status and new approaches for harvesting of chamomile flowers (*Matricaria recutita* L.). *Zeitschrift für Arzenei & Gewürzpflanzen* 16(3): 111–118.
El-Shamy, H. A. and Gendy, A. S. H. 2013. Improved productivity of German chamomile by foliar application of complete fertilizer and salicylic acid. *Bulletin of Faculty of Agriculture* 64(1): 67–77.
Emongor, V. E. and Chweya, J. A. 1992. Effect of nitrogen and variety on essential oil yield and composition from chamomile flowers. *Tropical Agriculture* 69(3): 290–292.
Emongor, V. E., Chweya, J. A., Kaya, S. O., and Munavu, R. M. 1990. The effect of nitrogen and phophorus on the essential oil yield. *East African Agricultural and Forestry Journal* 55(4). http://www.eaafj.or.ke/index.php/path/article/view/112/0 (Accessed July 19, 2013).
Emongor, V. E., Chweya, J. A., Keya, S. O., and Munavu, R. M. 1991. Effect of nitrogen and phosphorus on the essential oil yield and quality of chamomile (*Matricaria chamomilla* L.) flowers. In: *Traditional Medicinal Plants*. Tanzania: Dar Es Salaam University Press–Ministry of Health, p. 391.

Fahmi, T. 2005. Cultivation experiences in Egypt. In: R. Franke and H. Schilcher (eds.) *Chamomile: Industrial Profiles*. Boca Raton, FL: CRC Press, pp. 156–165.

Falistocco, E., Menghini, A., and Veronesi, F. 1994. Osservazioni cariologiche in *Chamomilla recutita* (L.) Rauschert [Karyological observations on *Chamomilla recutita* (L.) Rauschert]. In: *Atti del convegno internazionale: Coltivazione miglioramento di piante officinalo*. Trento, Italy, giugno 2–3.

Fanid, B. H., Far, M. S., Valouzi, H., Hashemi, S., Khorie, M. M. A., and Nezamoleslami, M. 2012. Survey the effect of on medicinal plants cultivation on rural economic in Iran. Poster Information Technology, Automation and Precision Farming presented at International Conference on Agricultural Engineering (CIGR-AgEng 2012), Agriculture and Engineering for a Healthier Life, Valencia, Spain, July 8–12, 2012, p. P-1337.

Farkoosh, S. S., Ardakani, M. R., Rejali, F., Darzi, M. T., and Faregh, A. H. 2011. Effect of mycorrhizal symbiosis and *Bacillus coagolance* on qualitative and quantitative traits of *Matricaria chamomilla* under different levels of phosphorus. *Middle East Journal of Scientific Research* 8(1): 1–9.

Fatma, R., Shahira, T., Abdel-Rahim, E. A., Afify, A. S., and Ayads, H. S. 1999. Effect of low temperature on growth, some biochemical constituents and essential oil chamomile. *Annals of Agricultural Science* 44(2): 741–760.

Felkova, M., Jasicova, M., Trnkova, L., and Ciutti, P. 1981. Effect of mineral nutrients on the yield and quality of chamomile flowers. *Acta Facultatis Pharmaceuticae Universitatis Commenianae* 36: 69–102.

Filipović, V., Kisgeci, J., and Salamon, I. 2007. The qualitative and quantitative characteristic of chamomile, *Matricaria recutita* L., from experimental cultivation in different areas of South Banat. ISHS *Acta Horticulturae* 749: 141–148.

Fogola, N. 2005. Experiences with the cultivation of chamomile in Argentina. In: R. Franke and H. Schilcher (eds.) *Chamomile: Industrial Profiles*. Boca Raton, FL: CRC Press, pp. 141–150.

Franke, R. 2005. Cultivation. In: R. Franke and H. Schilcher (eds.) *Chamomile: Industrial Profiles*. Boca Raton, FL: CRC Press, pp. 76–108.

Franke, R. and Hannig, H. J. 2005. Cultivation in Germany. In: R. Franke and H. Schilcher (eds.) *Chamomile: Industrial Profiles*. Boca Raton, FL: CRC Press, pp. 162–165.

Franz, C. 1980. Content and composition of the essential oil in flower heads of *Matricaria chamomilla* L. during its ontogenetical development. ISHS *Acta Horticulturae* 96: 317–321.

Franz, Ch., Müller, E., Pelzmann, H., Hårdh, K., Hälvä, S., and Ceylan, A. 1986. Influence of ecological factors on yield and essential oil of chamomile (*Chamomilla recutita* (L.) Rauschert syn *Matricaria chamomilla* L.). ISHS *Acta Horticulturae* 188: 157–162.

Galambosi, B., Holm, Y., Szeberi–Galambosi, Z., Repcak, M., and Cernaj, P. 1991. The effect of spring sowing times and spacing on the yield and essential oil of chamomile [*Chamomilla recutita* (L.) Rauschert] CV 'Bona' grown in Finland. *Herba Hungarica* 30: 47–53.

Gasic, O., Lukic, V., Adamovic, R., and Durkovic, R. 1989. Variability of content and composition of essential oil in various chamomile cultivars. *Herba Hungarica* 28: 21–28.

Gasic, O., Lukic, V., and Adamovic, D. 1991. The influence of sowing and harvest time on the essential oils of *Chamomilla recutita*. *Journal of Essential Oil Research* 3: 295–302.

Ghasemi, N. D., Asghari, G., Mostajeran, A., and Najafabadi, A. M. 2013. Evaluating the composition of *Matricaria recutita* L. flowers essential oil in hydroponic culture. *Journal of Current Chemical and Pharmaceutical Sciences* 3(1): 54–59.

Gosztola, B., Sárosi, S., and Németh, É. 2010. Variability of the essential oil content and composition of chamomile (*Matricaria recutita* L.) affected by weather conditions. *Natural Product Communications* 5(3): 456–470.

Gowda, T. N. V., Farooqui, A. A., Subbaiah, T., and Raju, B. 1991. Influence of plant density, nitrogen and phosphorous on growth, yield and essential oil content of chamomile. *Indian Perfumer* 35: 168–172.

Hadi, M. R. H. S. 2010. Effects of vermicompost and amino acids on the flower yield and essential oil production from *Matricaria chamomile* L. *Journal of Medicinal Plants Research* 5(23): 5611–5617.

Hadi, M. R. H. S., Darz, M. T., Ghandehari, Z., and Riazi, G. 2011. Effects of vermicompost and amino acids on the flower yield and essential oil production from *Matricaria chamomile* L. *Journal of Medicinal Plants Research* 5(23): 5611–5617.

Hadi, M. H. S., Noormohammadi, G., Sinaki, J. M., Khodabandeh, N., Yasa, N., and Darzi, M. T. 2004. Effects of planting time and plant density on the flower yield and active substance of chamomile (*Matricaria chamomilla* L.) In: T. Fischer et al. New directions for a diverse planet: *Proceedings for the 4th International Crop Science Congress*, Brisbane, Australia, September 26–October 1, 2004. http://www.cropscience.org.au/icsc2004/poster/3/5/280_hadim.htm (Accessed July 05, 2013).

Haikal, M., and Badr, M. 1982. Effect of some GA, and CCC treatments on growth and oil quantity and quality of chamomile. *Egypt Journal of Horticulture* 9: 117–123.

Holm, L., Doll, G., Holm, E., Pancho, J., and Herberger, J. 1997. *Matricaria chamomilla* L. Asteraceae (compostiae), Aster family. In: *Weeds: Natural Histories and Distribution*. New York, NY: John Wiley and Sons.

Holubář, J. 2005. Growing varieties of chamomile in the Czech Republic. In: R. Franke and H. Schilcher (eds.) *Chamomile: Industrial Profiles*. Boca Raton, FL: CRC Press, pp. 139–141.

Hussain, A., Virmani, O. P., Sharma, A., Kumar, A., and Misra, L. N. 1988. *Chamomilla* oil (German). In: *Major Essential Oil Bearing Plants of India*. Lucknow, India: CIMAP, pp. 45–48.

Ivanovic, S., Pajic, M., and Ivanovic, L. 2007. Choosing type of chamomile harvester based on current value of usage costs. ISHS *Acta Horticulturae* 749: 259–264.

Johri, A. K., Srivastava, L. J., Singh, J. M., and Rana, R. C. 1991. Effect of row spacings and nitrogen levels of flower and essential oil yield in German chamomile. *Indian Perfumer* 35: 93–96.

Juárez-Rosete, C. R., Rodríguez-Mendoza, M. N., Trejo-Téllez, L. I., Aguilar-Castillo, J. A., Gómez-Merino, F. C., Trejo-Téllez, L. I., and Rodríguez-Mendoza, M. N. 2012. Inorganic and organic fertilization in biomass and essential oil production of *Matricaria recutita* L. ISHS *Acta Horticulturae* 947: 307–311.

Katušin-Ražem, B., Dvornik, I., and Matić, S. 1983. Radiation treatment of herb for the reduction of microbial contamination (Flores Chamomillae). *Radiation Physics and Chemistry* 22: 707–713.

Katušin-Ražem, B., Ražem, D., Dvornik, I., Matić, S., and Mihoković, V. 1985. Radiation decontamination of dry chamomile flowers and chamomile extracts. In: *Proceedings of International Symposium on Food Irradiation Processing*. Washington, DC: International Atomic Energy Agency and Food and Agricultural Organization UN, March 4–8.

Katušin-Ražem, B., Matić, S., Ražem, D., and Mihoković, V. 1988. Radiation decontamination of tea herbs. *Journal of Food Science* 53(4): 1120–1126.

Krstic-Pavlovic, N. and Dzamic, R. 1984. Contribution to the investigation of the influence of fertilization on the yield and quality of cultivated chamomile (*Matricaria chamomilla* L.) in the region of Northern Banat (Yugoslavia, stimulant plant and crops). *Agrohemija* 3: 207–215.

Letchamo, W. 1992. A comparative study of chamomile yield, essential oil and flavonoids content under two sowing seasons and nitrogen levels. ISHS *Acta Horticulturae* 306: 375–384.

Letchamo, W. 1996. Development and seasonal variation in flavonoids of diploid and tetraploid chamomile ligulate florets. *Journal of Plant Physiology* 148: 645–651.

Letchamo, W. and Marquard, R. 1993. The pattern of active substances accumulation in chamomile genotypes under different growing conditions and harvesting frequencies. ISHS *Acta Horticulturae* 331: 357–364.

López Hernández, O. D., Menéndez Castillo, R. A., García Peña, C. M., González Sanabia, M. L., and Nogueira Mendoza, A. 2010. Spray drying study of *Plectranthus amboinicus*, *Ocimum tenuiflorum*, *Passiflora incarnata*, *Matricaria recutita* and *Melissa officinalis*. *Boletín Latinoamericano y del Caribe de Plantas Medicinales y Aromáticas* 9(3): 216–220.

Mackova, A. and Helemikova, A. 1992. Testing of sensitivity of *Tripleurospermum inodorum* and *Matricaria recutita* to dichlorprop. *Zahradnictvi-UVTIZ* 19(2): 101–108.

Marczal, G. and Petri, G.1988. Quantitative variations of some volatile compounds in various ontogenetical phases of *Matricaria chamomilla* L. *Chamomilla recutita* L. Rauschert grown in phytotron. *Acta Agronomica Hungarica* 37: 197–208.

Mashkani, M. R. D., Badi, H. N., Darzi, M. T., Mehrafarin, A., Rezazadeh, S., and Kadkhoda, Z. 2011. The effect of biological and chemical fertilizers on quantitative and qualitative yield of Shirazian Babooneh (*Matricaria recutita* L.). *Journal of Medicinal Plants* 10(38): Pe35–Pe48, Pe187.

Meawad, A. A., Awad, A. E., and Afify, A. 1984. The combined effect of N-fertilization and some growth regulators on chamomile plants. ISHS *Acta Horticulturae* 144: 123–134.

Mohammadreza, N., Mohammad, M. S., Houseyn, Z., and Bahari, B. 2012. Effects of different levels of nitrogen, phosphorus and potassium fertilizers on some agromorphological and biochemical traits of German chamomile (*Matricaria chamomilla* L.). *Journal of Medicinal Plants Research* 6(2): 277–283.

Nikitushkin, M. F. and Martynov, Y. F. 1978. Mechanized harvesting of wild chamomile. *Khim Farm Zh* 12: 89–92.

Nikolova, A., Kozhuharova, K., Zheljazkov, V. D., and Craker, L. E. 1999. Mineral nutrition of chamomile (*Chamomilla recutita* (L.) K. ISHS *Acta Horticulturae* 502: 203–208.

Ohe, C., Sugino, M., Minami, M., Hasegawa, C., Ashida, K., Ogaki, K., and Kanamori, H. 1995. Studies on cultivation and evaluation of chamomile flos. Seasonal variation in the production of the head (capitula) and accumulation of glycosides in the capitula of *Matricaria chamomilla* L. *Yakugaku Zasshi* 115: 130–135.

Oravec, Sr., V., Repcak, M., and Cernaj, P. 1993. Production technology of *Chamomilla recutita*. ISHS *Acta Horticulturae* 331: 85–87.

Oravec, V., Oravec, Jr., V., Repčák, M., Šebo, Ĺ., Jedinak, D., and Varga, I. 2005. Cultivation experiences in Slovakia. In: R. Franke and H. Schilcher (eds.) *Chamomile: Industrial Profiles*. Boca Raton, FL: CRC Press, pp. 121–141.

Pajic, M., Raicevic, D., Ercegovic, D., Mileusnic, Z., Oljaca, M., and Radojevic, R. 2007b. Influence of exploitation characteristics of harvester "NB 2003" on chamomile harvesting quality. ISHS *Acta Horticulturae* 749: 253–258.

Pajic, M., Raicevic, D., Miodragovic, R., Ivanovic, S., Gligorevic, K., and Jevdjovič, R. 2007a. The comparative analysis of basic working parameters for different chamomile harvesters. ISHS *Acta Horticulturae* 749: 245–251.

Patra, D. D. and Singh, D. V. 1995. Utilization of salt affected soil and saline/sodic irrigation water for cultivation of medicinal and aromatic plants. *Current Research on Medicinal and Aromatic Plants* 17: 378–381.

Pirbalouti, A. G., Bahrami, M., Golparvar, A. R., and Abdollahi, K. 2011. GIS based land suitability assessment for German chamomile production. *Bulgarian Journal of Agricultural Sciences* 17(1): 93–98.

Pirzad, A. 2012. Effect of water stress on essential oil yield and storage capability of *Matricaria chamomilla* L. *Journal of Medicinal Plants Research* 6(27): 4394–4400.

Pirzad, A., Shakiba, M. R., Zehtab-Salmasi, S., Mohammadi, S. A., Sharifi, R. S., and Hassani, A. 2011. Effects of irrigation regime and plant density on essential oil composition of German chamomile (*Matricaria chamomilla*). *Journal of Herbs, Spices & Medicinal Plants* 17(2): 107–118.

Prasad, A., Patra, D. D., Anwar, M., and Singh, D. V. 1997. Interactive effects of salinity and nitrogen on mineral N status in soil and growth and yield of German chamomile (*Matricaria chamomilla* [*Chamomilla recutita*]). *Journal of India Social Soil Science* 45: 537–541.

Radojević, R., Pavlekić, S., Raičević, D., Ercegović, Đ., and Oljača, M. 2000. Mechanized, harvesting flower of chamomile. In: M.S. Ristić (ed.) *Proceedings of First Conference on Medicinal and Aromatic Plants of Southeast European Countries & VI Meeting "Days of Medicinal Plants."* Arandjelovac (Federal Republic of Yugoslavia), May 29–June 3, 2000. http://www.amapseec.org/cmapseec.1/Abstracts/abs_p050.htm (Accessed July 29, 2013).

Ram, M., Gupta, M. M., and Kumar, S. 1997. Chamomile and its Cultivation in India. In: *Farm Bulletin Number 002*. Lucknow, India: Central Institute of Medicinal and Aromatic Plants.

Ram, M., and Kumar, S. 1996. The production and economic potential of cropping sequences with medicinal and aromatic crops in a subtropical environment. *Journal of Herbs, Spices and Medicinal Plants* 4: 23–29.

Reichling, J. 1980. Herbicides in chamomile cultivation. ISHS *Acta Horticulturae* 96: 277–292.

Reichling, J., Voemel, A., and Becker, H. 1980. Application herbicides in chamomile. Part-IV. Effect of MPSS and K, 50/w on the yield and essential oil. *Herba Hungarica* 19: 73–85.

Reinhold, C., Seidel, F., and Franz, C. 1991. Investigation into seed germination of *Chamomilla recutita* (L.). *Rausch Angewandte Botanik* 65: 1–8.

Rojas-Valencia, M. N., de Velásquez, M. T. O., and Franco, V. 2011. Urban agriculture, using sustainable practices that involve the reuse of wastewater and solid waste. *Agricultural Water Management* 98(9): 1388–1394.

Salamon, I. 1993a. Production of chamomile, *Chamomilla recutita* (L.) Rauschert and its production ecology. Faculty of Medicine PhD Thesis, Comenius University in Bratislava, Slovakia.

Salamon, I. 1993b. Chamomile. *The Modern Phytotherapist, Mediherb.* 13–16.

Salamon, I. 2007. Large scale production of chamomile in Streda Nad Bodrogom (Slovakia). ISHS *Acta Horticulturae* 749: 121–126.

Saleh, M. 1971. The effect of air temperature and thermoperiod on the quality and quantity of *Matricaria chamomilla* L. oil. *Mededelingen Landbouwhogeschool Wageningen* 70: 1–17.

Santucci, F. M., Cardone, L., and Mostafa, M. S. M. 2013. Labour requirements and profitability of chamomile (*Matricaria chamomilla* L.) in Egypt. *Journal of Agricultural and Biological Science* 8(5): 373–379.

Schilcher, H. 1978. Influence of herbicides on growth of *Matricaria chamomilla* and the biosynthesis of essential oils. ISHS *Acta Horticulturae* 73: 339–341.

Shams, A., Abadian, H., Akbari, G., Koliai, A., and Zeinali, H. 2012. Effect of organic and chemical fertilizers on amount of essence, biological yield and harvest index of *Matricaria chamomilla*. *Annals of Biological Research* 3(8): 3856–3860.

Singh, L. B. 1970. Utilisation of saline-alkaline soils for agro-industry without prior reclamation. *Economic Botany* 24(4): 439–442.

Solís-Pacheco, J. R., Villanueva-Tiburcio, J. E., Peña-Eguiluz, R., González-Reynoso, O., Cabrera-Díaz, E., González-Álvarez, V., and Aguilar-Uscanga, B. R. 2013. Effect of plasma energy on the antioxidant activity, total polyphenols and fungal viability in chamomile (*Matricaria chamomilla*) and cinnamon (*Cinnamomum zeylanicum*). *Journal of Microbiology, Biotechnology and Food Sciences* 2(5): 2318–2322.

Stevanovič, Z. D., Vrbničanin, S., and Jevdjovič, R. 2007. Weeding of cultivated chamomile in Serbia. ISHS *Acta Horticulturae* 749: 149–155.

Stričík, M. and Salamon, I. 2007. Investment rating with a combine harvester acquisition for chamomile flower picking. ISHS *Acta Horticulturae* 749: 265–268.

Sváb, J. 1983. Results of chamomile cultivation in large-scale production. ISHS *Acta Horticulturae* 132: 43–47.

Szabó, K., Németh, É., Sárosi, Sz., and Czirbus, Z. 2010. Essential oil content of chamomile during primary processing. *Z Arznei-Gewurzpfla* 15(2): 63–68.

Tamizkar, A. and Khoshouei, Z. 2011. Fertilizer Management in Chamomile for achieve to the Sustainable Agriculture at Iran. In: *Proceedings of the International Conference on Technology and Business Management*. Dubai, March 28–30, 2011, pp. 10–14. www.trikal.org/ictbm11/pdf/agriculture/D1239-done.pdf (Accessed July 17, 2013).

Tilebeni, H. G. and Sadeghi, H. 2011. Effect of priming on biochemical regeneration of chamomile (*Matricaria recutita, Chamaemelum nobile*) deteriorative seeds. *American-Eurasian Journal of Agricultural and Environmental Sciences* 10(6): 954–961.

Urcoviche, R. C., Volpini, A. F. N., Dias, D. C., Lopes, A. R., Zaghi Jr., L. L., de Souza, S. G. H., and Alberton, O. 2012. Mycorrhization and microbial activity from a soil cultivated with coriander and chamomile. *Arquivos de Ciências Veterinárias e Zoologia da UNIPAR* 15(2): 121–125.

Vaverkova, S. and Herichova, A. 1980. Qualitative characteristics of secondary metabolites of *Matricaria chamomilla* L. after pyrimidine application. *Physiologia Plantarum* 17: 47–54.

Vomel, A., Reichling, J., Becker, H., and Drager, P. 1977. Herbicides in chamomile culture. Ist report. The influence of herbicides on flower production and the weed stand. Residue investigations. *Planta Medica* 31: 378–389.

Vukmanovic, L. and Stepanovic, B. 1987. Contribution to the study of cultivating medicinal and aromatic herbs in mountain areas (Yugoslavia). *Ekonomika poljoprivrede* 34(7–8): 431–439.

Wagner, T. 1993. Chamomile production in Slovenia. ISHS *Acta Horticulturae* 344: 476–478.

Wali, A. and Kawy, A. S. 1982. Yield and volatile oil content of chamomile as effected by water supply. *Herba Hungarica* 19: 65–75.

Walker, J. J. 1995. Garden herbs as hosts for Southern root knot nematode [*Meloidogyne incognita* (Kofoid and White) Chitwood, race 3]. *HortScience* 30: 292–293.

Wegner, T. 1993. Chamomile production in Slovenia. ISHS *Acta Horticulturae* 344: 476–478.

Weldt, S. E. 2005. Chamomile in Chile. In: R. Franke and H. Schilcher (eds.) *Chamomile: Industrial Profiles*. Boca Raton, FL: CRC Press, pp. 150–155.

Yaskonis, Y. A. 1978. Seed germination in chamomile. Effect of fertilizers on its growth and essential oil content in its organs in Lithuanian SSR. *Liet TSR Mokslu Akacf Darb Ser C Biol Mokslai* 2: 51–58.

Zalecki, R. 1972a. Cultivation and fertilizing of the tetraploid *Matricaria chamomilla* L. I. The sowing time. *Herba Polonica* 17: 367–375.

Zalecki, R. 1972b. Cultivation and fertilizing of the tetraploidal form of *Matricaria chamomilla* L.II. Spacing and density of sowing. *Herba Polonica* 18: 70–80.

7 Chamomile
Patents and Products

7.1 CHAMOMILE PATENTS

Numerous patents have been granted to chamomile to date all over the world. These patents are in areas as diverse as oral and topical medicines, medical items such as plasters, skin, hair and oral care, cosmetics, beverages, herbicides and pesticides, veterinary medicines and feed, and flower varieties and flower harvesters. As per the available information, these patents have been granted since the 1800s. As the number of patents on chamomile runs into several hundreds, it is not possible to enumerate all the patents in this chapter. Also, the patents on the individual chemical constituents of chamomile, such as α-bisabolol, chamazulene, or apigenin, are numerous and not included here. However, an attempt has been made here to present information about a few selected patents that are representative of the various areas of uses. Most of the patents are U.S. patents. Other patents, such as the European, Japanese, Korean, Russian, Indian, and those through the Patent Cooperation Treaty (PCT) have also been granted to many inventors.

7.1.1 Oral Medicine Containing Chamomile

The earliest patents for chamomile-based medicines that could be taken orally were granted mostly to tonics. These medicines and tonics were prepared as decoctions of several herbs including chamomile. These medicines and tonics were prescribed for treating a variety of ailments.

As early as 1866, a U.S. patent on chamomile was granted to J. Levy for an improved "medicine for treating abdominal diseases and female irregularities" (Levi 1866). This medicine was essentially a decoction made from various fruit juices and essential oils mixed in prescribed proportions. In 1868, a U.S. patent was granted to G.V. Rambaut for inventing a *medicine* containing several plants including chamomile flowers. The method of preparing the mixture was prescribed as steeping the ingredients in alcohol (Rambaut 1868). A U.S. patent was granted in 1868 to Rachel Feibelman for inventing a mixture for treating "felons and other inflammatory sores." The mixture contained chamomile water decoction along with the extracts of other plants (Feibelman 1868). Another U.S. patent was granted in the same year to Theodore A. Barry and Benjamin Patten for a tonic containing chamomile and 16 other herbs that were steeped in alcohol and water (Barry and Patten 1868).

In 1870, a U.S. patent was granted to Michele Ferro for a medicinal compound containing chamomile. The formulation was made by macerating chamomile flowers, rhubarb, and aloe in gin for 6 days and filtered. It was claimed that the medicine cured fever, ague, cholera, diarrhea, and colic (Ferro 1870). In 1878, a patent was granted to Amanda Owen for a *medical compound*, which contained chamomile, for the cure of cough, colds, and general debility (Owen 1878). In the following years, several tonics, called bitters, based on chamomile were patented (Kieffer 1884; Seibert 1884; Haubert and Haubert 1869; Llado 1869). It may be noted that the preparations were mostly made from a combination of several herbs, including chamomile flowers, and several methods were used to create the mixtures. The methods used were chiefly extraction through the use of hot water or maceration in alcohol. Much later, with the invention of new extraction techniques, patents were granted for chamomile medicinal compositions that used these techniques. In 2002, a U.S. patent was granted to Thomas Newmark and Paul Schulick for a method using supercritical fluid extraction to create a herbal composition containing chamomile and seven other herbs (Newmark and Schulick 2002).

7.1.2 Topical Use of Chamomile

Several chamomile medicinal products have been invented for topical use and patents have been granted to these products. The topical products range from creams, ointments, lotions to sprays.

In 1888, a U.S. patent was awarded to George H.A. Williams for a poultice containing chamomile. It was prepared by boiling lard, tallow, bark of sweet elder, chamomile flowers, borax, sulfur, and almond oil for 5 hours and then sieving (Williams 1888). The preparation was claimed to cure boils, burns, and various inflammations when applied externally as salve or poultice.

In 1990, a U.S. patent was granted to Michael R. Schinitsky and Lorraine F. Meisner for inventing a composition for treating aging skin. The composition contained chamomile extract along with other ingredients, such as ascorbic acid, glycerol, and mineral oil. This composition could be used with a cream base (Schinitsky and Meisner 1990).

In 1994, Bozo Tasevski was granted a patent through the PCT international application for inventing a skin treatment containing chamomile extract and linden tree (Tasevski 1994).

In 1998, a U.S. patent was awarded to Lydia Robertson and Edward M. Harris for inventing a composition suppressing or eliminating snoring. The composition contained chamomile, *Dioscorea*, and *Zingiber*. It was claimed that the composition acted as a vasoconstrictor and reduced the swelling of the nasal mucus membranes and nasal passages. It could be administered as a throat spray, nose drop, nasal inhalant, mouthwash, swab, or gargle (Robertson and Harris 1998).

A liquid formulation containing chamomile and capsaicin along with other chemical ingredients developed by Timothy R. Laughlin and Stephen D. Holt was granted a U.S. patent in 1998. It claimed to provide relief against pain and itching, when applied topically (Laughlin and Holt 1998). Another U.S. patent was granted to them in 1999 for a separate composition containing chamomile and capsaicin that could be used to neutralize the discomfort due to various types of pain (Laughlin and Holt 1999).

In 1999, a Japanese patent was granted to Akihiro Ishino for inventing a composition containing chamomile and other herbal ingredients that activated the growth of hair papilla (Ishino 1999).

A formulation was developed by Brian M. Albert and R. Richard Riso containing chamomile, plantain, and jewelweed sap, and other chemicals such as zinc oxide, menthol, and manganese gluconate for curing dermatitis. In 1999, a U.S. patent was granted to this formulation (Albert and Riso 1999/406).

A PCT patent was granted in 1999 to Vsevolod Nikolaevich Rudin and colleagues for inventing a composition containing chamomile that could be used for combating caries, parodenitis, and paradentosis (Rudin et al. 1999). The composition could be used in toothpastes, tooth creams and gels, and mouthwashes. It could also be used in chewing gums.

A composition to provide protection against lumpiness, edema, and other effects of liposuction and cosmetic surgery was invented by Jack Mausner. This invention was a skin cream containing chamomile and other complex molecules used in cosmetics, such as a long-chain fatty acid ester of ascorbic acid and a glyceryl ester complex. A U.S. patent was granted in 1999 (Mausner 1999).

In 2000, a PCT patent was granted to Jacqueline Vijgen and her colleagues for inventing a nasal spray containing chamomile and xylometazoline. HCl, among others (Vijgen et al. 2000).

A formulation containing chamomile, curcuma extract, and natural gum was invented by Rattan Lal Bindra and colleagues to cure cracked skin. In 2000, a U.S. patent was granted to this formulation (Bindra et al. 2000).

In 2001, a U.S. patent was granted to Murray Borod for inventing a formulation that treated hemorrhoids and the associated pain, swelling, and discomfort, when applied topically. The formulation contained chamomile essential oil, lavender oil, extracts of grape seed, gotu kola, horse chestnut, and aloe vera (Borod 2001).

In 2002, a U.S. patent was granted to Elliott Farber for inventing a topical allantoin composition containing chamomile, and other chemicals, for the treatment of skin inflammatory diseases (Farber 2002a).

An astringent composition was invented to deliver salicylic acid into the skin by Geraldine Watson. This astringent composition contained chamomile extract as one of the ingredients. Watson was granted a U.S. patent for this in 2002 (Watson 2002).

In 2003, Mario Baraldi was awarded a PCT patent for inventing a composition containing chamomile extract and other anti-inflammatory herbal plants for the treatment of skin conditions such as inflammations and tissue damage (Baraldi 2003).

In 2003, a PCT patent was granted to Karoly Lukacs and colleagues for inventing a composition containing chamomile essential oil and other antibacterial and fungicidal compounds to be used topically as antimicrobial pharmaceutical (Lukacs et al. 2003).

In 2003, Bruce Marshall Frome was granted a U.S. patent for inventing composition containing chamomile oil, and other essential oils, that could be used to treat various neuralgias and other conditions, such as viral pharyngitis, insect bites, sunburn, facial paralysis due to Bell's palsy, pain of postherpetic neuralgia, reflex sympathetic dystrophy, terminal neuralgia, sciatica, and tic douloureux (Frome 2003).

In 2003, a European patent was granted to Mishal Hamid Al-Sari for inventing a composition to be used as an injection containing chamomile flower extract, vitamins, and insulin to treat several disorders, such as genetic disorders, skin diseases, cancer, and viral infections (Al-Sari 2003).

In 2003, a U.S. patent was granted to Tom C. Veach and colleagues for inventing a composition to provide iodine to mucosal areas of the human body. This composition contains chamomile glycolic extract, sodium percarbonate, butylenes glycol, and others, and exerts antimicrobial effect, when applied topically (Veach et al. 2003).

In 2003, an Australian patent was granted to David McIntyre and colleagues for inventing an antibacterial composition containing essential oils of chamomile, clary sage, sandalwood, basil, bergamot, birch, eucalyptus, fir, geranium, lavender, pine, and tea tree (McIntyre et al. 2003).

In 2004, David M. Stoll was granted a U.S. patent for inventing a muscle stimulant containing chamomile essential oil along with other essential oils and creatinine (Stoll 2004).

In 2004, David Nakar was granted a PCT patent for inventing a composition to be used as an ear gel to treat ear disorders. The composition contained chamomile and about 12 other ingredients (Nakar 2004).

In 2004, Maria Villani was granted a U.S. patent for inventing a composition containing chamomile extract and Porifera for treating various skin disorders including acne, dermatitis, eczema, psoriasis, photoaging, and skin resurfacing (Villani 2004).

In 2005, Patricia R. Springstead was granted a U.S. patent for inventing a composition for treating skin disorders, such as dermatitis, dry skin, rough skin, cracking, itching, and psoriasis. To be used as a lotion, lotion bar, or soap, this composition contains chamomile essential oil along with about 13 other ingredients (Springstead 2005/280).

Natalia Dmitrivna Oltarzhevskaya and colleagues invented a composition that had hemostatic properties, that is, the ability to reduce or stop bleeding. In addition, it had anti-inflammatory properties. The composition contained chamomile and *Urtica dioica*. The composition could be used topically or applied in dressing material. A PCT patent was granted to this invention in 2006 (Oltarzhevskaya et al. 2006).

Another topical formulation to treat inflammatory conditions, such as gynoid lipodystrophy, was invented by Henry Okigami. The formulation contained chamomile, passion fruit, and artichoke. A PCT patent was granted to this invention in 2006 (Okigami 2006/247).

In 2008, a PCT patent was granted to George Botos for inventing an antimicrobial composition containing chamomile extract and other chemicals (Botos 2008).

In 2008, a U.S. patent was granted to Wanda Fontaine and colleagues for inventing a composition containing chamomile, silicone, and an analgesic, among others, for treating burns. The composition is to be applied topically (Fontaine et al. 2008).

Kaira Rouda was granted a U.S. patent in 2009 for inventing a composition that promoted sleep. It contained hydroalcoholic extracts of chamomile, lavender, poppy, valerian root, sage, gingko, and/or peppermint (Rouda 2009).

Chamomile: Patents and Products

In 2010, a European patent was awarded to Vincenzo Russo for inventing a composition containing chamomile extract and hyaluronic acid for the treatment of pathologies of the respiratory tract (Russo 2010).

Christine Parkes and Mervyn Keynes invented a formulation containing chamomile and lavender oils to treat skin disorders such as dermatographia. In 2011 a U.K. patent was granted to this invention (Parkes and Keynes 2011).

In 2012, a European patent was granted to Jianying Yam and Matthias-Heinrich Kreuter for inventing a composition containing chamomile extract and vitamin A for the treatment of cancer (Yam and Kreuter 2012).

7.1.3 Medical Devices

Medical devices, such as liniments, patches, wipes, and garments, that have chamomile-based medicines adsorbed on them have been granted patents. Other devices such as dental prosthesis have also been patented.

Patrick Bonjour invented a liniment containing chamomile and other herbal extracts for applying on the eye to relieve external eye irritation and ocular fatigue. This was granted a European patent in 1990 (Bonjour 1990).

Monika Van de Loecht-Blasberg and Rainer Knollman invented an adhesive for dental prostheses in the form of a patch containing chamomile extract. A PCT patent was granted in 1996 to this invention (Van de Loecht-Blasberg and Knollman 1996).

Medicated feet wipes containing chamomile essential oil, along with the essential oils of other plants, were invented by Anna Kavaya. In 1998, a U.K. patent was granted to the invention (Kavaya 1998).

In 2002, a U.S. patent was granted to Elliot Farber for inventing a flexible applicator that was in the form of a bandage or wipe, and an allantoin containing chamomile and other compounds that could be applied using the applicator (Farber 2002b).

Carol L. Erdman invented an absorbent garment incorporated with skin wellness ingredients containing chamomile and other ingredients. The composition can be applied to the adhesive that can be placed in those parts of the absorbent garment where the benefits of the composition are desired. In 2003, a U.S. patent was granted to this invention (Erdman 2003).

Rainer Lange invented a feminine hygiene wipe containing chamomile. The wipes were strips of spunlace and the composition containing chamomile and other ingredients was sprayed on to the wipes for use. A U.S. patent was granted to this invention in 2003 (Lange 2003).

In 2006, Joanna Bader was granted a U.K. patent for inventing a foot therapy pad made of latex foam covered with muslin and infused with the essential oils of chamomile and other oils (Bader 2006).

In 2009, a PCT patent was granted to Mindee Doney and Julie Pickens for inventing a nose wipe containing chamomile extract and other ingredients. This wipe was for mucus removal and skin moisturization (Doney and Pickens 2009).

In 2009, Matthias-Heinrich Kreuter was granted a PCT patent for treating viral infection. Chamomile flowers were extracted through a specific method and mixed with ricinus oil to prepare the formulation (Kreuter 2009).

7.1.4 ORAL HYGIENE

Oral hygiene products that contain chamomile have been developed and patented. These products are in the form of toothpastes, mouthwashes, dentifrices, chewing gums, and so on.

Mihaly Videki and colleagues invented a composition containing chamomile extract and other herbal extracts with antimicrobial properties for preventing paradentosis and inflammation of the periodontium. In 1991, a U.S. patent was granted to the invention (Videki et al. 1991).

In 2001, a Korean patent was granted to Sei Goon Oh for inventing a toothpaste containing chamomile and other essential oils for the maintenance of oral hygiene (Oh 2001).

An oral hygiene formulation comprising chamomile and several herbal ingredients was invented by Joseph Lesky and colleagues. The formulation could be used as an oral rinse, gel, or ointments and has a synergistic effect on the improvement and maintenance of gum and oral health. It also prevented sores and tooth decay. In 2002, a U.S. patent was granted to it (Lesky et al. 2002).

In 2004, a PCT patent was granted to Antonio Carlos Pereira Coelho for inventing a composition containing chamomile and other herbal extracts for curing halitosis and other forms of oral problems. The composition can be used in a dentifrice gel, chewing gum, sweet or confetti, and liquid or spray (Coelho 2004).

A composition containing chamomile oil and other herbal and chemical ingredients for reducing oral malodor was invented by John Martin Behan and colleagues. In 2004, a PCT patent was granted to it (Behan et al. 2004a). Another PCT patent was granted to them in the same year for inventing a composition containing chamomile oil and other herbal extracts to cure dental caries (Behan et al. 2004b). A third patent was granted to them in the same year for inventing a composition containing chamomile oil and other herbal and chemical ingredients for curing gum disease. It could be used in toothpaste (Behan et al. 2004c).

To provide relief of the symptoms of teething in children, Rachel Wilson invented a homeopathic composition that contained chamomile, belladonna, and coffee extracts. The solution could be applied orally on the gums of the child. It was granted a U.S. patent in 2006 (Wilson 2006).

In 2008, a U.S. patent was granted to Mein Laali for inventing a composition containing chamomile oil and other herbal oils for treating oral calculus (Laali 2008).

7.1.5 DIAPERS

A disposable baby diaper containing chamomile extract was invented by Juan. The diaper could ameliorate the problems of graze, rash, and pruritus. A Mexican patent was granted to it in 2002 (Giacoman 2002).

7.1.6 FEMININE HYGIENE

Chamomile-based products for feminine hygiene have been developed and patented.

In 2003, a U.S. patent was granted to Rainer Lange for inventing a composition containing chamomile for feminine hygiene wipes (Lange 2003).

Another U.S. patent was awarded in 2003 to Kimberly Marie Geiser for inventing a composition containing chamomile to be used in tampons, liners, or sanitary pads (Geiser 2003).

7.1.7 Food

Chamomile extracts and essential oil have been used in many food products, such as candies and jellies. These special products have been patented.

7.1.7.1 Candies

In 2001, a Korean patent was granted to In Soo Bang for inventing a Korean traditional candy containing chamomile and other ingredients (Bang 2001).

In 2004, three Russian patents were granted to O.I. Kvasenkov for inventing jelly and jelly-fruit paste containing chamomile, peppermint leaves, yarrow, marigold herb, and other herbs (Kvasenkov 2004a–c).

In the same year four more Russian patents were granted to O.I. Kvasenkov and his colleagues for inventing a jelly candy containing chamomile and other herbal extracts (Dobrovol'ski and Kvasenkov 2004; Rodionova et al. 2004; Zhivagina et al. 2004).

A Chinese patent was granted in 2010 to Ying Xu for inventing a throat-soothing chewing gum containing chamomile and Chinese medicinal plant extracts (Xu 2010).

Steven T. George was granted a U.S. patent in 2013 for inventing a formulation containing chamomile extract, kava root extract, and lemon balm that could be hard chewable candies, a beverage, an effervescent powder, or a dietary supplement (George 2013).

7.1.7.2 Flavoring

In 2010, a Russian patent was granted to N.I. Shchetkina for inventing a flavoring for food using chamomile extract and other micro- and macroelements (Shchetkina 2002).

Lars Månsson and Senanayake were granted a U.S. patent in 2012 for inventing a food emulsion containing chamomile extract to be used to flavor food. It can be used as a salad dressing (Månsson and Senanayake 2012).

7.1.8 Beverages

Chamomile extract has been used to flavor beverages. Several beverages containing chamomile have been patented.

A beverage containing brandy, chamomile extract, and fruit juices was invented by Vida-Bronislava Semeniene and Valentas Semenas. In 1994, a Lithuanian patent was granted to it (Semeniene and Semenas 1994).

In 1995, a Latvian patent was granted to Dukulis Leons and colleagues for inventing a beverage containing alcohol, chamomile extract, *Sedum roseum* extract, and cherry leaf extract (Leons et al. 1995).

A Japanese patent was granted in 1998 to Junji Asari for inventing an awamori-containing chamomile extract. Dried chamomile flowers were soaked in the awamori to yield a light yellow awamori. (Asari 1998).

Kyoji Kiyama and Kazuo Kawaguchi invented a fermented tea containing chamomile extract, olive leaves, and other herbal material. In 1999, a Japanese patent was granted to this beverage (Kiyama and Kawaguchi 1999).

In 1999, a Japanese patent was granted to Akinobu Ono for inventing a beverage containing chamomile extract, lavender, lemon grass oil, rose, and other herbal material that imparted a refreshing feeling (Ono 1999).

In 2000, a Latvian patent was granted to Janis Mazis and colleagues for inventing a liqueur containing alcohol, chamomile extract, thyme, cranberry leaves, marsh tea sprouts, and other herbal material (Mazis et al. 2000).

A Korean patent was granted to Ji Seong Yoo in 2001, for inventing a herbal drink containing chamomile extract and *Ziziphus jujuba* (Yoo 2010).

In 2001, a Spanish patent was granted to Carlos Gonzalez Garcia for inventing a liqueur containing grape rum, chamomile, and orange (Gonzalez 2001).

Dorothy J. Blount invented a herbal drink containing chamomile, catnip, mullein, pennyroyal, and other herbal extracts. In 2001, a U.S. patent was granted to this drink (Blount 2001).

In 2001, a Korean patent was granted to Yeong Kim for inventing a herbal drink containing chamomile, green tea, St. John's wort, catnip, and other herbal extracts (Kim 2001).

Jong Soo Lee was granted a Korean patent in 2002 for inventing a beverage containing alcohol and chamomile (Lee 2002).

A Japanese patent was granted to Tatsuyuki Kudo in 2002 for inventing a beverage containing chamomile extract, lindane, passionflower, catnip, and so on (Kudo 2002).

Scott Vincent Borba invented a composition to be used as a nutritional supplement. The composition contained chamomile extract and an antioxidant. It could be used as a beverage, paste, powder, gel, or encapsulated. Its uses included improving skin moisture, elasticity, fine lines, dryness, roughness, firmness, complexion breakouts, hydration, texture, and redness. In 2006, a PCT patent was granted to this composition (Borba 2006).

Allen Griffiths and colleagues were granted a PCT patent in 2007 for inventing a beverage precursor containing chamomile and one insoluble material such as strawberry, and lemon slice. This precursor was used in the preparation of tea, coffee, hot chocolate, and other beverages (Griffiths et al. 2007).

Andrew K. Benson and colleagues were granted a PCT patent in 2007 for inventing a probiotic composition containing a bacterial strain, a yeast strain, and chamomile. The composition can be used to treat a variety of gastric disorders such as pouchitis, ulcerative colitis, Crohn's disease, and diarrhea (Benson et al. 2007).

In 2010, a Korean patent was granted to Jong Cheon Ham and colleagues for inventing a beverage containing chamomile extract and other herbs for protecting vocal cords (Ham et al. 2010).

A Russian patent was granted to O. Ya Mezenova and O.V. Skapets in 2011 for inventing a whey-based beverage containing chamomile extract and other additives (Mezenova and Skapets 2011).

Chamomile: Patents and Products   277

Xishou Yu was granted a Chinese patent in 2012 for inventing a *Matricaria recutita* nectar containing chamomile, honey, maltose, rock sugar, and water (Xishou 2012).

In 2012, a Hungarian patent was granted to Jozsef Papp for inventing a beverage containing chamomile fruit brandy and other herbs (Papp 2012).

A Korean patent was granted in 2013 to Bong Seon Choi for inventing a beverage containing chamomile and other herbs for adjusting the acidity of stomach (Choi 2013).

7.1.9 Perfume

Chamomile essential oil and extracts have been extensively used in perfumery. Several formulations containing chamomile have been patented. These formulations can be used in soaps, shampoos, and perfumes.

A European patent was granted to Leenert Maarten van der Linde and colleagues in 1986 for inventing a perfume composition containing chamomile and other essential oils. This composition could be used in soaps (van der Linde et al. 1986).

Lajos Szente and colleagues invented an air-scenting composition containing chamomile oil and other herbal oils and chemical compounds. They also invented an inhalant powder containing chamomile oil and additional compounds. In 1988, a PCT patent was granted (Szente et al. 1988).

In 1999, a Japanese patent was granted to Masako Konishi for inventing a perfume composition containing chamomile extract and other herbal extracts. This composition could be used to mask the smell of a lower alcohol (Konishi 1999).

In 2002, a perfume formulation invented by Richard A. Weiss and colleagues was granted patent. The formulation contained chamomile, citronellol, geraniol, and 3-methyl octahydro-2H-1-benzopyran-2-one. This formulation can be used to impart aroma to different articles (Weiss et al. 2002).

Yoon I. Kim was granted a Korean patent in 2003 for inventing a perfume for undergarments. The perfume was composed of chamomile oil, ylang ylang, and other natural essential oils (Kim 2003).

A Korean patent was granted to Yeong Han in 2003 for inventing a perfume for enhancing dopamine levels in human blood. The perfume was composed of chamomile, ylang ylang, bergamot, lemon, and other natural aromatic ingredients (Han 2003).

In 2008, Pam Asplund was granted a PCT patent for inventing a perfume composition containing chamomile oil and other herbal oils. The composition could be used to improve sleep as well (Asplund 2008).

7.1.10 Cosmetics

Patents on products containing chamomile have been granted to hair care (Deane 2001), shampoos and conditioners (Korkis 1976), cleansers (Konishi 1999), skin whiteners (Yoshida 2002), masks (Doi 2000), creams (Posner 1998), moisturizers

(Mausner 1993) and lotions (Hayase 1998), aftershaves (Minetti 1988), and soaps (Park 2002).

In 1968, a U.S. patent was granted to Madeline V. Knudsen for inventing a composition for strengthening the nails. The composition contained chamomile extract, formaldehyde, eucalyptus oil, and other chemicals. The composition is claimed to fortify the nails (Knudsen 1968).

A U.K. patent was granted to Jeanette Hill in 1987 for inventing a hair treatment containing chamomile extract and bay rum or malt. The composition can be used for promoting hair growth as a treatment for alopecia (Hill 1987).

In 1992, Henri Pinchon and Alain Saboureau were granted a European patent for inventing a composition using chamomile and several other herbs to be used for making cosmetics, such as facial mask, antiwrinkle agent, keratinocyte-separating agent, bathing agent, toothpaste, and soap (Pinchon and Saboureau 1992).

In 1995, a U.K. patent was granted to Jean Reidy for inventing a herbal sachet containing chamomile and more than 23 ingredients. The sachet is water-permeable and can be used for tired eyes, conditioning hair, and soothing and relaxing the body (Reidy 1995).

Robert Bell and Denman Gray invented a sunscreen containing chamomile extract and other ingredients. A U.S. patent was granted to the product in 1999 (Bell and Gray 1999).

A U.S. patent was granted to Jeremy T. Miller and colleagues in 2003 for inventing a conditioning shampoo containing chamomile extract and other ingredients (Miller et al. 2003).

In 2003, a U.S. patent was granted to Neil D. Scancarella and colleagues for inventing a composition that ameliorated the adverse effects of the skin and hair from aging. The composition contains chamomile and several other ingredients and can be used as an exfoliating agent to ameliorate blotchy skin, skin laxity, skin blemishes and skin yellowing (Scancarella et al. 2003).

Catherine A. Piterski was awarded a PCT patent in 2003 for inventing a lotion containing chamomile oil and other ingredients (Piterski 2003).

An Indian patent was granted in 2004 to Neena Sharma for inventing a herbal talc containing chamomile and other herbs (Sharma 2004).

A U.S. patent was granted to Christine Marie Cahen in 2006 for inventing a composition with hyposensitive properties with chamomile oil and other essential oils. The composition is gentle to the skin (Cahen 2006).

In 2007, Wendy L. Hill was granted a U.S. patent for a massage emollient containing chamomile and other herb oils. The composition can be used as an emollient for softening the skin (Hill 2007).

In 2007, Paula Dorf was granted a U.S. patent for inventing an emollient for the area around the eyes. The composition contained chamomile extract and other chemical ingredients (Dorf 2007).

Kenneth S. Day and George A. Hansen invented a composition to promote hair growth. The composition contained chamomile extract, sage leaf extract, and other ingredients. A U.S. patent was granted in 2007 (Day and Hansen 2007).

Chamomile: Patents and Products

In 2008, Olivier Courtin was granted a U.S. patent for inventing a composition for treating hyperpigmentation of the skin. The composition contained chamomile extract along with parsley extract and *Alchemilla vulgaris* extract, among other ingredients (Courtin 2008).

In 2011, Shannon Elizabeth Klingman was granted a U.S. patent for inventing an antiperspirant and deodorant containing chamomile extract and other ingredients (Klingman 2011).

In 2012, Landra Booker Johnson was granted a U.S. patent for inventing a composition for a shampoo and bodywash. It was a tear-free composition and contained chamomile and other ingredients (Johnson 2012).

A PCT patent was granted in 2012 to Aleksandar Pavlov and Jovanka Bubnjevic-Pavlov for inventing a herbal cosmetic composition containing chamomile and several other herbal and chemical ingredients. The composition can be used for treating disorders of facial and body skin, blood vessels, muscle–bone system, reproductive system, and thyroid gland (Pavlov and Bubnjevic-Pavlov 2012).

7.1.11 INSECTICIDE

Chamomile could be used as an insecticide or pesticide. In 1881, a U.S. patent was granted to Ferdinand H. Renz for inventing a *medicated fluid* to be used as an insecticide. The composition contained chamomile extract along with other ingredients (Renz 1881).

7.1.12 TRICHOMONADICIDAL

The trichomonadicidal activity of chamomile is known. In 1985, a European patent was granted to Aurelia Tubaro and colleagues for inventing a composition with trichomonadicidal properties. The composition contained hydroalcoholic extracts of chamomile (Tubaro et al. 1985).

7.1.13 VETERINARY MEDICINE

Chamomile formulations have been developed to be used as veterinary medicine and these have been patented. In 1999, a U.S. patent was granted to Anthony Costa for inventing a formulation containing chamomile extract and valerian root extract for reducing the stress levels in the fish in aquariums (Costa 1999).

In 2002, a U.S. patent was granted to Thomas Newmark for inventing a herbal composition containing chamomile that reduced inflammation in animals (Newmark 2002).

A U.S. patent was granted to Ralph S. Robinson in 2007 for inventing a veterinary pharmaceutical composition containing chamomile and six other herbs. This can be used for treating gastric ulcers in horses (Robinson 2007).

In 2012, a PCT patent was granted to Somabhai Dhulabhai Parmar for inventing a pharmaceutical composition containing chamomile and three other herbs for treating bloating in animals (Parmar 2012).

7.1.14 ANIMAL FEED SUPPLEMENT

The flowers and herb of chamomile can be used as animal feed.

Karoly Pelek invented a fodder substitute, which contained chamomile flowers, powdered bone, washed chalk, lemon oil, charcoal powder and so on. A Hungarian patent was granted to it in 1948 (Pelek 1948).

In 1998, a U.K. patent was granted to Phelim Greene and Terence Jennings for inventing a food supplement with herbs. The composition contained chamomile, molasses, and calcium oxide (Greene and Jennings 1998).

Hisashi Watanabe invented a feed for hens that contained chamomile. The feed was claimed to produce eggs with superior flavor and taste. In 2002, a Japanese patent was granted to the product (Watanabe 2002).

7.1.15 CHAMOMILE VARIETY

Several varieties of chamomile have been developed and patented worldwide. In 1986, a patent was granted in Spain to Degussa on a chamomile variety with high contents of chamazulene and bisabolol. It was developed from the variety H-29 (Degussa 1986).

7.1.16 EQUIPMENT

A PCT patent was granted to Federico Bonomelli of Italy in 1992 for inventing a batcher that is essentially a sorting machine to sort and pack chamomile flowers. The machine consists of seven batching units placed side by side. The sorted chamomile flowers are finally packed in filter bags (Bonomelli 1992).

A device was invented by Andrea Romagnoli for feeding particulate products such as chamomile tea in a user machine that could pack the tea into bags (Romagnoli 2003).

7.2 CHAMOMILE PRODUCTS

There is a wide range of products based on chamomile, such as food, beverages, cosmetics, and hygiene products in the international market. Some products have been selected and listed in Table 7.1. The aim is to acquaint the reader with the available range of chamomile-based products. The current market price of the product is also mentioned.

The information on the patents and products based on chamomile is expected to motivate more entrepreneurs to take up chamomile cultivating and/or produce value-added chamomile products in the future.

Chamomile: Patents and Products 281

TABLE 7.1
Selected Chamomile Products and Their Current Cost in the International Market

S. No.	Category	Product	Price	Reference
1.	Oil	Chamomile oil (30 mL)	US$70	http://www.biosourcenaturals.com/chamomile-essential-oil.htm
2.	Oil	Chamomile oil (05 mL)	US$48	http://www.youngliving.com/en_US/products/essential-oils/singles/german-chamomile-essential-oil
3.	Hydrosol	Chamomile hydrosol (113 mL)	US$16.50	http://www.biosourcenaturals.com/hydrosols-floral-waters.htm
4.	Tea bags	Chamomile (tea pack/25 bags)	US$5.67	http://www.officeworks.com.au/retail/products/Catering-and-Cleaning/Tea/Fruit-and-Herbal-Tea/LE03566006
5.	Tea bags	Organic chamomile: box of 15 tea sachets	US$7.95	http://www.twoleavestea.com/organic-chamomile
6.	Tea bags	Chamomile: 1/4 pound loose tea sleeve	US$16	http://www.twoleavestea.com/loose-tea/loose-tea-herbal/organic-chamomile-1-4-pound-loose-tea-sleeve
7.	Tea bags	Chamomile herbal tea bags (50 g)	US$9.50	http://brewspace.com.au/serenitea-infusions-chamomile-herbal-loose-leaf-tea-50g
8.	Cosmetics	Chamomile and green tea eye makeup remover, fragrance free (4 fl. oz.)	US$6.79	http://www.drugstore.com/earth-science-chamomile-and-green-tea-eye-makeup-remover-fragrance-free/qxp141271
9.	Cosmetics	Shampoo for sensitive hair and scalp, fragrance-free 12 fl. oz. (355 mL)	US$7.99	http://www.drugstore.com/earth-science-shampoo-for-sensitive-hair-and-scalp-fragrance-free/qxp141262?catid = 183491
10.	Cosmetics	Chamomile hair lightener (6.8 fl. oz.)	US$9.26	http://www.shopping.com/herbatint-herbatint-chamomile-hair-lightener-6-8-fl-oz/info
11.	Cosmetics	Chamomile conditioner (15 fl. oz. liquid)	US$6.69	http://www.swansonvitamins.com/pure-life-soap-chamomile-conditioner-15-fl-oz-liquid?SourceCode = INTL401&CA_6C15C = 5300246000001569
12.	Cosmetics	Turquoise chamomile wax (16 oz.)	US$19.95	http://www.bodyworkmall.com/home/amber-turquoise-chamomile-wax-16-oz/?cid = shopping.com&CA_6C15C = 1498814584
13.	Cosmetics	Chamomile soap bar (8 oz.)	US$2.28	http://www.shopping.com/Kiss-My-Face-Camomile-Soap-Bar-8-oz-From-Kiss-My-Face-Kiss-My-Face/info
14.	Cosmetics	Body deodorant spray chamomile and green tea (4 fl. oz. liquid)	US$3.59	http://www.swansonvitamins.com/crystal-body-deodorant-spray-chamomile-green-tea-4-fl-oz-liquid?SourceCode = INTL401&CA_6C15C = 1588525237

(Continued)

TABLE 7.1 (Continued)
Selected Chamomile Products and Their Current Cost in the International Market

S. No.	Category	Product	Price	Reference
15.	Cosmetics	Calm skin chamomile moisturizer (sensitive skin) (60 mL/2 oz.)	US$59.50	http://us.strawberrynet.com/skincare/eminence/calm-skin-chamomile-moisturize/140217/?trackid = 4459
16.	Cosmetics	CoQ10 cream, 97% natural 2 fl. oz. (59 mL) cream	US$5.99	http://www.swansonvitamins.com/swanson-premium-coq10-cream-97-natural-2-fl-oz-59-ml-cream
17.	Cosmetics	Itch relief lotion with tea tree, E, and chamomile (6 fl. oz.) liquid	US$9.62	http://www.swansonvitamins.com/derma-e-itch-relief-lotion-tea-tree-e-chamomile-6-fl-oz-liquid
18.	Cosmetics	Chamomile soothing mist (100 mL)	US$34.00	http://www.jurlique.com/products/csm/chamomile-soothing-mist
19.	Oral care	Spry oral rinse blue (16 fl. oz.) liquid	US$4.76	http://www.swansonvitamins.com/xlear-spry-oral-rinse-blue-16-fl-oz-liquid
20.	Baby products	Raw shea chamomile argan oil baby head-to-toe ointment (4 oz.)	US$8.49	http://www.target.com/p/sheamoisture-raw-shea-chamomile-argan-oil-baby-head-to-toe-ointment-4-oz/-/A-13690528 = $15.00
21.	Baby products	Baby powder pure cornstarch lavender and chamomile (22 oz.)	US$15.00	http://www.amazon.com/Johnsons-Powder-Cornstarch-Lavender-Chamomile/dp/B000VEBG26
22.	Baby products	Pampers ThickCare touch of chamomile wipes refill (7x)	US$27.77	http://www.amazon.com/Pampers-ThickCare-Touch-Chamomile-Refill/dp/B003ART3H6
23.	Baby products	Baby oil with sunflower, jojoba, and chamomile (8.4 fl. oz.) liquid	US$6.74	http://www.swansonvitamins.com/bentley-organic-baby-oil-sunflower-jojoba-chamomile-8-4-fl-oz-liquid
24.	Baby products	Bubble bath chamomile (12.5 fl. oz.) liquid	US$8.69	http://www.swansonvitamins.com/earth-friendly-baby-bubble-bath-chamomile-12-5-fl-oz-liquid
25.	Feminine hygiene	Natural feminine wash: Chamomile fresh (200 mL)	PHP 149.75	https://humanheartnature.com/buy/index.php/checkout/cart/
26.	Feminine hygiene	Naturella	177 Rouble	http://foodle.ru/en/taxonomy/term/401/all?page = 1
27.	Detergent and fabric softener	Procter & Gamble	£14.99	http://www.dooyoo.co.uk/household-products/bold-2-in1-powder-lavender-and-camomile/

REFERENCES

Albert, B. M. and Riso, R. R. 1999. US 5888515 A 19990330. USA.
Al-Sari, M. H. 2003. EP 1366768 A1 20031203. European Patent Office.
Asari, J. 1998. JP 10215855. Japan.
Asplund, P. 2008. WO 2008044046 A1 20080417. PCT.
Bader, J. 2006. GB 2416664 A 20060208. Great Britain.
Bang In Soo. 2001. KR 1020010044759. Korea.
Baraldi, M. 2003. WO 2003033007 A1 20030424. PCT.
Barry, T. A. and Patten, B. A. 1868. US 75837 A 18680324. USA.
Behan, J. M., Bradshaw, D. J., Richards, J., Munroe, M. J., Minhas, T., and Cawkill, P. M. 2004a. WO 2004073668 A1 20040902. PCT.
Behan, J. M., Bradshaw, D. J., Richards, J., and Munroe, M. J. 2004b. WO 2004073670 A1 20040902. PCT.
Behan, J. M., Bradshaw, D. J., Richards, J., and Munroe, M. J. 2004c. WO 2004073672 A1 20040902. PCT.
Bell, R. and Gray, D. 1999. US 6007797 A 19991228. USA.
Benson, A. K., Hoerr, R. A., and Bostwick, E. F. 2007. WO 2007136553 A2 20071129. PCT.
Bindra, R. L., Gupta, R., Shukla, Y. N., Dwivedi, S., and Kumar, S. 2000. US 6126950 A 20001003. USA.
Blount, D. 2001. US 6287567. USA.
Bonjour, P. 1990. EP 365427 A1 19900425. European Patent Office.
Bonomelli, F. 1992. PCT/IT/1992/000056.
Borba, S. V. 2006. WO 2006055550 A2 20060526. PCT.
Borod, M. 2001. US 6228387 B1 20010508. USA.
Botos, G. 2008. WO 2008061375 A1 20080529. PCT.
Cahen, C. M. 2006. US 20060002876 A1 20060105. USA.
Choi, B. S. 2013. KR 2013002613. Korea.
Coelho, A. C. P. 2004. WO 2004009105 A1 20040129. PCT.
Costa, A. 1999. US 5942232 A 19990824. USA.
Courtin, O. T. 2008. US 20080020074 A1 20080124. USA.
Day, K. S. and Hansen, G. A. 2007. US 20070212434 A1 20070913. USA.
Deane, J. A. 2001. US 6312675. USA.
Degussa. 1986. ES 8705741. Spain.
Dobrovol'ski, V. F. and Kvasenkov, O. I. 2004. RU 2233598. Russia.
Doi, N. and Watanabe, K. 2000. JP 2000169321. Japan.
Doney, M. and Pickens, J. 2009. WO 2009023192 A1 20090219. PCT.
Dorf, P. 2007. US 20070154439 A1 20070705. USA.
Erdman, C. L. 2003. US 20030100877 A1 20030529. USA.
Farber, E. 2002a. US 20020055531 A1 20020509. USA.
Farber, E. 2002b. US 20020102288 A1 20020801. USA.
Feibelman, R. 1868. US 81152 A 18680818. USA.
Ferro, M. 1870. US 107024 A 18700906. USA.
Fontaine, W., Madfes, D., Palefsky, I., and Wilson, N. 2008. US 20080206371 A1 20080828. USA.
Frome, B. M. 2003. US 20030224072 A1 20031204. USA.
Geiser, K. M., Koenig, D. W., Minerath, B. J., Dvoracek, B. J., Tyrrell, D. J., and Krzysik, D. G. 2003. US 20030120224. USA.
George, S. T. 2013. US 8383169. USA.
Giacoman, J. G. M. 2002. MX PA/a/1999/011514. Mexico.
Gonzalez Garcia, C. 2001. ES 2152896. Spain.

Greene, P. and Jennings, T. 1998. GB 2321583 A 19980805. Great Britain.
Griffiths, A., Ormerod, A. P., Russell, A. L., and Wantling, S. D. 2007. WO 2007009600 A1 20070125. PCT.
Ham, J. C., Choi, S. Y., and Ham, S. M. 2010. KR 2010028186. Korea.
Han, Y. 2003. KR 1020030074525. Korea.
Haubert, W. P. and Haubert, J. 1869. US 87169 A 18690223. USA.
Hayase, M. 1998. JP 10152428. Japan.
Hill, J. 1987. GB 2179551 A 19870311. Great Britain.
Hill, W. L. 2007. US 7205012 B1 20070417. USA.
Ishino, A. 1999. JP 11240823. Japan.
Johnson, L. B. 2012. US 2012/0065115A1. USA.
Kavaya, A. 1998. GB 2321587 A 19980805. Great Britain.
Kieffer, N. 1884. US 308900 A 18841209. USA.
Kim, Y. I. 2003. KR 1020030050722. Korea.
Kim, Y. 2001. KR 102001006227. Korea.
Kiyama, K. and Kawaguchi, K. 1999. JP 11262378. Japan.
Klingman, S. E. 2011. US 20110256082 A1 20111020. USA.
Knudsen, M. V. 1968. US 3382151 A 19680507. USA.
Konishi, M. 1999. JP 11079967. Japan.
Korkis, G. N. 1976. US 3984538. USA.
Kreuter, M. H. 2009. WO 2009138860 A1 20091119. PCT.
Kudo, T. 2002. JP 20022306142. Japan.
Kvasenkov, O. I. 2004a. RU 02227635. Russia.
Kvasenkov, O. I. 2004b. RU 02227633. Russia.
Kvasenkov, O. I. 2004c. RU 2226876. Russia.
Laali, M. 2008. US 20080031831 A1 20080207. USA.
Lange, R. 2003. US 20030114324 A1 20030619. USA.
Laughlin, T. R. and Holt, S. D. 1998. US 5854291 A 19981229. USA.
Laughlin, T. R. and Holt, S. D. 1999. US 5856361 A 19990105. USA.
Lee, J. S. 2002. KR 1020020035685. Korea.
Leons, D., Janis, Mazis, V., M., Tamara, P., Elza, Z., Livija, V., and Helena, K. 1995. LV 10641. Latvia.
Lesky, J., Lesky, J., and Lesky, T. 2002. US 20020054858 A1 20020509. USA.
Levy, J. 1866. US 57155 A 18660814. USA.
Llado, J. 1869. US 93209 A 18690803. USA.
Lukacs, K., Lukacs, K., Dajka, L., and Mueller, J. 2003. WO 2003039559 A1 20030515.
Månsson, L. and Senanayake, S. P. J. N. 2012. US 20120263850 A1 20121018. USA.
Mausner, J. 1993. US 5215759. USA.
Mausner, J. 1999. US 5922331 A 19990713. USA.
Mazis, J., Eliasa, B., Ponomarjova, T., Kaulina, H., Zdanovska, E., and Vaivode, L. 2000. LV 12385. Latvia.
McIntyre, D., Planken, E., and Compton, S. 2003. AU 2003200927 A1 20030925. Australia.
Mezenova, O. Ya. and Skapets, O. V. (2011). RU 2432768. Russia.
Miller, J. T., Maningat, C. C., Bassi, S., Makwana, D., and Chinnaswamy, R. 2003. US 20030138391 A1 20030724. USA.
Minetti, D. C. 1988. US 4758599. USA.
Nakar, D. 2004. WO 2004035071 A1 20040429. PCT.
Newmark, T. 2002. US 6391346.
Newmark, T. and Schulick, P. 2002. US 6391346 B1 20020521. USA.
Oh, S. G. 2001. KR 1020010020067. Korea.
Okigami, H. 2006. WO 2006063422 A1 20060622. PCT.

Oltarzhevskaya, N. D., Savilova, L. B., and Krichevsky, G. E. 2006. WO 2006067549 A1 20060629. PCT.
Ono, A. 1999. JP 11215973. Japan.
Owen, A. 1878. US 200331 A 18780212. USA.
Papp, J. 2012. HU 2011000094. Hungary.
Park, H. H. 2002. KR1020010007428. Republic of Korea.
Parkes, C. and Keynes, M. 2011. GB 2473735 A 20110323. Great Britain.
Parmar, S. D. 2012. WO 2012131728 A2 20121004. PCT.
Pavlov, A. and Bubnjevic-Pavlov, J. 2012. WO 2012033422 A1 20120315. PCT.
Pelek, K. 1948. HU 13424419480316. Hungary.
Pinchon, H. and Saboureau, A. 1992. EP 519777 A1 19921223. European Patent Office.
Piterski, C. A. 2003. WO 2003077856 A2 20030925. PCT.
Posner, R. M. 1998. US 5738859. USA.
Rambaut, G. V. 1868. US 74940 A 18680225. USA.
Reidy, J. 1995. GB 2281730 A 19950315. Great Britain.
Renz, F. H. 1881. US 244932 A 18810726. USA.
Robertson, L. and Harris, E. M. 1998. US 5804211 A 19980908. USA.
Robinson, R. S. 2007. US 20070092589 A1 20070426. USA.
Rodionova, L. Ya., Kvasenkov O. I., Donchenko, L. V., and Zhivagina, I. S. 2004. RU 2227598. Russia.
Romagnoli, A. 2003. US 20030159975. USA
Rouda, K. 2009. US 20090061029 A1 20090305.
Rudin, V. N., Zuev, V. P., Komarov, V. F., Melikhov, I. V., Minaev, V. V., Orlov, A. Y., Mishin, A. A., and Bozhevolnov, V. E. 1999. WO 9920237 A1 19990429. PCT.
Russo, V. 2010. EP 2251016 A2 20101117. European Patent Office.
Scancarella, N. D., Reinhart, G. M., and Russ, J. G. 2003. US 20030190300 A1 20031009. USA.
Schinitsky, M. R. and Meisner, L. F. 1990. US 4938969 A 19900703. USA.
Seibert, G. 1884. US 294687 A 18840304. USA.
Semeniene, V. B. and Semenas, V. 1994. LT 3105. Lithuania.
Sharma, N. 2004. IN 192288 A1 20040327. India.
Shchetkina, N. I. 2002. RU 2396031. Russia.
Stoll, D. M. 2004. US 6730331 B1 20040504. USA.
Szente, L., Szejtli, J., Magisztrak, H., and Horvath, E. 1988. WO 8808304 A1 19881103.
Tasevski, B. 1994. PCT Int. Appl., WO 9401082 A1 1994012. PCT.
Tubaro, A., Della Loggia, R., Banfi, E., Cinco, M., and Redaelli, C. 1985. EP 143141 A1 19850605. European Patent Office.
Van de Loecht-Blasberg, M. and Knollman, R. 1996. WO 9626704 A1 19960906. PCT.
van der Linde, L. M., van Lier, F. P., and van der Weerdt, A. J. A. 1986. EP 173395 A1 19860305. European Patent Office.
Veach, T. C., Lewis, P., and Siegel, P. B. 2003. US 20030211173 A1 20031113. USA.
Videki, M., Varadi, J., Mozsgai, K., Kiss, N. V. Z., Malasics, G., Puskas, I., Sipos, M., and Budavari, O. 1991. US 5043153 A 19910827. USA.
Vijgen, J. L. J., Admiraal, S. M., and Titulaer, H. A. C. 2000. WO 2000041709 A1 20000720. PCT.
Villani, M. 2004. US 20040109872 A1 20040610. USA.
Watanabe, H. 2002. JP 2002095425. Japan.
Watson, G. 2002. EP 1210946 A1 20020605. European Patent Office.
Weiss, R. A., Sprecker, M. A., Hanna, M. R., Beck, C. E. J., and Jackson, H. W. 2002. US 6462015 B1 20021008. USA.
Williams, G. H. A. 1888. US 377592 A 18880207. USA.

Wilson, R. 2006. US 20060280778 A1 20061214. USA.
Xishou, Y. 2012. CN 102754761. China.
Xu, Y. 2010. CN 101912025. China.
Yam, J. and Kreuter, M. H. 2012. EP 2420228 A1 20120222. European Patent Office.
Yoo, J. S. 2010. KR 2010026487. Korea.
Yoshida, K. 2002. JP 202284644. Japan.
Zhivagina, I. S., Rodionova, L. Ya., Donchenko, L. V., and Kvasenkov O. I. 2004. RU 2227540. Russia.

Index

A

Azotobacter chrocooccum, 254
Abstinence, effect of chamomile extract on, 125
Azulene, 103
Acetate–mevalonate pathway works, 79
Acetylcholine, 130

B

Acidic soil, 245
Acute allergic contact dermatitis, 106
Bacillus coagulans, 74
Adonis aestevis L., 50
Baraldi, Mario, 271
Adonis flammea Jacq, 50
Bath, essential oils of chamomile, 6
Adulteration
Belgium, effect of inorganic and organic
    chamomile homeopathic formulations,
        fertilization, 252
        quality issues of, 29
Benzodiazepines, 121
    essential oil, 16, 19
Beverages, chamomile extract, 275–277
    flowers, 18–19
Biofertilizers, 252
Agrobacterium rhizogenes, 233
Biopesticides, effect of, 75
Alchemilla vulgaris extract, 279
Biotic stress, 75–78
Alkaline soil, 245
α-Bisabolol, 66, 120
Anacyclus clavatus, 52
(–)-α-Bisabolol, 115–116
Angiospermae, 60
    antioxidant property of chamomile, 112
Animal feed supplement, 280
α-Bisabolol oxide A, 214
Animal husbandry, chamomile use in poultry
α-Bisabolol oxide B, 214
    and, 136–137
Bisabolol oxides, 164–165
Anisakiasis, chamomile as potential cure for, 135
Bisabolols, 164–165
Anisakis simplex, 135
Botanical polysaccharides, 117
Anthemidae, 61
β-Bourbonene, 172
Anthemis arvensis, 52
Brazil, greenhouse pot experiment, 252
Anthemis cotula, 52
Breeding
Anticancer property of chamomile, 119–120
    aim of, 223
Anti-inflammatory properties, 18
    different varieties, 225, 226
Antioxidant property of chamomile, 111–113
    diploid and tetraploid varieties, 224
Anxiolytic properties of chamomile, 126
British Homeopathic Pharmacopoeia, 27–28
Anxiolytic-reducing properties of chamomile,
Bulgaria, tetraploid variety, 225
    122–123
Apigenin, 104, 185, 189

C

    anti-inflammatory effects of, 106
Apis cerana, 64
τ-Cadinol, 172
Apis florae, 64
Cadmium toxicity, effect of calcium on, 74
Argentina, chamomile cultivation
Calcium, effect of, 68
    herbicide Treflan, 255
Callus cultures, constituents of chamomile oil
    sowing, 247
        distilled from flowers and, 178–180
Aromatase converts, 128
CAM, *see* Complementary and alternative
Aromatherapy, 6
    systems of medicine
Artemisia pollen, 138
Camphene, 170–171
Aspergillus niger, 135
Candida albicans, 112, 117, 134
Asteraceae, 61
Candies, 275
Asterales, 60–61
Capillary electrochromatography (CEC), 198
Astringent composition, 271
Capsaicin, 270
Autumn sowing, 247
Carbon dioxide, 194
Autumn-sown crops, 244
Cardiovascular disease, chamomile and, 128–129
Azospirillum lipoferum, 254
Caspases, 120
C-banding techniques, 65

CBP, *see* CREB Binding Protein
CC, *see* Column chromatography
CCA, *see* Chamazulene carboxylic acid
CEC, *see* Capillary electrochromatography
Cedrelanol, 173
Cell viability, 120
Central glutamatergic system, 124
Central Institute of Medicinal Plants (CIMAP), 81
Central nervous system, effect of chamomile essential oil on, 127
Ceraceous substances, 191
Chamaemelum nobile, 51
Chamazulene, 66, 135, 156
    anti-inflammatory effects of, 102–103
    antioxidant property of chamomile, 111–112
    content, 252
    precursors, 163–164
    synthesis, 79
Chamazulene carboxylic acid (CCA), 103, 164
Chamomile
    anatomical characteristics of
        inflorescence/capitulum, 63
        leaf, 62
        stem, 61
    against anisakiasis, effect of, 135
    antibiotic effect of, 133–135
    anticancer effect of, 119–120
    anticulex effect of, 135–136
    anti-inflammatory effects of, 100–102
    antileishmania effect of, 133
    antimutagenic effect of, 118–119
    antinociceptive effect of, 113–114
    as antioxidant, 111–113
    anxiolytic effect of, 125–126
    biochemistry, 78–80
    (–)-α-bisabolol, anti-inflammatory effects of, 103–104
    botanical classification, 50–61
        differences in, 57
        Linnaeus, 52
        nomenclature, 58
        taxonomic tools, 61
        taxonomic key, 60
    and cardiovascular disease, 128–129
    chamazulene, anti-inflammatory effects of, 102–103
    chemical constituents of, 269
    common names of, 49
    compounds, toxicity and allergenicity of, 138–139
    cytology and ploidy effects, 64–66
    on dermatological problems, effect of, 127
    descriptors of, 80–86
    flavonoids, anti-inflammatory effects of, 104–111
    hepatoprotective effect of, 131–132
    as immunomodulator, 117
    introduction, 99
    on morphine withdrawal syndrome, effect of, 124–125
    nephroprotective effect of, 132–133
    neuroprotective effect of, 123–124
    and oral health, 129–130
    pain-relieving properties of, 113–114
    patents
        animal feed supplement, 280
        beverages, 275–277
        cosmetics, 277–279
        diapers, 274
        feminine hygiene, 274–275
        food, 275
        insecticide, 279
        medical devices, 273
        oral medicine containing, 269–270
        perfume, 277
        topical use of, 270–273
        veterinary medicine, 279–280
    and pest management, 136
    physiology
        biotic stress, 75–78
        calcium, sulfur, and magnesium effect, 68
        heat stress, 72
        indole acetic acid, gibberellin, and kinetin, 69
        inorganic and organic fertilizers and mycorrhiza, 74
        salt stress, 70
        weedicide, effect of, 75
    phytoestrogenic effect of, 128
    and pregnancy, 136
    products, 280
        and cost in international market, 281, 282
    psychopharmacological effect of, 126–127
    for relieving stress, 122–123
    reproductive biology, 64
    spasmolytic effect of, 130–131
    trans-EN-YN-dicycloether, anti-inflammatory effects of, 104
    for treating insomnia, 120–122
    treatment of colicky infants with, 130
    use in poultry and animal husbandry, 136–137
    as veterinary medicine, 137–138
    wound healing and ulcer-protective property of, 114–117
Chamomile-based medicines, 269
Chamomile Capitula, flavonoids and coumarins, 186
Chamomile chromosomes, karyotype of, 65
Chamomile cultivation
    fertilization, *see* Fertilization, chamomile cultivation
    habitat and climatic conditions, 243–245

harvesting, 256–258
herbicide resistance, 256
hydroponic culture, 261
irrigation, 254
pest control, 256
postharvest treatment, 258–261
seeds, 246–247
soil conditions, 245–246
sowing, 247–249
weeding, 254–256
yield, 258, 259
Chamomile flowers, 273
  extract, chemical composition, 183–185
    major compounds, 185–189
    minor compounds, 190
Chamomile herbal formulations
  GMP, guidelines for, 20–21
  national regulations for, 21, 22
  quality, safety, and efficacy issues of, 16–18
Chamomile oil, 116, 127, 155
  chemical composition
    chamomile flower extract, 183–199
    flower essential oil, 160–172
    herb essential oil, 156–158
    root essential oil, 156
  comparative account, essential oils of, 174
  distilled from flowers and callus
    cultures, 178–180
  extraction methods, 191–196
  identification methods, 196–199
  localization and content, 156–198
  physical properties of flower essential oil, 173
Chamomile plant
  essential oil compounds, various organs of, 175, 176
  oil content and oil color, 159
Chamomile population, variability in
  chemical types of, 215–217
  chemotypes, chamomile flowers, 217
  environmental conditions, 218–219
  morphological characteristics of, 210
  oil composition, 211–215
  oil content, 211
  ontogeny, 218
Chamomile seeds, germination of, 67
Chamomilla recutita, 139
Chamomillol, 170, 171
Chemotypes, based on flower extract, 217
Chile, chamomile cultivation, 250
  weed control in, 255
Chronic inflammation, 100
Chrysanthemum, 50
CIMAP, *see* Central Institute of Medicinal Plants
Clayey lime soils, 245
ColiMil, 131
Column chromatography (CC), 19

Combined fertilizers, chamomile cultivation, 252–253
Complementary and alternative systems of medicine (CAM), 2
Compositae-allergic patients, 138
Compress, chamomile oil, 7
Convolvulus arvensis, 256
Copper chelates, effect of, 74
Corticosteroids, 100
Cosmetic ingredient, chamomile, 4
Cosmetics, 277–279
Coumarin, 190
  biosynthesis of, 80
  solvent extracts of chamomile capitula, 183, 186
CREB Binding Protein (CBP), 109
Crop descriptors, 81
Crotalaria spectabilis, 252
Cryopreservation, 233
Culex, inhibitory effect of chamomile against, 136
Culex mosquito, 135
Culex quinquefasciatus, 136
Cultivars, chemical constituents in, 161
Cultivation of chamomile, *see* Chamomile cultivation
Cultures, essential oil in, 232–233
Czech Republic, chamomile cultivation, 224

D

Delicate chamomile, 51
Dexamethasone, 115
Diapers, 274
Dichlorprop, 75
Dicotyledones, 60
Dioscorea, 270
Dioscorides, 50
2,2-Diphenyl-1-picrylhydrazyl (DPPH), 111
Diploid chamomile plants, morphological characteristics of, 210
Diploids, identification of, 65
Diploid varieties *vs.* tetraploid varieties
  flower extracts, compounds in, 230
  morphological characters, 227–228
  ploidy levels, determination of, 228–229
Direct sowing method of chamomile cultivation, 248
Diurnal rhythms, effect of, 68
DNA markers, 66
Dried chamomile flowers, 258
Drying techniques, 260

E

Egypt, chamomile cultivation, 250
  nursery beds, 247
  weeds, 255

Egyptian oil, 66
Electron microscope, 135
β-Elemene, 171
Environmental conditions
   cultivation practices, 218
   phytotron conditions, 219
Enzymes, 78
Eranthemon, 50
ERK, *see* Extracellular-regulated kinase
Escherichia coli, 134
Essential oil, 66, 155, 251
   discovery of, compounds, 159–160
   descriptor of, 87, 88
   flowers, content in, 156–158
Europe
   chamomile cultivation, 255
   homeopathy system of medicine, national regulations, 32
European patent, 272
European Pharmacopoeia, 10–12, 121
European Scientific Cooperative on Phytotherapy (ESCOP) monograph, 13–14
Exoborneol, *see* Isoborneol
Extracellular-regulated kinase (ERK), 104
Extraction methods, 191–196

F

Farmyard manure (FYM), 247
(*E*)-β-Farnesene, 165
Farnesol, 166–167
Feminine hygiene, 274–275
Fertilization, chamomile cultivation
   combined fertilizers, 252–253
   inorganic fertilizers, 249–251
   physiological treatments, 253
   rhizobacteria, 254
   vascular arbuscular mycorrhizal fungi, 253–254
   vermicompost, 251–252
Field preparation, chamomile cultivation, 249
Flavone, 128
Flavonoids
   anti-inflammatory effects of, 104–111
   in chamomile flower extract, 183
Flavoring for food, using chamomile extract, 275
Flora Europea (Tutin), 59
Flora of Great Britain and Ireland (Sell and Murrell), 60
Flora Suecica (First Edition) (Linnaeus), 55
Flora Suecica (Second Edition) (Linnaeus), 56
Flower essential oil
   chemical composition, 160–163
      major compounds, 163–166
      minor compounds, 166–172
   physical properties of, 173

Flower essential oil quality determining compounds, descriptors of, 88, 89
Flowering, stages of, 64
fMLP, *see* N-Formyl-methionyl-leucyl-phenylalanine
Food, 275
   chamomile, 3–4
N-Formyl-methionyl-leucyl-phenylalanine (fMLP), 112
Frome, Bruce Marshall, 271
Fusarium pallidoroseum, 134
FYM, *see* Farmyard manure

G

Gas chromatography-mass spectrometry (GC-MS), 198–199
G-banding techniques, 65
GC-MS, *see* Gas chromatography-mass spectrometry
Genetics
   chamazulene and bisaboloids, 220–221
   molecular techniques, 221–222
   patents on cloned genes, 222–223
Geographic information system (GIS), 246
Germacrene D, 166, 167
German Commission E Monograph, 12–13
Germany, chamomile cultivation, 224, 255
Germination
   of chamomile seeds, 67
   chamomile cultivation, 249
Gibberellin, effect of, 69
GIS, *see* Geographic information system
Glandular hairs, 62
Granulocytes, 102
Greenhouse pot experiment, 252

H

Hamilton anxiety rating (HAM-A), 126
Harvesting of flower, 256–258
Heat stress, 72
Heavy metals, 17
   chamomile homeopathic formulations, quality issues of, 29
   detection of, 19
Helicobacter pylori, 108, 134
Helminthosporium sativum, 134
Herbaceous annual chamomile plant, 49
Herbal medicines, manufacture of, 20
Herb essential oil, chemical composition of, 173
Herbicide resistance, 256
Herbicides, 255
Herbicide Treflan, 255
Herniarin, 185–194
Hexaconazole, 70
Hexadec-11-yn-13,15-diene, 172

Index

n-Hexanol, 167
Highly discriminating descriptors for international harmonization, 81
High-performance liquid chromatography (HPLC), 197–198
High-performance thin-layer chromatography (HPTLC), 197
Histamine, 130
HMC-1, *see* Human mast cells
Homeopathic drugs, methods to improve, 29–32
Homeopathic Pharmacopoeia of the United States (HPUS), 26–27
Homeopathy
   chamomile homeopathic formulations, quality issues of, 28–29
   disease conditions, 24
   formulations in, 25–26
   homeopathic drugs, improve quality, safety, and efficacy of, 29–32
   Homeopathic Pharmacopoeia, 26–28
   medicine, Unani system of, 35
   national regulations, 32–35
Hortus Cliffortianus (Linnaeus), 54–55
HPLC, *see* High-performance liquid chromatography
HPTLC, *see* High-performance thin-layer chromatography
HPUS, *see* Homeopathic Pharmacopoeia of the United States
Human mast cells (HMC-1), 107
α-Humulene, 171
Humus-rich soils, 245
Hungary, breeding experiments, 224
Hydroponic cultures, 261
   of chamomile, 74
Hydroxycoumarin, 190

I

ICBN, *see* International Code of Botanical Nomenclature
Immunomodulating effect of chamomile, 117
Impurities, 17
India
   breeding experiment, 225
   homeopathy system of medicine, national regulations, 33–35
Indirect sowing method of chamomile cultivation, 248
Indole acetic acid, effect of, 69
Indomethacin, 115
Inducible nitric oxide synthase (iNOS) genes, 103
Infusion chamomile flowers, 191
Inorganic fertilizers, 74
   chamomile cultivation, 249–251
iNOS genes, *see* Inducible nitric oxide synthase genes

Insecticide, 279
International Bureau of Plant Genetic Resources in the 1970s, 80
International Code of Botanical Nomenclature (ICBN), 58
Inter simple sequence repeat (ISSR), 222
IPP, *see* Isopentenyl diphosphate
Iran, chamomile cultivation in, 244, 248
Irrigation, 158
   during cropping season, 254
Isoborneol, 168
Isoflavone phytoestrogen, 128
Isolated guinea pig ileum, 127
Isopentenyl diphosphate (IPP), 79
ISSR, *see* Inter simple sequence repeat

K

Kenya, chamomile cultivation, 250
Kinetin, 253
   effect of, 69

L

Labeling, information, quality of, 18
Leaf, chamomile, 62
Leishmania, inhibitory effect of chamomile on, 133
Leishmania mexicana, 133
Leishmaniasis, 133
Leucanthemum dioscoridis, 55
Leukocytes, 114
Levomenol, *see* Bisabolols
Light, effect of, 67
Limonene, 169
Linuron, 254
Lipidic substance, 191
Lonicera japonica, 110
Luteolin, 104, 185, 189
   anti-inflammatory effects of, 109–111

M

Magnesium, effect of, 68
Manzanilla, 225
Massage, chamomile, essential oils of, 6
Matricaria, 1, 52
   types of, 54
Matricaria argentea, 55
Matricaria aurea, 52
Matricaria chamomilla, 52, 55, 56, 58
Matricaria inodora, 56–57
Matricaria parthenium, 55
Matricaria recutita, 50, 56, 58
Matricaria recutita L., 58
Matricaria suaveolens, 52, 57
Matricin, 163, 164

Medical compound, 270
Medical devices, 273
Medicinal uses, chamomile
  disease conditions, 7
  essential oil of, 5–6
  herbal formulations of, 10, 11
  pharmacopoeias and monographs, 10–15
  southwestern Serbia, 2
  tincture and extract, preparation of, 5
Medicine, Unani system of
  chamomile, formulations of, 35–38
  India, National Legislation in, 41
  Unani formulations, 39–41
  Unani pharmacopoeia, 38–39
Meloidogyne incognita, 136
Metachlamydeae, 60
N-Methyl-D-aspartic acid (NDMA), 124
Methyl jasmonate, 71
Microbial contamination, 17
  chamomile homeopathic formulations, quality issues of, 29
Microorganisms, detection of, 19
Micropropagation, of chamomile
  cryopreservation, 233
  cultures, essential oil in, 232–233
MicroRNAs (miRNAs), 108
Minimum descriptors, 81
miRNAs, *see* MicroRNAs
Molecular markers, 66
Monograph
  Canada, compendium of, 15
  WHO, 14–15
Monoterpene, 164
Morphine, 124
  dependence, effect of chamomile extract on, 125
Morphological descriptors of chamomile, 82–86
Mucuna aterrima, 252
Mutagenesis, 118
Mycorrhiza, 74

N

Naloxone, 125
Naval Medical Research Institute (NMRI) mice, 114
NDMA, *see* *N*-methyl-D-aspartic acid
Nephrotoxicity, effect of chamomile on, 132–133
Nerol, 168, 169
(E)-Nerolidol, 166, 167
Neuroprotective properties of chamomile, 123–124
Nitrogen, effect of, 68
NMRI mice, *see* Naval Medical Research Institute mice
NMR spectroscopy, *see* Nuclear magnetic resonance spectroscopy

Nociceptive pain, 113
Non-benzodiazepines, 121
Nonmevalonate pathway works, 80
Nuclear magnetic resonance (NMR) spectroscopy, 199

O

Ontogenetic variation, essential oil, content and composition of, 218
OPLC, *see* Overpressured-layer chromatography
Oral health, chamomile and, 129–130
Oral hygiene, 274
OR-banding techniques, 65
Organic fertilizer, 74, 252
Organ-specific essential oil, variation in composition of, 158
Overpressured-layer chromatography (OPLC), 198
Oxidative damage, 108

P

Packaging, 17
PAL, *see* Phenylalanine ammonia lyase
Papaverine, 131
Parasites, on chamomile plants, 76–78
PCT patent, 271
PDE10A, *see* Phosphodiesterase enzyme
Pectic acid, 79
Peduncle, chamomile, 63
Perfume, chamomile essential oil and extracts, 277
Peroxidase activities, 71
Peruviol, *see (E)*-Nerolidol
Pest control, 256
Pesticide, 279
Pesticide residues, 16
  chamomile homeopathic formulations, quality issues of, 29
  detection of, 19
Pest management, chamomile and, 136
Pharmacologically active compounds, levels of, 211
Phenolic enzymes, activities of, 71
Phenylalanine ammonia lyase (PAL), 68
  activity, 73
Phenyl carboxylic acids, 191
Phorbol 12-myristate 13-acetate (PMA), 102
Phosphatidylserine, 120
Phosphodiesterase enzyme (PDE10A), 130
Phosphorus, effect of, 68
Photoperiod, effect of, 68
Photosystem II (PSII) reaction, 70
Phytoestrogens, chamomile as, 128
Phytol, 169
Plant-based drugs, 119

Index

Plasma energy, 259
PMA, *see* Phorbol 12-myristate 13-acetate
Poland, breeding experiments, 224
Polysaccharides, 79, 191
    antioxidant property of chamomile, 113
Postoperative sore throat (POST), 103
Potassium, effect of, 68
Poultry, and animal husbandry, chamomile use in, 136–137
Prostaglandins, 100
Pseudocercospora griseola, 135
Pseudomonas aeruginosa, 134
PSII reaction, *see* Photosystem II reaction
Psychostimulants, 126
Pulvis, 247, 257
Pyrimidines, 253

Q

Quercetin, 188–190
    anti-inflammatory effects of, 106–109

R

Random amplified polymorphic DNA (RAPD), 222
Rhizobacteria, 254
Romania, tetraploid varieties, 224
Root essential oil, chemical composition of, 174
Roots, chamomile, 63

S

SA, *see* Salicylic acid
Safety, 18
Salicylic acid (SA), 69
    on chamomile plants, 72
Saline-alkaline soils of Banthra, Uttar Pradesh, 245
Salt stress, 70
Scentless chamomile, 51
SCFC, *see* Supercritical carbon dioxide fluid extraction
Seeds, 246–247
Self-inspection, 21
Self-sowing method of chamomile cultivation, 248–249
Sesquiterpenes, 79, 166
Shigella sonnei, 134
Slovakia
    breeding experiments, 224
    chamomile cultivation, 250
        herbicide Hedonal DP (dichlorprop) on chamomile plants, effect of, 256
        sorting of flowers, 260
Soil conditions for chamomile cultivation, 245–246

Solar power, drying method, 260
Solid-phase extraction (SPE), 194
Solid-phase microextraction (SPME), 195
Solvent extraction, 192–193
    flavonoids and coumarins identified in, 186
Sorghum halepenses, 256
Sowing, 247–249
Spasms, 130
Spathulenol, 167, 168
SPE, *see* Solid-phase extraction
Species Plantarum (First Edition) (Linnaeus), 55
Species Plantarum (Second Edition) (Linnaeus), 57
Spectrocolorimeter, 197
Spermidine, 71
Spiroethers, 165–166
SPME, *see* Solid-phase microextraction
Spray drying method of chamomile extracts, 260
Spring sowing, 247
Steam distillation, 191–192
Stem, chamomile, 61
Stinking chamomile, 51
Stress-reducing properties of chamomile, 122–123
Sulfur, effect of, 68
Supercritical carbon dioxide fluid extraction (SCFC), 194

T

Tabernaemontanus, 51
Tamoxifen, 128
Tasevski, Bozo, 270
Taxonomic key, of chamomile, 60
Temperature effect, 67
α-Terpineol, 170
Tetraploid chamomile plants, morphological characteristics of, 210
Tetraploidization, 65
Tetraploids, identification of, 65
Tetraploid varieties of chamomile, and diploid varieties
    characteristics of, 228, 229
    flower extracts, compounds in, 230
    morphological characters, 227–228
    ploidy levels, determination of, 228–229
Tezontle, 252
Thin-layer chromatography (TLC), 196
Thin-layer chromatography–ultra violet (TLC-UV) spectrometry, 196
Tincture/extract
    characteristics of, 6
    preparation of, 5
TLC, *see* Thin-layer chromatography
TLC-UV spectrometry, *see* Thin-layer chromatography–ultra violet spectrometry
Toxic metals, 72

TPP, *see* Triose pyruvate pathway
Trans-en-yn-dicycloether, anti-inflammatory effects of, 104
Trans-β-farnesene, *see (E)*-β-Farnesene
Transplantation, chamomile cultivation, 249
trans-β-farnesene, *see (E)*-β-Farnesene
Tribolium castaneum, 136
Trichomonadicidal activity of chamomile, 279
Trichomonas, 133
Trigona iridipennis, 64
Triose pyruvate pathway (TPP), 80
Tripleurospermum inodorum, 52

U

Ulcer-protective property of chamomile, wound healing and, 114–117
Ultra-performance liquid chromatography (UPLC), 198
Umbelliferone, 66, 190
United States, homeopathy system of medicine, national regulations, 32–33
Unweighted pair group method with arithmetic mean (UPGMA), 222
UPLC, *see* Ultra-performance liquid chromatography
Urtica dioica, 272
Usar soils of north India, 245
U.S. patent, 269–271

V

Vacuum headspace, 195
Vanillosmopsis erythropappa, 199
Vascular arbuscular mycorrhizal fungi, 253–254
Vermicompost, 251–252
Veterinary medicine, 279–280

W

Water stress, 70
Weedicide, effect of, 75
Weeding, chamomile crop, 254–256
Williams, George H.A., 270
Wound healing, and ulcer-protective property of chamomile, 114–117

Y

Yield, descriptor of, 87, 88
Yield of chamomile flowers, 258, 259
Yugoslavia
    breeding programs, 225
    chamomile cultivation, 248

Z

Zingiber, 270

Life Science

Volume 13 of the book series
Traditional Herbal Medicines for Modern Times

Series Editor: *Roland Hardman*

In use as a medicinal plant since time immemorial in Europe and the Middle East, chamomile is gaining popularity in the Americas, Australia, and Asia. The spectrum of disease conditions in which it is used in traditional medicine systems is, quite simply, mind boggling. There is, without a doubt, a growing demand for this plant and therefore a growing need for an updated ready reference for the researchers, cultivators, and entrepreneurs who wish to work with chamomile. **Chamomile: Medicinal, Biochemical, and Agricultural Aspects** is just that.

Based on extensive research, this book provides the latest information on the medicinal, aromatic, and cultivation aspects of chamomile. It covers chamomile's geographical distribution, taxonomy, chemistry, pharmacology, genetics, biochemistry, breeding, and cultivation. The book also discusses the profiles of the several medicinally active compounds of the oil and extracts and how their levels could be increased through breeding.

The author highlights several potentially useful compounds discovered in the chamomile oil and extracts and discusses the cultivation and postharvest technology aspects of the plant in different agroclimatic zones including that of India. She presents guidelines on the good manufacturing practices laid out in different systems of medicine and provides an overview of the patents and products of chamomile especially important to researchers and entrepreneurs.

Although there is a plethora of information available on chamomile, the challenge has been finding a compendium that covers all aspects of the plant. Some books provide general coverage, others focus on only pharmacological uses, and many are outdated. This book examines all aspects from cultivation and harvesting, to essential oil content and profile as well as pharmacology and biotechnology. It is a reference for current information, an entry point for further study, a resource for using oils and extracts in product development, and a guide for following the best agronomic practices.

ISBN 978-0-367-37855-4